OCEAN ENGINEERING SYSTEMS

OCEAN ENGINEERING SYSTEMS

John P. Craven

Compiled and edited by
T. Gray Curtis
John R. Mittleman
James M. Patell

The M.I.T. Press
Cambridge, Massachusetts, and London, England

ISBN 0 262 53021 X (paperback)

Sea Grant Report No. MITSG 71-6
Index No. 71-106-Not

Library of Congress catalog card number: 70-165160

Printed in United States of America

CONTENTS

PUBLISHER'S FOREWORD

The aim of this format is to close the time gap between the preparation of a monographic work and its publication in book form. A large number of significant though specialized manuscripts make the transition to formal publication either after a considerable delay or not at all. The time and expense of detailed text editing and composition in print may act to prevent publication or so to delay it that currency of content is affected.

The text of this book has been photographed directly from the author's typescript. It is edited to a satifactory level of completeness and comprehensibility though not necessarily to the standard of consistency of minor editorial detail present in typeset books issued under our imprint.

The MIT Press

ACKNOWLEDGMENT

This text was prepared under the auspices of the Department of Naval Architecture and Marine Engineering at the Massachusetts Institute of Technology. The preparation was supported in part by a grant made to M.I.T. by the National Sea Grant Program, GH88, in part by a grant from the Henry L. and Grace Doherty Charitable Foundation, Inc., and in part by a sponsored research project funded by Sperry Rand Corporation.

This publication is a substantial expansion of material found in Chapter VI, "Sea Systems," which the author, together with Paul R. Stang, prepared for the book Hydronautics (edited by H. E. Sheets and V. T. Boatwright, Academic Press, 1970). The material presented in this text was developed for a course entitled "Ocean Engineering Systems" offered in 1969-1970 by Dr. John P. Craven as part of the Ocean Engineering Program of the Department of Naval Architecture and Marine Engineering. The case studies appearing in the text were prepared by M.I.T. students as part of the course requirements.

April 1971 — Alfred H. Keil, Chairman

AUTHOR'S ACKNOWLEDGMENTS

EPIGRAMS

The quality of an executive is limited by the quality of his secretary.

The quality of an executive can be improved by the quality of his secretary.

Some executives can even play golf.

Students tackle assignments that no sane practitioner would undertake.

Students do not know their own limitations.

No other group can accomplish so much in so little time.

Some students can even go skiing.

These epigrams acknowledge in some small manner the dedication and persistence of Mr. T. Gray Curtis, Mr. John R. Mittleman and Mr. James M. Patell in compiling this text, securing additional information, pre-drafting chapters from class notes and editing case studies.

They acknowledge the enthusiastic assistance of Mr. Joseph Lassiter in carrying the task to completion.

They cannot adequately acknowledge the skill of Miss Eleanor Baker in the preparation of this manuscript. Not only the format and setting of type script are her handiwork, but through her nimble fingers numerous unintelligible, ungrammatical sentences were instantly converted into impeccable and lucid prose.

John P. Craven
Honolulu, Hawaii
April 15, 1971

OCEAN ENGINEERING SYSTEMS

CHAPTER I

OCEAN ENGINEERING SYSTEMS

EPIGRAMS

A true system has components which exceed the comprehension of any single man.

System design is very different from component design--most engineers are trained to design components.

The greatest system catastrophes occur when one sybsystem fails to operate at all.

The greatest system costs are incurred when one subsystem is not delivered on time.

If a thing is worth doing, it is worth doing badly.

If a thing is not worth doing, it is not worth doing well.

Ocean Engineering Systems

A subject as broad as ocean engineering systems must have a focus and structure if generalized principles of ocean systems design and management are to be adduced. At the outset, some generalized understanding should be reached of what is meant by an ocean engineering system and what makes it distinct and unique from other systems. The concepts of system, subsystem and component need also to be initially defined in a formal and precise manner. Generalized parameters such as "complexity" and "magnitude" are certainly two qualitative characteristics which most people would identify as being common to a system, but generalized criteria, while permitting us to identify some systems which have already been established or which are contemplated, will not suffice for a more rigorous treatment.

The differentiation between ocean systems and land systems is also a difficult one to make since almost all of the major logistic systems of man involve both land- and marine-oriented activities. For example, international air transportation is heavily dependent upon the movement of aviation fuels by tanker, since the movement of cargo by air is essentially a tradeoff between high cost per unit commerce items of trade, such as people, and low-cost bulk products, such as petroleum. This tradeoff utilizes the slow movement of petroleum to permit the subsequent rapid movement of high-value items of commerce. Our differentiation between land and sea systems is, therefore, somewhat arbitrary and will depend heavily upon the extent to which the design of the total system is constrained by the characteristics of the sea, the sea/air interface, and the sea/land interface. For illustrative purposes, a number of real or possible ocean systems are here enumerated. This enumeration is made pedagogically in order to permit students to have an array of alternatives and choices for subsequent study and to demonstrate the kind of subjective human purposes which are, or can be, fulfilled by a sea-based system. Thirty-six sea systems are chosen for exposition, as follows:

1. Fisheries System A. Objective: search, location, retrieval, processing and delivery to market of specific fish products garnered from the open ocean beyond the twelve-mile limit.
2. Fisheries System B. Objective: search, location, retrieval, processing and delivery to market of demersal or sedentary marine life garnered from the open ocean within the twelve-mile limit.

3. Fisheries System C. Objective: the selective breeding, tracking, trailing, and protection of fish species which spawn and return to coastal waters and which grow on the high seas together with the harvest and processing of such species.

4. Fisheries System D. Objective: the mariculture of sedentary or demersal marine life, the harvest, processing, and delivery, on the continental shelf.

5. Desalininization Systems. Objective: the collection, desalinization, and distribution of fresh water, and disposal of brine ocean water, in sufficient quantities that the total local ecological balance is altered.

6. Hard Mineral Recovery System A. Objective: the recovery, processing and delivery to market of specific minerals from geologic deposits on the continental shelf in selected areas of the world, for selected United States or world markets.

7. Hard Mineral Recovery System B. Objective: recovery, processing and delivery to market of specific minerals from geologic deposits in the deep ocean in selected areas of the world for selected United States or world markets.

8. Extraction of Minerals from Brine. Objective: the collection, extraction, and distribution of minerals from salts dissolved in sea water.

9. Oil Recovery System A. Objective: the recovery, processing and delivery to market of offshore oil located in the geologic structures of the continental shelf in selected areas of the world for selected United States or world markets.

10. <u>Oil Recovery System B</u>. Objective: the recovery, processing and delivery to market of offshore oil located in the geologic structures of the deep ocean (greater than 6000 feet) in selected areas of the world for selected United States markets.

11. <u>Bulk Transportation</u>. Objective: the collection, transport and distribution of diversified bulk cargoes in either international or domestic trade.

12. <u>Containerized Transport Systems</u>. Objective: the collection, transport and distribution of containerized cargoes in either international or domestic trade.

13. <u>Air Transportation System</u>. Objective: the construction and deployment of airfields in the ocean in the vicinity of urban centers, together with means for the support of aircraft and receipt and delivery of passengers to appropriate destinations.

14. <u>International Aid System A</u>. Objective: the collection and distribution of wheat, grains, proteins, or other products required on a fluctuating basis for purposes of international aid.

15. <u>International Aid System B</u>. Objective: the use of sea-based educational facilities, hospital ships, marine academies, technological institutions for specialized education, in developing countries on a rotating or seriatim basis.

16. <u>Urban Renewal Systems</u>. Objective: a modification and utilization of port harbor facilities and adjacent waters for urban renewal for cities with populations of not less than one million.

17. Disaster Relief Systems. Objective: the identification of major disaster areas accessible to the sea and the deployment of sea-based task forces for temporary or long-term establishment of mechanisms for economic and social viability in the disaster areas.

18. Pollution Control System A. Objective: the identification of pollution sources from distressed ships or structures on the high seas and the deployment and operation of devices to contain the pollutants and to control the source.

19. Pollution Control System B. Objective: systems for collection, transport, and dispersal of specific wastes in the ocean without detriment to the ocean environment.

20. Harbor and Coastline Protection. Objective: the maintenance and construction of shoreline facilities to ensure the continued effective configuration of harbor and shoreline for purposes of commerce, conservation and recreation.

21. Weather Forecast and Control System. Objective: the acquisition and analysis of data and operational measures to forecast weather primarily generated by oceanic phenomena and to control such weather.

22. Recreation System A. Objective: the receipt, transport, and provision of facilities, instruction, demonstration, and assurance of safety for persons desiring to obtain recreation at the free surface, or within ten feet of the free surface, of the sea.

23. Recreation System B. Objective: the receipt, transport, and provision of facilities, instruction, demonstration, and assurance of safety for persons desiring to obtain recreation in atmospheric

pressure facilities below the free surface to depths of several thousand feet.

24. Recreation System C. Objective: the receipt, transport, and provision of facilities, instruction, demonstration, and assurance of safety for persons desiring to obtain recreation at hyperbaric pressures below the free surface to depths of several hundred feet.

25. Amphibious Assault Systems. Objective: the capability to deploy and support men and equipment, and transfer them across essentially arbitrary configurations of coastline in the face of active and hostile measures to prevent such transfer.

26. Fast Logistic Deployment. Objective: the capability to identify and respond to needs for deployment of specified complements of men and equipment from the continental United States to arbitrary locations in the world.

27. Antisubmarine Warfare System A. Objective: the location and defense against submarines engaged in active attack on merchant shipping and/or other military vehicles.

28. Antisubmarine Warfare System B. Objective: the search, location, and tracking of ballistic missile submarines.

29. Strategic Deterrence System A. Objective: the deployment of weapons on the surface of the ocean to provide credible threat of assured retaliation to some postulated aggressive action.

30. Strategic Deterrence System B. Objective: the deployment of weapons under the surface of the ocean, or near or on the bottom of the sea, to provide credible threat of assured retaliation to

some postulated aggressive action.

31. Intelligence Collections System. Objective: the capability to collect, monitor, and retransmit electronic intelligence, utilizing components which are sea-based.

32. Harbor Protection. Objective: the assurance that harbors and approaches to harbors are free from mines or other military obstructions, and that mines are detected, identified, and nullified as they are inserted or as insertion is attempted.

33. Search and Rescue System A. Objective: the search, location and recovery of mariners in distress on the surface of the ocean and the salvage of their equipment.

34. Search and Rescue System B. Objective: the search, location and recovery of mariners in distress in submersibles, and the salvage of their equipment.

35. Search and Rescue System C. Objective: the search, location and recovery of aquanauts, scuba divers, and swimmers in distress in water depths up to 1000 feet.

36. Arms Control Inspection System. Objective: the location, inspection, and confirmation of activities on the ocean bottom, within the ocean, or on the surface of the ocean, believed to be in violation of a specified arms control agreement.

This list, while not all-inclusive nor specified in other than the broadest terms, is an example of the spectrum of subjective human purposes for which major sea systems have been built or have been contemplated.

System Definition

In order for systems of scope and complexity to become realized fact, in one way or another, design teams consisting of individuals with integrated knowledge of the technology required for successful construction of the system must be assembled. These design teams must be coordinated and the results of their efforts must be transmitted to artisans, craftsmen, surveyors, lawyers, teachers, and other individuals concerned with the system who possess an actual responsibility for actions which must be taken before the sea system can become a reality. A major and central problem in the design, development and construction of a system is, therefore, the communication with, and coordination of, this tremendously large number of individuals and the efficient use of their skills and talents in order that the system be produced with a minimum expenditure of human resource, and in order that the system perform in the manner which is intended.

Many techniques have been devised for the management of large systems and many differing social structures have been involved. The most controlled and/or regimented approaches have been taken by the Department of Defense. The most permissive have probably been undertaken by the University. In retrospect, it must be said that the degree of success has been unpredictable. The author has been associated with a program that many have classified as successful (i.e., the Polaris-Fleet Ballistic Missile Program). An attempt is therefore made to distill from this program what may have been an ingredient of success. This ingredient is identified as the management of information transfer between the elements of the society which will design, produce and deploy the system.

For an otherwise perfectly organized community of individuals, the limiting factor in systems design and development capability is therefore deemed to arise from the limitation of a human being in processing or comprehending the total information which is required to design, produce, and operate the system. It is this recognition that leads to the definition chosen for an engineering system:

A. An engineering system is an identifiable set or collection of mechanical devices and trained individuals, designed and/or trained to operate in concert to fulfill a subjective human purpose, and of sufficient complexity that it is impossible or impractical to interrelate, cross-coordinate, or cross-correlate the design or training of subsets of the collection.

B. An engineering component is an identifiable set or collection of mechanical devices and trained individuals such that a single individual, or a single management unit of individuals having nearly continuous communication capability, can have, or does have, full knowledge of the design, fabrication, and operation of the collection of devices and/or the indoctrination, training, and employment of the individuals under his purview.

It should be noted that these definitions admit of a hierarchy of systems and subsystems and, in fact, hierarchies of components. Systems hierarchy results from subsystem characteristics; that is, the subsystems of a major system can be sufficiently complex that it is impossible or impractical to interrelate or cross-coordinate or cross-correlate the designing or training of sub-subsets of the collection.

Component hierarchy results from a different viewpoint, mainly that of utilizing "black boxes" or "service organizations" as part of the total component. If a "black box" can be obtained for which all of the relevant functional characteristics are known and all of the relevant interfaces are specified, then, regardless of the complexity of the contents of the "black box," a systems designer or manager may treat that "black box" as a component. This also can be applied to services of service organizations where the service to be rendered is clearly specified or specifiable in terms of function and in terms of interface with other elements of the collection. It must be emphasized that a "black box" or "service organization" cannot be treated as a component unless it has been previously produced by the society, and its characteristics clearly demonstrated, as well as its ability to be replicated.

Employing these definitions may result in some initial difficulty with the term "subjective human purpose." This difficulty generally stems from a failure to recognize that even systems whose major characteristics have been quantized and specified by detailed cost-effectiveness analysis are, in fact, chosen because they were most cost effective for some more generally specified subjective human purpose. For example, if one were designing a space system to make voyages to the moon, and only the moon, the most cost-effective system for this mission would be very different from the most cost-effective solution for a space system which was designed for the multiple missions of voyage to the moon with subsequent voyage to Mars.

Information Management

Of greater difficulty in utilizing the systematic definitions is the necessity for quantizing, at least generally, the actual limitations on human information processing, the scope of the manager's knowledge and the resulting determination that one is dealing with a system or with a component.

In an overly-simplified model, each individual is deemed to have a set of loosely-defined information which will be employed in issuing action items with respect to the design, construction, maintenance, and operation of the system. This information we will call "relevant knowledge." It will be convenient to further subdivide this set of "relevant knowledge" into input and output sets. The input set consists of the knowledge which the individual has received in the form of audio, visual, and tactile inputs. The output set consists of the knowledge which the individual has generated by logic processes in his brain and which is available for output in terms of verbal, graphic, or tactile presentation. Because of the characteristics of human beings, the output set for any individual will be very, very much larger than the set of information which he is, in fact, able to reveal in the form of verbal, graphic, or tactile outputs. This knowledge is, nevertheless, available in his brain in a "staff reservoir" from which specific output can be selected in an adaptive manner for the design, development, or operation of the system. The failure to recognize this fact has frequently resulted in the erroneous notion that all of the relevant information for the design, development, and maintenance of a complex system can be transferred and stored in computer memories, thereby relieving the individual of further need for adaptive response during the design, construction, and operation phases.

In order to identify and quantify this "relevant knowledge," the set of knowledge will be measured by an empirical measure which will be called the "page." The limits of human processing, in terms of this unit, are shown in Figure 1-1. Insofar as inputs are concerned, the human being receives his input either by reading pages of material, by listening to the spoken word, or by visual observation, or by some combination of the three, and, in rare instances, by tactile impressions. The spoken word can be reduced to a page equivalent, if it is presumed that it transmits information at approximately the same page rate if the words were transcribed by dictaphone and subsequently reassembled. Of greater difficulty is quantification of the observation, and it will be presumed that the amount of observational information which a human being can absorb in a given time period is roughly the same as the amount of information he would obtain if he were to leaf through photographs of the same situation and, further, that the quantity of information on a photograph which is "leafed through" is crudely equivalent to the amount of information on a printed page. As we shall see, the crudity of these empirical measures of "pages" will not affect the final outcome in the decision process as to whether one has, or has not, a system with which to deal. Indeed, it is for this reason that the "page" is chosen as the unit of knowledge, rather than some more esoteric, precise, and misleading (though technically correct) unit of information theory. Granting these crude equivalents, it is now possible to make estimates of the upper limits of human knowledge processing. Insofar as input is concerned, if a person sits in conference all day, the average daily transcript will run about three hundred pages. If, on the other hand, an individual spends the entire day reading material of modest or moderate intellectual content, then he will be able to absorb about one thousand pages as a reasonable upper limit. Speed readers may wish to quarrel with this estimate, but they will certainly not quarrel with the assumption that it is not off by more than one order of magnitude, which, on the scale of information required to build a system, becomes a very small change in level.

In a similar manner, one can estimate levels of output. If an individual utilized the entire day for purposes of lecture, the average daily transcript will again run about three hundred pages. If, on the other hand, an individual spends the entire day generating written material (and it is further presumed that he has the staff assistance of secretarial skills), then the nature of this information generation process is such that somewhere between ten and twenty pages of text represents a reasonable upper bound of daily output of knowledge, information, or instructions. Again many could or would quarrel with this figure, suggesting for technical information that this bound is greatly reduced to perhaps as little as one or two pages or, in the case of gifted reporters, to a level of perhaps as much as fifty pages of text or even higher, but they will certainly not quarrel with the assumption that it is not off by more than one order of magnitude, which, on the scale of information required to build a system, as before, becomes

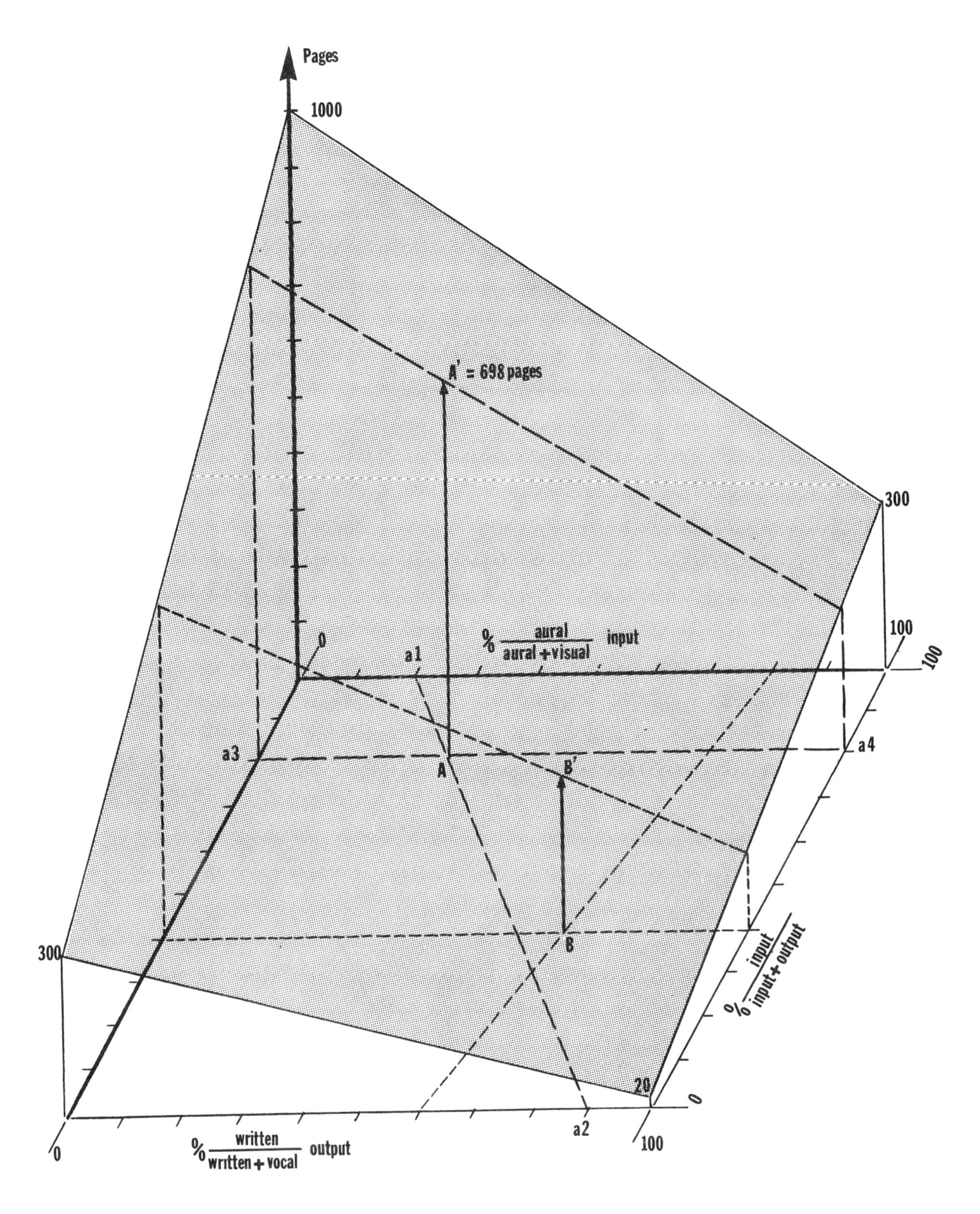

Figure 1-1. Upper Bound of Information Transfer.

Figure 1-1. Upper Bound of Information Transfer

Example A of Figure 1:

Based on an eight-hour day: line a3 - a4 indicates that 80% x 8 hours = 6.4 hours were spent for the input of relevant knowledge; 20% x 8 hours = 1.6 hours were spent for the output of relevant knowledge.

Point a1 indicates that 20% x 6.4 hours = 1.28 hours were spent listening and 80% x 6.4 hours = 5.12 hours were spent reading.

Point a2 indicates that 90% x 1.6 hours = 1.44 hours were spent writing and 10% x 1.6 hours = 0.16 hours were spent talking.

Point A graphically represents the combined effect of this information, and point A', which is on the surface, indicates the maximum amount of relevant knowledge which could be processed by one individual following the above time allotments.

a very small change in level. Individuals involved with the processing of knowledge utilized in system design, development, construction, and operation will, of course, be involved in a complex joint function of input and output. The total amount of knowledge which is processed per day, therefore, falls as an upper limit on the surface of Figure 1. If, as will be required, a substantial portion of the day is utilized for thinking processes, then the total amount of information or knowledge which is handled will be greatly reduced below this upper level.

With these definitions of relevant knowledge and with these gross estimates of magnitude, we can now establish a few characteristics of systems management, as follows:

1. The relevant knowledge in a system is a maximum when the intersection of knowledge sets is zero; that is, the relevant knowledge is a maximum when no one knows what anyone else knows. The ability to build a system under such conditions is, of course, essentially zero, since such a requirement results in no system coherence and a resulting random behavior of each individual in the system.

2. The relevant knowledge in a system is a minimum when the intersection of sets is unity; that is, the relevant knowledge is a minimum when everyone knows what everyone else knows. The building of a system is not possible in this instance, even though the coherence is exceedingly high (and we shall presume that it is unity), since the total amount of information or relevant knowledge which is available is not enough to cover the complexities of the total system.

3. It follows, therefore, that the management of a system is most

efficient from an information standpoint when the set intersections are at the minimum required for complete system definition. In the real world in which understanding and coherence are highly limited, a great deal of redundancy on set intersections will, of course, be required if compatible components are indeed to be designed, constructed, assembled, and operated.

<u>Information Transfer</u>

Before the decision-making structure, scope, and content of relevant knowledge which must be possessed by individuals in the decision-making structure can be more precisely defined, some further elaborations on types and varieties of information transfer must be made. A number of types of information transfer can be identified, having various efficiencies from a systems standpoint, as follows:

A. Person-to-person, including telephone communication

B. Conference

C. Lecture

D. Directives, announcements, messages, plans

E. Management information schemes, such as PERT, line of balance, and other computer-aided information processing systems.

Information flows most effectively when a person, or persons, is directly involved in a real-time feedback loop of the communication process. From the standpoint of the individual, person-to-person communication proves the most effective, but insofar as the system is concerned, it is a highly inefficient and painfully slow process, subject to the garble and distortion of successive repetition of the message content in going from person to person and subject to the long-time delays before a relevant message can be communicated to substantial segments of the system at large.

This limitation is somewhat alleviated by the conference. In a conference there develops a disproportionate balance between the time which is available for output and the time required for input, such that the average output time per individual is 1/n of the total time of the conference. As an empirical rule of thumb, if n is greater than 10, then a substantial number of participants at the conference can be no more than listeners and such representation is likely to be highly inefficient insofar as the flow of information is concerned.

When the conference becomes very large, or when feedback from conferees is not allowed, the situation develops into that of the lecture; that is, a great deal of efficiency is generated with respect to the output of a single individual at the expense of any significant reaction from the members of the audience. While some progress is being made in the form of automatic polling procedures or real-time indicators of agreement, disagreement or understanding by the members of the audience as an entity, the lecture is essentially a one-way communication channel.

An alternative to this is multiply-addressed messages which can be disseminated to as large a segment of the system as is desired. This communication system suffers from the drawback that the total response can be greater than can be assimilated by the message center, and that a high degree of logical inconsistency can result if all responses are to be believed.

These communication difficulties result in an inherent characteristic of systems management; there is an asymmetry between a system's capability to pass information down and the ability to pass information up the management chain. Passing information up the management chain, in general, floods the communication channels of the higher echelon, unless the information has been selectively reduced by predetermined logic patterns. Such management information schemes, if properly designed, can indeed achieve the required ingredient of reducing the flow of information up the system line within a structure which gives a high probability that information which does finally get through has a high degree of relevance.

The PERT system as originally formulated was one such management technique. Under the PERT program a network of events and milestones which must be accomplished in order to meet system goals is established. The network is sequential and is characteristically single-valved at the inception and completion points and multibranched and multipathed between events. A simplified diagram is shown in Figure 1-2. Assignments are made at the lowest level of management (preferably at the component working level) for responsibility to complete each event in the network. At periodic intervals (usually weekly) the individuals with the designated responsibility make a subjective estimate of the expected time to complete the event, the maximum expected time to complete the event and the minimum expected time to complete the event. (Unfortunately, this estimate generally requires the assumption that

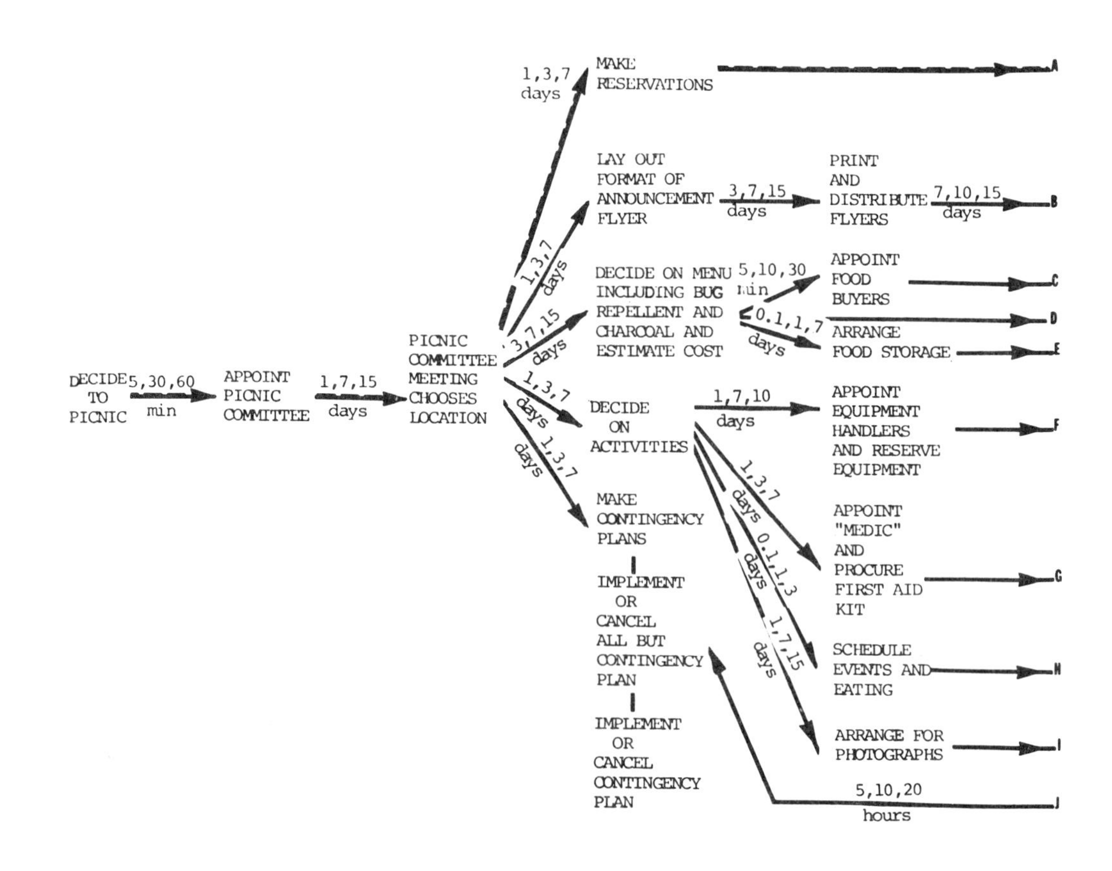

DECIDE TO PICNIC
5,30,60 min
APPOINT PICNIC COMMITTEE
1,7,15 days
PICNIC COMMITTEE MEETING CHOOSES LOCATION
1,3,7 days
MAKE RESERVATIONS
A
1,3,7 days
LAY OUT FORMAT OF ANNOUNCEMENT FLYER
3,7,15 days
PRINT AND DISTRIBUTE FLYERS
7,10,15 days
B
3,7,15 days
DECIDE ON MENU INCLUDING BUG REPELLENT AND CHARCOAL AND ESTIMATE COST
5,10,30 min
APPOINT FOOD BUYERS
C
D
0.1,1,7 days
ARRANGE FOOD STORAGE
E
1,3,7 days
DECIDE ON ACTIVITIES
1,7,10 days
APPOINT EQUIPMENT HANDLERS AND RESERVE EQUIPMENT
F
1,3,7 days
MAKE CONTINGENCY PLANS
IMPLEMENT OR CANCEL ALL BUT CONTINGENCY PLAN
IMPLEMENT OR CANCEL CONTINGENCY PLAN
1,3,7 days
APPOINT "MEDIC" AND PROCURE FIRST AID KIT
G
0.1,1,3 days
SCHEDULE EVENTS AND EATING
H
1,7,15 days
ARRANGE FOR PHOTOGRAPHS
I
5,10,20 hours
J

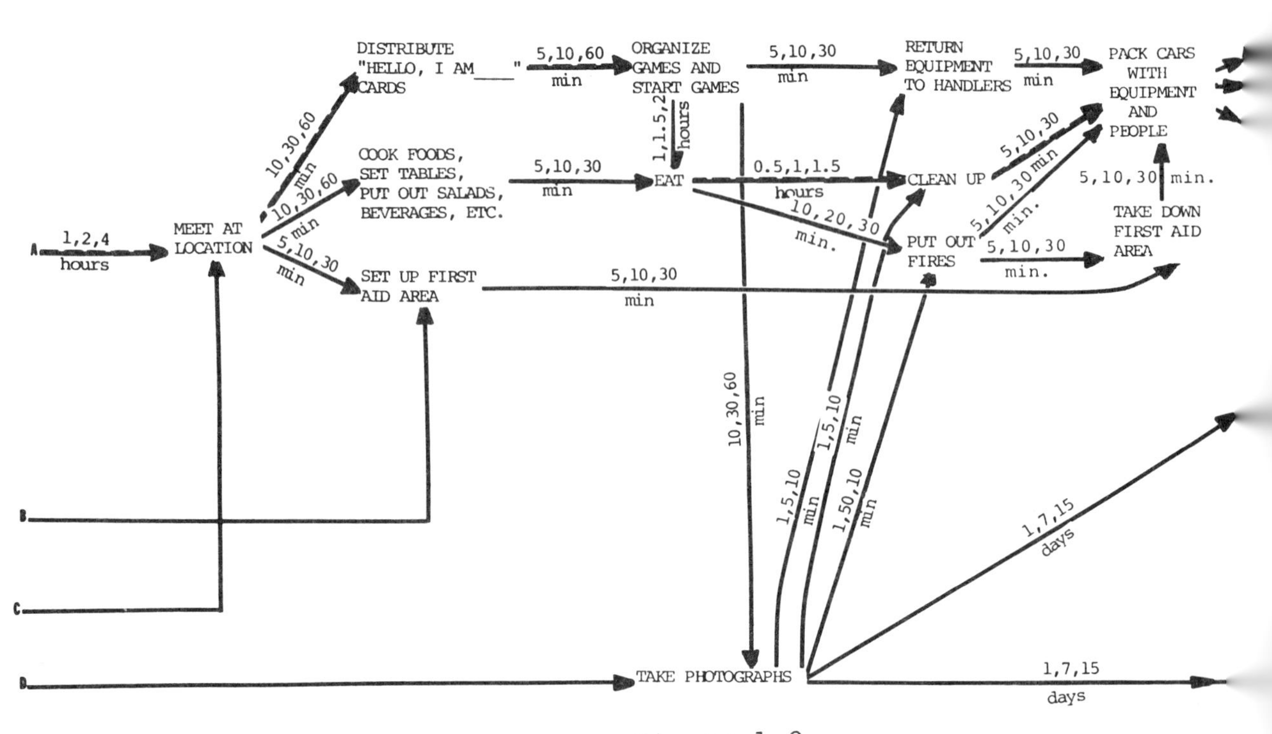

A
1,2,4 hours
MEET AT LOCATION
10,30,60 min
DISTRIBUTE "HELLO, I AM___" CARDS
5,10,60 min
ORGANIZE GAMES AND START GAMES
5,10,30 min
RETURN EQUIPMENT TO HANDLERS
5,10,30 min
PACK CARS WITH EQUIPMENT AND PEOPLE
10,30,60 min
COOK FOODS, SET TABLES, PUT OUT SALADS, BEVERAGES, ETC.
5,10,30 min
1,1.5,2 hours
EAT
0.5,1,1.5 hours
CLEAN UP
5,10,30 min
10,20,30 min.
PUT OUT FIRES
5,10,30 min.
5,10,30 min.
TAKE DOWN FIRST AID AREA
5,10,30 min.
5,10,30 min
SET UP FIRST AID AREA
5,10,30 min
10,30,60 min
1,5,10 min
1,5,10 min
1,50,10 min
1,7,15 days
B
C
D
TAKE PHOTOGRAPHS
1,7,15 days

Figure 1-2

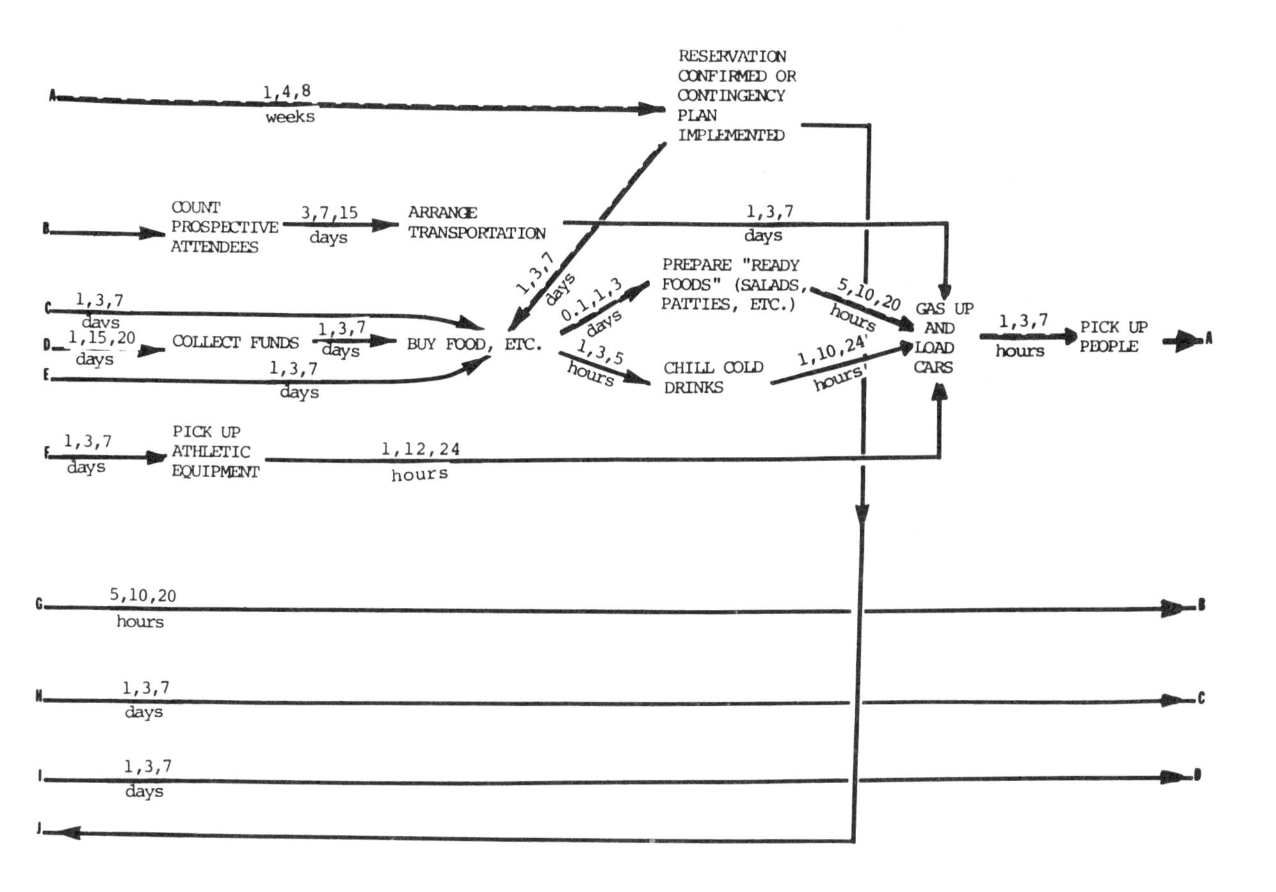

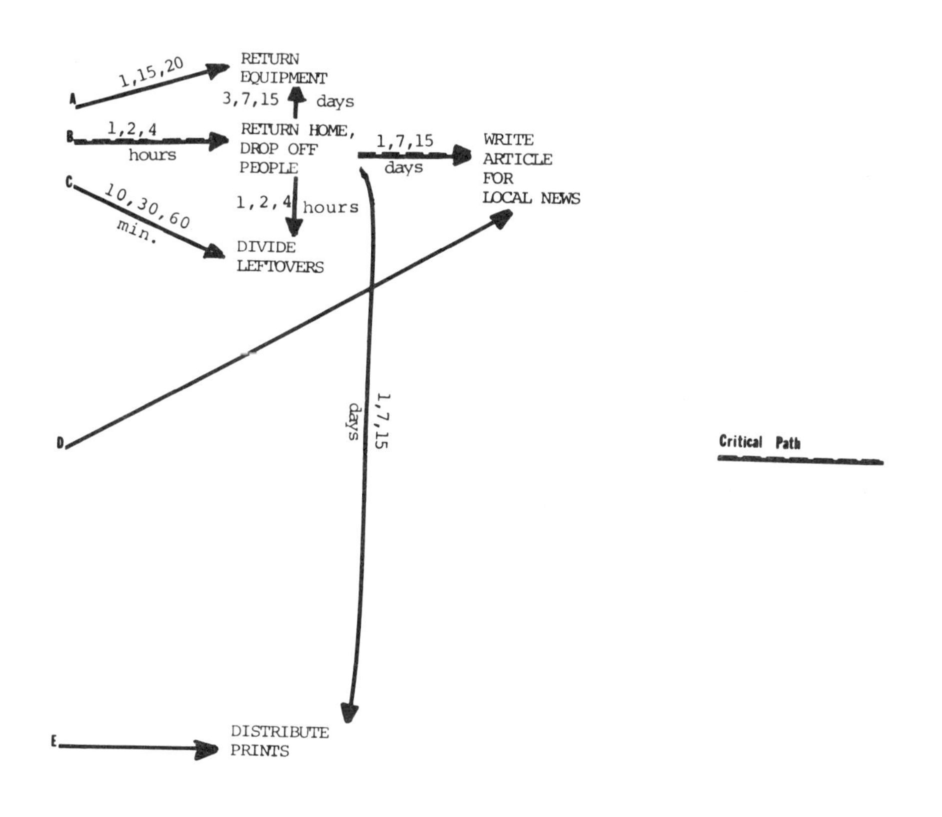

Critical Path

resources will be made available.) Utilizing these estimates, a computer can determine the "critical path," i.e., those sequences of events which govern the period of time between the inception and completion of the network. Management attention is thus directed solely to the critical path and/or those which are just below it. In this sense, the system automatically ignores information inputs from people who are in "good shape" and also ignores people who are in "bad shape" but for which this otherwise unhappy condition is irrelevant. It should be noted that, when dealing with systems, the completion of a subsystem on time is highly related to economic efficiency since large cost penalties are paid if the remainder of the system is forced to mark time.

Variants of the PERT system have been introduced which include estimates of cost and time in the network. It is the author's opinion that many of these have been counterproductive, since they generate additional information and often generate the need for separate PERT-COST organizations who prepare the estimates although divorced from the "knowledge reservoir" of the engineer on the production floor.

The characterization of information having relevance is again a central problem of systems management and, for purposes of this exposition, three classes of relevant knowledge or information will be defined, as follows:

1. Category A Knowledge

This is knowledge which is translated or which will be translated into action which affects the state of the system. Category A Information or Knowledge includes design drawings, instructional materials, shop drawings, certain aspects of contract directives, specifications, and alterations. Category A Information also includes knowledge possessed by skilled welders, machinists, electronics technicians, engineers, and scientists, which makes them uniquely qualified to perform actions which affect the system state. It is a major task of systems management to insure that Category A Information is generated in a timely manner and of such completeness that the entire system changes its state in accordance with the goals of design, development, production, operation, and maintenance.

2. Category B Knowledge

This is knowledge which is convertible into Category A Information and which forms a basis for Category A choices. This knowledge consists of alternatives which may or may not, in fact, be implemented to change the state of the system. Such knowledge includes all the designs, cost-effectiveness studies, contract definition studies, systems tradeoffs and other information which may be generated in order to assist in choice of Category A Information.

3. Category C Knowledge

This is knowledge which is required for administrative purposes only, such as payrolls, university class schedules, availability of life support services for the administrator or the designers, and other information which is relevant for support of the individuals who are designing, developing, operating, or maintaining the system, but which is otherwise irrelevant to the changes of system state. Although Category C Knowledge is an irrelevant category for further discussion, the burdening of executives involved with systems management with this technological alternatives, and to ensure that the system is produced within a specified time envelope, a specified cost envelope, and a specified performance envelope. In addition, the organization of a system must recognize that the subsystem of a system may itself be a system. This then requires a replicating structure to satisfy all system level requirements, i.e., the information transfer and technological decision structure must be compatible with the requirement that the system itself is a subsystem of a similar higher level structure.

Subsystem Interfaces

One solution to this problem has been the grouping of individuals and organizations into closed communications structures in which these collections of individuals and organizations are assigned precisely-defined functional requirements within precisely-defined physical limits. The precisely-defined limits are referred to as interfaces. Except for communications concerning the interfaces themselves, official communications (i.e., communications affecting the state of the system) across subsystem interfaces are specifically denied. While this has the disadvantage of being incapable of eliminating "ideal world" redundancies or inefficiencies, it has the overwhelming advantage that the total systems task can be accomplished in a finite period of time and in such a manner that the total system works as required.

At each level of a systems organization, the number of subsystems into which a system can be divided is again limited by human communication. Seven to ten appears to be an optimum number in order that the type of information is a serious problem. The filling out of travel forms, of requisitions, of telephone usage forms, and other bureaucratic minutiae is a type of time pollutant which destroys the capability of the executive to absorb Category B Information and disseminate Category A Information. Zeal in the development and enforcement of Category C Information requirements can render a management system almost helpless.

Having categorized these three types of information, a number of observations can now be made, as follows:

a. The volume of Category B Knowledge which can be converted into Category A Knowledge is limited by the communication capacity of the individual charged with responsibility for generating the Category A Knowledge.

b. The volume of Category B Knowledge which can be generated is infinitely expandable once its generation is detached from the possessors of Category A responsibility and capability.

c. Since this is the case, the accomplishment of systems goals will therefore require on many, if not most, occasions the rejection of Category B Information which would otherwise be relevant to the systems decision, but which cannot be communicated to the Category A generator because of time constraints.

The organization of the design and management team to design, develop, produce, deploy and operate an ocean system requires, therefore, a mechanism to ensure a balance between Category A Information and Category B Information throughout the system, to ensure that the functional intent of the system is maintained, to ensure freedom for span of control be within the communication capability of a single system manager. In assessing the question of span and control, it is highly desirable that each executive in the system prepare for himself an "information budget" with careful schedules in order to satisfy himself that the different systems levels can be directed in a manner which does not delay systems progress and, in particular, does not cause large numbers of people to "mark time."

Information Budget

The information budget is defined as the allocation of time between superiors, staff, peers, subordinates, self, libraries, and information storehouses as a function of the types and varieties of input-output and output-input communication processes. A sample information budget is shown in Figure 1-3. It should be noted in this information budget that a certain amount of time should be devoted to acquisition of knowledge outside of the system in order that the total knowledge in the system be increased. If such an allocation is not made, then the individual in the system is no more than a mechanism for reordering information which already exists. While such reordering is, of course, required for system coherency, it contributes little to innovation.

The information budget does not, of course, duplicate itself on a day-to-day or even a week-to-week basis. Time cycles must be accounted for which correspond to the time cycles to which our society is accustomed. Such cycles are generally annual, semiannual, bimonthly, monthly, weekly, or daily. Information budgets for systems planning,

TIME SPENT IN HOURS PER WEEK		Boss	Peers 1 2 . . .	Staff 1 2 . . .	Subordi- nates 1 2 . . .	Others (in the system) 1 2 . . .	Others (out of the system) 1 2 . . . i.e., Friends Books Study Movies . . .
INPUT	Aural Reports	1	4	4	1/2	1	2
	Written Reports	1/2	1/2	4	1/2	1-1/2	1
	Observed			1/2	1/2	2	2
	. . .						
INPUT/OUTPUT	Telephone		1-1/2	1-1/2		2	1
	Conferences	1	2	3	1	6	1
	Interviews					1-1/2	1-1/2
	Conversations	1	1	1	1/2	3	3
OUTPUT	Oral Reports	1-1/2	1-1/2	2	4	2	2
	Written Reports	1-1/2	1-1/2	4	4	2	1

Figure 1-3. Sample Information Budget.

WEEK'S SCHEDULE FOR

SUBSYSTEM MANAGER

MONDAY

0800-1000	Monday Management Review (boss and peers) (Categories A and B)
1000-1015	Interim coffee, informal discussion (Category B)
1015-1115	Special presentation (Category B)
1115-1230	Problems meeting on the morning's discussion (Category B)
1230-1300	Lunch at desk, read mail (Category B)
1300-1400	Generate replies to correspondence (Category A)
1400-1500	Plan next day's meeting (Category B)
1500-1600	Staff conferences (Category B)
1600-1700	Prepare reports and memoranda (Categories A, B and C)
1700-1800	Issue line directives to subordinates (Category A)
1800-2300	Homework, dinner (Category B)
2300	Telephoned field report (Categories A and B)

TUESDAY

0800-1800	Replicate Monday with staff and subordinates instead of peers
1800	Depart for airport for one-day field trip

WEDNESDAY

0800	Arrive at destination
0900-1000	Review and briefing (Category B)
1000-1200	On-site inspection (Category B)
1200-1300	Lunch with corporation (Category B)
1300-1500	Problems conference and problem-solving (Categories A and B)
1500-1700	Ancillary inspections and visits (Category B)
1700	Depart for airport (Category C)

THURSDAY

0800-0900	Mail and correspondence (Categories A, B and C)
0900-1200	Quarterly Budget Review (Category A)
1200-1300	Lunch with Budget Review authorities (Category B)

1300-1400	Prepare material for boss to use for his Friday morning meeting (Category B)
1400-1500	Meeting with boss and peers (Category B)
1500-1600	Prepare alternatives resulting from Budget Review Category B)
1600-1700	Brief subordinates on alternatives from Budget Review and on field trip (Categories A and B)
1700-1800	Visitors (Category B)
2300	Telephoned field report (Category A)

FRIDAY

0800-0900	Correspondence (Categories A, B and C)
0900-1030	Lecture to management interns (Category B)
1030-1100	Free reading or visitors (Category B)
1100-1130	Awards ceremony (Category C)
1130-1200	Visitors connected with ceremony (Category C)
1200-1400	Ceremonial luncheon and speech (Categories B and C)
1400-1500	Congressional hearing (Category B)
1500-1800	Review and initial plans specifications and action documents (Category A)
1800-2300	Dinner, theatre or a play (Category B)
2300	Telephoned field report (Category A)

SATURDAY

0800-1000	Staff meeting in preparation for Monday morning (Category B)
1000-1100	Meeting with Graphic Arts for presentation material (Category C)
1100-1300	Tie up Irish pennants (e.g., unfinished odds and ends) (Category C)
1300	Family and home life (Category C)

SUNDAY

Morning	Religion or golf (Category C)
Afternoon	Emergency Crisis Conference (Category A)
Evening	Type C information (bills, income tax, etc.)

design, manufacture, deployment, and operation must take such annual cycles into account. Typical cycles include annual program plans, annual program change proposals, annual budget proposals, annual budget reviews, semiannual program reviews, semiannual budget reviews, bimonthly systems appraisals, bimonthly systems technology reviews, monthly system sampling reviews, and weekly subsystem progress and problem reviews. The total information budget must be prepared not only with these reviews in mind, but with the recognition that the total expected number of hours from an individual with action responsibilities reaches an upper limit of approximately seventy hours a week for fifty weeks a year. This limit assumes that he adjusts his schedule so that all travel is accomplished between the hours of 1700 and 0900. Very few individuals are capable of sustained operation at this level of activity and it is highly desirable to note that such executive performance requirements must permit occasional sabbaticals or detachments from executive duty for appropriate human maintenance.

Subjective Human Purpose

Within these highly limited constraints of human information processing, the systems development process must take place. Recognizing these limitations, we may now turn to the problems of subsystem definition and to the initial hardware identification. At the outset, the subjective human purpose for which the system is to be built must be established. This must include the approximate time and cost constraints which govern the accomplishment of this subjective human purpose and the assurance that the society will provide the resources necessary to meet this subjective human purpose. All too often these hard questions are avoided by systems managers on the presumption that some "higher authority" in the society is charged with this responsibility or, even more dangerously, on their presumption that there exist some economic-deterministic processes in the society which dictate the tasks to be done and the choices of the alternatives in performing these tasks. It is therefore worthwhile to digress at this point in the text for a very brief discussion of the identification of subjective human purpose in the society and the assessment that resources will indeed be available to meet these purposes.

In order to survive, a society, at the bare minimum, must provide food, clothing, shelter, and defense. If a modest level of technological sophistication is introduced, it is necessary to introduce transportation and distribution for the food, clothing, and shelter and, concomitantly, the acquisition of minerals and manufacturing processes in order to ensure the existence of a clothing, shelter and transportation system. When there is a surplus of income beyond these basic necessities, then the society has an option of directing its surplus into such areas as health, recreation, environment, and luxuries. The luxuries include: non-defense activities of space, pure science, luxury foods, luxury clothing, luxury shelter, luxury transportation, and the generation of Type B and Type C information in the form of nonfiction, news communications, etc.

In making an assessment of the society's need for a specific functional capability, the system planner must assess whether the government at the federal, state or local level will appropriate the necessary resources to meet the function, or whether the capital will be available from private industry for investment on the basis of assured or expected profit return, or whether some eleemosynary or charitable section of the society will contribute the resources to meet the specified capability.

In the first instance, the system planner must determine the governmental requirements or anticipate the governmental requirements as they will be assessed by the Congress. In the military sphere this assessment will be aided by a threat analysis in which the potential harm to national security is measured as the result of the known or projected developments on the part of other nations of the society. The result of such threat analysis will in general indicate the magnitude of the effort which is required in order to meet such a threat, but it is by no means automatic that the Congress will approve the resources which are required in order to meet such hypothetical threats. Indeed, a highly subjective judgment is generally employed in determining what will be a minimum response to the real or projected threat, what will be a moderate response to the real or projected threat, and what would be a maximal response to the real or projected threat. It is only when such order of magnitude estimates of the nature of the threat and the nature of the response are determined that one is faced with the more easily ascertained question of alternatives and the cost effectivity of each alternative to meet the postulated requirement. Even then a great deal of difficulty results since a specific hardware configuration utilized by the military will generally have some multipurpose application and, when comparing two alternatives, one is inevitably cast in the role of recognizing that each alternative, while being adequate for the prime mission of concern, will have a different set and spectrum of alternate mission capabilities. To this extent, cost effectiveness is but one measure that is employed in a highly subjective decision-making process which eventually results in a recommendation to the Congress to proceed with the design, development, test, evaluation, and deployment of a specific military capability together with a recommendation as to the force levels, the rate at which the capability is acquired, the locale of deployment, and the total magnitude of resources which will probably be required in order to meet the postulated capability.

Where the subjective human purpose is one which requires investment by a profit-making industry on the basis of a sure or expected profit return, the calculus of choice is equally, or perhaps more, complex. But a major failing on the part of many engineers is to presume that the most cost-effective system, from an engineering point of view, which can be produced to meet some need of the society will, in fact, be the most profitable. Nothing could be further from the truth. The current structure of our society in terms of taxes, subsidies, labor union constraints, psychological aspects of customer preference, etc., are such that it may be safe to say that very rarely in the society is the most cost-effective system the most profitable. Just a brief statement about each of the major categories will provide sufficient examples to illustrate the point.

In the basic food industry, competition between marine foods and agriculturally-developed foods is such that food which is grown on the land enjoys a substantial competitive advantage as a result of agricultural supports and agricultural programs. For example, in 1962 the net farm income of the United States was 12.8 billion dollars and the U.S. farm support program was for that same year at the level of seven billion dollars. While it is probably quite true that the set of individuals who received the 12.8 billion dollars is not all-inclusive of the set of individuals who received seven billion dollars, it is probably true that net farm income would have been six billion dollars or less were it not for the farm support program. In assessing the desirability of entering into such businesses as fishing for menhaden as a food supplement in the production of hogs, the profit-minded entrepreneur must compare the cost of producing menhaden in a relatively unsupported economy with the cost of producing soy beans in a highly-supported economy.

The luxury food market, on the other hand, is one in which the basic needs of the society play a very small role in determining the profit and loss margin. Indeed, in the luxury foods, as in most luxuries, the profit margin is highly conditioned by the advertising program which is utilized by the purveyor of the commodity. Thus, when one contemplates such marine industries as the mariculture of lobster or of shrimp, the entrepreneur can assume that the market is far from saturated and the society will tolerate differentials in price several dollars a pound above the cost of acquisition. In such types of markets, the profit-making entrepreneur is much more interested in rapid and immediate capability than he is in optimizing the process by which the product is brought to market.

In the recovery of minerals from the ocean--oil, gas, extractive minerals, or hard minerals--a very complex set of artificially-developed social relations has been imposed which, in some instances, favor the acquisition of ocean minerals if profit, rather than engineering cost effectivity, is used as the measurement criterion. It is well known and universally acknowledged that the import quotas on oil greatly affect the ratio of domestic to imported crude which is utilized in the United States and, as a result, has made viable the option of recovery of oil from the Gulf of Mexico, from the offshore continental slope in the area of California, and even from the northern slopes of Alaska. In this instance, the differential in cost for a barrel of oil delivered in a port such as Philadelphia between oil which is acquired from off-shore oil reserves and oil which is imported from foreign fields is at least $1 per barrel.

Other artifacts of this industry include the depletion allowance and its effect on the total organization of the industry in order to optimize profits within the constraints and opportunities afforded by this allowance. Some very subtle differences must be noted. For example, in the extraction of minerals from brackish waters no depletion allowance is permissible when extracting minerals from the sea. On the other hand, the extraction of the same minerals from brackish waters which are found inland is entitled to depletion allowances of varying amounts, depending upon a rather detailed and complex schedule. Such a differential in allowance, of course, has an effect on the relative choice between extraction from sea water and extraction from brackish sources on land.

Significant social regulations also exist in the transportation industry. In this instance, for example, the requirement that the gasoline tax be earmarked for roads and highways and specifically precluded for use in alternative means of transportation results in an effective differential subsidy in favor of the road and highway system. This particular tax law may be conceived either as placing the highway transportation industry into the status of a tax-exempt industry or it may be conceived as putting it into the status of a normally-taxed industry in which roads and arteries are provided free of charge. Although there are subsidy provisions with respect to marine transportation, the nature of these provisions is such that the more common use of marine transportation utilizes the foreign flag and, as a result, a much more completely laissez-faire form of competition. When a choice is to be made between a sea transportation system or the alternative use of road transportation, the system planner must recognize this significant differential.

When the subjective human purpose revolves around health, recreation, environment, and other public services and public works, a most difficult assessment must be made as to the magnitude of resources which the society will desire to expend in each of these areas. The estimate is as much dependent upon estimate of political processes as it is upon any cost-benefit analysis. Indeed, the major problem in this type of analysis is the attachment of a value to such nonquantifiables as health, recreation, or environmental quality. Economists are now wrestling with this problem, but as yet no firm monetary equivalent of social or psychic income has been agreed upon.

Techniques for factoring into tradeoff analyses the effect of such subjective choices have been attempted. The approach which is most attractive to the author is that of Raiffa.[1] In this approach the analyst does indeed make a subjective estimate of the value of various nonquantifiables and, in addition, makes probabilistic estimates of the willingness of the society to pay for such nonquantifiables as a function of other variables in the problem.

The approach by Raiffa recognizes that future actions and their outcomes can legitimately be represented by subjectively-determined probabilities and that the value of the outcomes can be appropriately represented by subjective values. The introduction of these new concepts permits the determination of a spectrum of probable outcomes for a given development policy together with a spectrum of subjective costs which would be expended to reach these outcomes. Quite obviously, approaches of this type lead to the generation of alternate scenarios and spectrums of outcomes which are extremely difficult, if not impossible, to enumerate even with the aid of the digital computer. The judicious introduction of simplification is therefore required. The approach which the systems manager must take, therefore, is one in which he believes the society wishes to be fulfilled in order to discharge his responsibilities as a public servant or as a profit maker or as an eleemosynary executive. The political, social and environmental constraints which will be imposed upon the carrying out of the subjective human purpose should then be enumerated. These should be enumerated in terms of the broad generic type of hardware which could be employed to meet the subjective human purpose, the broad general

---

[1] Raiffa, Howard, "Decision Analysis," Addison-Wesley, 1968.

resource level at which the subjective human purpose can be or could be fulfilled, some subjective measure of the payoff for each level and type of system, and some assessment of the probabilities that the society will make the resources available for each of the broadly-specified alternatives. Such analysis, whether it is carried out rigorously or heuristically, will lead the systems designer to the most realistic initial formulation of the subjective human purpose which his system might ultimately fulfill. Such qualitative analyses are probably best carried out with small concept-formulation teams, con sisting of a mix of expertise from segments of the society desiring the system, segments of the society who will design and manufacture the system, segments of the society who will deploy and operate the system, and segments of the society which will provide the resources for the system. It is only after such initial formulation that one can seriously enter into more formalistic cost-effectiveness tradeoffs between specific hardware alternatives. Even then, continuous attention must be paid to the political, social and cultural acceptability of the configuration which is otherwise most "cost effective."

Recognizing now the limits on information and its rate of transfer, the large amount of accumulated information in the range of trained individuals which is incapable of transfer except in limited selectively-adapted situations, the highly subjective aspects of determining the society requirements and willingness to support the investment in the system, it must be realized that the evolution from initial identification of the system to its final specification and configuration is an iterative process. This iteration is most effectively accomplished initially by the collection of an identified set of experts who are maintained in seclusion for periods of time ranging from one to three months, and who are provided with real-time operational analysis inputs, as requested. The effectiveness of such a technique depends on the initial presumption that the individuals chosen indeed have the appropriate technical and management expertise and this selection process is critical to the initiation of a socially-useful system. If, as indicated, systems development is an iterative process, then we should expect that this group of experts will at one level of abstraction carry through the same general logic processes which will be reiterated at more detailed levels of abstraction by larger and larger sets of individuals as the development moves more toward design and the ultimate realization of specific hardware or specific training programs. We shall therefore carry through the generalized process which must be engaged in by such a collection of individuals, recognizing that this process must replicate itself in a more formalized structure.

The initial task is a first estimate of the magnitude of the parameters which delineate the subjective human purpose. This statement must be made in a probabilistic way since systems development philosophy and systems development cost will depend very highly on the expected performance and reliability of the system.

The generalized statement of the subjective human purpose, therefore, is as follows: It is desired that a system be designed, constructed and deployed and that this system be capable of operating in a specified manner (herein specify the manner) on "x" elements of the society, human or physical, (herein specify the elements) and change these elements into a new state or configuration (herein specify the configuration or state) and that such operations will be successfully performed "z" percent of the time (herein specify the value of "z") with conditional probabilities "p" (herein specify the value of the conditional probabilities and the conditions) that reliability "z" will be obtained. This generalized statement can be made more meaningful by a few examples, as follows:

a) Oil recovery system A, subjective human purpose: The recovery, processing and delivery to market of 10 million barrels per year of oil located on the continental shelf of California with delivery to the market at a cost not greater than $1.50 per barrel with a systems reliability such that 99 years out of 100 the quantity of oil will be delivered at the price (price adjusted to 1970 monetary value) with a sufficient demonstration of system performance that the probability of attaining 99 percent reliability (as measured on an annual basis) is 75 percent.

b) Containerized transport systems, subjective human purpose: The collection, transport and distribution of 500,000 containers of cargo between five European ports and five east coast Atlantic ports with the assurance that this rate can be maintained for at least nine out of ten years on the conditional presumption that this assessment is made independently of strike, war, or major natural catastrophe and that the performance of the system be so demonstrated that a probability of 80 percent can be assigned to the achievement of this reliability.

c) Pollution control system A, subjective human purpose: A system of surveillance, classification, communication and containment such that pollution sources from distressed ships or structures on the high seas which are generated on an average rate of not more than one per month, when averaged over a one-year time base, with the further presumption that twelve occurrences will occur randomly throughout the year. They will be detected within twelve hours after their occurrence and will be contained within 48 hours after their occurrence with a systems performance and reliability such that 95 percent of all such occurrences will be detected and contained within this time interval and with a demonstration of systems performance to the extent that it can be estimated that the system has a probability of 50 percent of meeting the systems goal.

These examples are, of course, purely arbitrary, have not been developed on the basis of the kind of reasoning which goes into the development of a subjective human purpose, and are very brief and incomplete, as compared with the kind of specification that will be required for any major system. However, it is important to note several characteristics. One is the generality of specification of the system in terms of hardware. In other words, a minimum of constraints on the type and form of hardware which will be constructed and designed must be placed by the specification of the subjective human purpose in order that the systems designer have widest range in selecting equipment to meet the desired goal. On the other hand, the early quantification of the design reliability of the system and the tests and evaluation which must be conducted to demonstrate the reliability and the quantity of elements to be handled by the system must be specified quite precisely at the outset, since the entire approach to the systems development and the magnitude of the cost are critically dependent upon the assigned values of these quantities.

Basic Sea Systems

Once the subjective human purpose has been established, the next task is to prepare a mission analysis as a basis for broad hardware choices. In order to accomplish this effectively, we may first note that most sea systems fall in the broad category of transportation systems where the subjective human purpose is the collection of an object or objects, the operation on, or processing of, the object or objects, and the delivery of the object or objects to specified locations in the face of obstacles, environmental and/or man-made. In order to prepare a mission analysis and mission profile for the generic class of transportation systems, the following quantizations must take place:

a) The quantity and type of objects to be collected;
b) The location of the objects, the precision of location and the type of location which must be made before the objects can be retrieved;
c) Constraints on retrieval (possible damage to object, environmental constraints which must be met);
d) Obstacles to retrieval (sea state, environment, predators, man-made interferences);
e) Time limits on collection, location and retrieval;
f) The quantity and type of processing of the retrieved objects which must take place;

g) The time, location and environmental limits on the processing which must take place;

h) The constraints on processing (permissible damage to objects, etc.);

i) Obstacles to processing which must be overcome;

j) The time constraints on transportation;

k) Obstacles to transportation, natural and man-made;

l) The routing and intermediate stops which may be required by the mission;

m) The quantity and type of objects to be delivered;

n) The addressees and the precision of delivery and the type of delivery;

o) The constraints on delivery;

p) The obstacles to delivery;

q) The time limits on delivery;

r) The probability of systems success, reliability of the system together with the demonstrated probability that the reliability has been achieved.

A second generic class of sea systems are transportation-interdiction systems where the subjective human purpose is to prevent the delivery of an object or objects to a specified address or addressees. For such a system the following quantization must take place before the broad outline of the system can be ascertained:

a) The quantity and type of objects which are expected to be interdicted or diverted;

b) The source of the object or objects, the location of the source, and the specification of location uncertainties which are required for the object to be interdicted at the source;

c) Requirements which must be met (damage, containment, modification or impediment) in order to interdict or divert the object at the source;

d) Obstacles to interdiction at the source;

e) Time limits on location or diversion or interdiction;

f) The character of the dissemination, transport and intermediate stages of processing between source and delivery;

g) The time of dissemination and transport and temporal limits on the processing which takes place;

h) Requirements which must be met to interdict or divert the object or objects during dissemination and transport phase;

i) Constraints on interdiction or diversion of the object or objects during this phase;

j) Obstacles to interdiction or diversion of the object or objects during this phase.

k) The terminal phases of the delivery process, the location of the terminals, the precision of location of terminals, and the specificity of location of terminals which is required if the object is to be interdicted during the terminal delivery phase;

l) Requirements which must be met (damage, containment, modification or impediment) in order to interdict the object during the terminal phase;

m) Obstacles to interdiction during the terminal phase;

n) Time limits on interdiction during the terminal phase;

o) The probability of system success and reliability of the system together with an estimate of the probability that the reliability has indeed been achieved.

A third generic class of systems falls into the category of information and surveillance, where the subjective human purpose is to collect, analyze and disseminate information about the state of some specified ocean phenomena (man-made or natural or some combination thereof). For such systems the following quantizations must take place before the broad outlines of the system can be specified.

a) The purpose for which information must be obtained;

b) The required success probability and time limits which are required to meet the objective;

c) The temporal requirements for the synthesized information (continuous, intermittent, cyclic, acyclic) together with time intervals;

d) The parameters which must be measured and the synthesized information which is required to meet the objective;

e) The precision and accuracy with which the parameters must be measured;

f) The spatial and temporal location of the phenomena to be measured and the spatial and temporal uncertainties which are permissible in parameter measurement;

g) The constraints on measurement of the parameters;

h) The obstacles to measurement of the parameters;

i) Information processing which is required for each parameter with limits on accuracy, precision and specification of the temporal nature thereof;

j) The requirements on information processing imposed by the synthesis. This includes simultaneity of measurement of the different parameters, relative spatial relationships between measurement, locales of the different parameters, etc.

Nearly all of the systems which have been described will fall into one of these three categories or will be comprised of subsystems which fall into one of these three categories. The delineation of the mission profile can therefore follow these generic models.

Subsystem Identification

The delineation of each of these requirements will result in a number of limitations on the choice of system, the definition of a number of system tradeoffs which will be required, and the identification of a number of options which may be left open. As a basic rule of systems efficiency, maximum flexibility should be a goal in making subsystem choices; that is, the choice of subsystem should be primarily functional and should provide freedom for the subsystem designer to optimize the subsystem with his own choices insofar as he is capable. It is then a question as to whether subsystem identification or specification of mission profile should be the next sequence in the systems process. In any event, the process is iterative, requiring an initial approximation of one or the other with progressive modification in order to reach compatible mission profiles and subsystem identification.

In this text, the choice of subsystem identification is therefore arbitrarily considered next.

No fixed rule can or ought to be given to subsystem identification other than the fact that it must provide the basis for "a clearly definable set of objects and trained individuals designed to operate in concert to perform a specific subsystem function." Since the subsystem's function must be carved out of the total system function, the basis for such carve-out is to a first order arbitrary. Subsystems may be chosen primarily on the basis of subsystem function, physical location, technology, or any recognizable classification, including the personality of management, sociological factors, economic factors, etc. There is no reason why hybrid classification may not be used as the basis for subsystem identification, and there is no reason why all subsystems should be of equal size or even of approximately equal size. The primary care that must be taken is the recognition of the point at which a subsystem becomes a component and the point, therefore, at which the principles of design and management indicated in this text are no longer applicable.

For pragmatic reasons the writer's preference for subsystem identification is a combination of function and technology. If a subsystem is chosen which has a diversity of technologies (e.g., heavy construction, electronics, and precision mechanisms) and if such diverse technologies are assigned to a management group specializing primarily in one technological discipline, then unsophisticated and archaic elements are liable to appear in the subsystem product in those areas of technology with which the group is unfamiliar.

On this basis, technologies which are deemed to be markedly different are as follows:

a) Heavy construction - low tolerance

b) Heavy construction - high tolerance

c) High performance structures

d) Electrical power - heavy

e) Electrical power - light

f) Chemical power (heavy, low performance)

g) Chemical power (high performance)

h) Nuclear power

i) Precision mechanisms

j) Machinery - heavy

k) Machinery - light

l) Physiology and psychology

m) Instrumentation - high performance

n) Electronics - computations

o) Electronics - communications

p) Hydrodynamics - seaworthiness and seakeeping

q) Hydrodynamics - stability and control

r) Aerodynamics - low speed

s) Aerodynamics - high speed

The list of basic technologies can be almost endless, and the aforementioned are given only as examples, but it should be noted that very frequently subsystems can be chosen on the basis of such characteristics. For example, it is somewhat classical in aircraft design to subdivide the airplane between Avionics, consisting primarily of sophisticated electrohydraulic systems, Air Frames and Structure, involving forces and moments on extremely high performance structures in aerodynamic environment, and Power and Propulsion, involving the design of complex high-performance chemical power machinery. When applying the same criteria to sea systems and after a review of a number of sea systems (see original list), the writer has concluded that one set of generic subsystems can be identified which fairly completely encompasses sea systems currently extant. These subsystems are as follows:

a) Platform subsystem or subsystems

b) Life support

c) Environmental sensing

d) Navigation

e) Communication

f) Object delivery and retrieval

g) Object processing

h) Command and control

i) Test and training

j) Terminals

k) Construction, maintenance and logistic support

The first of these, the platform, is of such importance that it is important to now identify three basic sub-subsystems generally associated with platform choice, as follows:

a1) Hull and structure

a2) Power and propulsion

a3) Stability

Mission Profile

After an initial specification of subsystems, the test of completeness of subsystem identification is most generally uncovered in the specification of the total mission profile or scenario. It is therefore vital in systems identification and design to specify such profile in meticulous detail. The mission profile or scenario is written in two phases: the operational and tactical phase, and the life cycle phase which includes construction, test, training, overhaul and maintenance. The scenario, therefore, provides a first cut on the minimum number of operational units which are required to complete the system. It also provides a first cut on the minimum number of operational units which are required to operate the system to meet the subjective human purpose. The operational scenario should detail the sequence and times which are involved. The life cycle phase will require an initial estimate of maintenance and overhaul philosophy. Tradeoffs in this regard are ordinarily undertaken since the total cost of the system depends heavily upon such choices.

An overly-simplified example of the mission profile is presented here for illustrative purposes. The example chosen is that of submarine rescue. The subjective human purpose for rescue is to provide relief to a distressed submarine rapidly, at least in a time scale shorter than the survival time of personnel on board. The distressed submarine may be at, or close to, its collapse depth, may be at extreme angles of repose on the ocean floor, and may have an elevated atmosphere as the result of partial flooding or inability to recompress the added gases associated with life-support or pneumatic systems.

Analysis of the geographic distribution of depths from which a rescue is possible and analyses of submarine deployments reveal that very many rescue vehicles deployed at a large number of bases would be required to ensure a short time scale unless the rescue-vehicle system were air transportable. Air transportability thus dictates the maximum size, maximum weight, and limits on mass distribution for major assemblies of the system. Air transportability further requires that the system be road mobile and that the equipments for road mobility are themselves capable of air transport.

Further analysis of sea state and ice in submarine-operation areas dictates that the rescue system have an all-weather capability. This is best achieved by use of a nuclear submarine as a mother craft. Therefore, submerged rapid-mating capability is another design requirement.

Since the size of the vehicle is limited by air-transport requirements, it is not possible to transfer even a major portion of the crew of a modern military submarine in a single transit, so the vehicle must be designed for rapid turnaround and maximum number of rescuees per excursion.

In summary, the rescue mission design requires all-weather capability to effectuate, with high probability of success, rescue from military submarines in normal operating areas, at collapse depth, at extreme angles of repose, at elevated pressure, and in a time scale which provides a high probability of survival.

The system required leads to the following operational mission profile:

1. Alert the rescue system;
2. Transfer the ready vehicle and support equipment from surface-ship tender to road-mobile transport;
3. Transport to predesignated airfield for loading aboard predesignated mission aircraft;
4. Load and fly to predesignated airfield proximate to the nearest port of embarkation;
5. Transport from airfield to port of embarkation;
6. Prepare nuclear submarine to receive rescue vehicle; assemble rescue system and piggy-back unload vehicle to submarine;
7. Travel to rescue site;
8. Establish navigation system and locate distressed submarine with respect thereto;
9. Travel to distressed submarine;
10. Inspect, clear hatch, communicate with, and investigate internal conditions of submarine;
11. Mate and transfer;
12. Repeat transfer until complete.

It should be realized that the mission profile has imposed the following constraints on the submarine rescue vehicle:

Depth	5000 ft
Length Overall	49 ft 4 in.
Maximum Beam	8 ft
Maximum Height	12 ft
Air Weight	62,000 lb
Maximum Speed	5 knots
Power	118 kwh

Endurance	3 knots, 12 hours
Rescuees	24
Maneuverability	Unaided mating in 1-knot current; mechanically aided mating in 3-knot current
Life Support	78 man hours in control sphere; 184 man hours each in mid and aft spheres

The life cycle mission profile would be as follows:

1. Design and construction of initial units;
2. Test evaluation of initial units;
3. Attainment of initial operational capability;
4. Scheduled maintenance periods for each individual rescue submersible;
5. Scheduled maintenance periods for the submersible support ship;
6. Scheduled training periods for the operational system;
7. Scheduled guaranteed operational availability period;
8. Scheduled major overhaul periods for support ship and operational vehicles;
9. Scheduled decommissioning and replacement periods for elements of the system;
10. Scheduled periods for system expansion and augmentation.

This very brief example is presented for purposes of completeness in this introductory chapter. A more complete example follows on page 56 as the first of the case studies which were conducted by students in the M.I.T. course on ocean systems.

Subsystem Interfaces

The identification of subsystems does not complete the system manager's initial responsibility. The extremely important task of definition of subsystem interfaces must now commence, and must continue until these interfaces are resolved in the most minute detail. The most egregious mistake which can be, and often is, made by system managers is the failure to properly and appropriately identify interfaces, thereby disregarding their importance. By the very philosophy of systems management espoused here, the system manager will not pay attention to the details of subsystem design, but he must pay meticulous attention to the specific hardware details associated with the interface. The student who is not familiar with interfaces will best understand this if he recognizes that the standard 110-volt home and office electrical wiring system constitutes a subsystem in which the interface with large numbers of unknown appliances takes place at the wall plug and wall socket. The definition of the spacing of tines, the tolerances on the plug, the voltage and amperage levels which are expected is one such set of clearly- and precisely-defined interfaces which permits manufacturers of electrical equipments of wide variety and scope to design their equipments with the full assurance that they will be able to operate on house or office electrical supply systems without the need for adapter modifications or modification of the power system. Similarly, the system manager must identify interfaces between all of his subsystems. While it is true that this is also an iterative process during the design and development phase, it is *not* true that the process shall remain flexible and delayed as in the case of subsystem design. *The earliest convergence on rigid interface specifications should be pursued with diligence and vigor*. These interfaces are initially defined by the parameters of:

a. Location

b. Weight

c. Volume

d. Forces

e. Moments

f. Electromagnetics (the nature of the electrical power and signals which flow across the interface; voltage levels, variations in voltage level, types of current, current levels, etc.)

g. Environment (the types and varieties of atmospheres, toxic elements and other physical vapors and materials which may or may not flow across the interface)

h. Heat (the heat flow across the interface, and the specification of the allowable interface temperature range)

i. Vibrations and Kinetics (the motions and vibrations to which the subsystem will be subjected at the interface)

It is a good management practice, which appears wasteful only to the naive and uninitiated, to employ three interface contractors at each interface. The system manager will or should employ his own interface monitor or contractor whose task is to detail the interfaces, to identify incompatibilities between subsystems at the interface, and to recommend resolutions to these incompatibilities. It should be the system manager's task to resolve these incompatibilities, going as they will to the heart of the systems operation, and to specify the general form of the interface. Each subsystem manager should be allowed and encouraged to have his own interface manager whose function is to examine the interface from the standpoint of the subsystem manager, to propose interfaces, to propose modifications of interfaces, and to ensure that the subsystem manager does not violate interfaces inadventently as the result of the design of the subsystem. This tripartite scheme of adversary-adjudicatory proceeding will do much to ensure that all interface problems are appropriately resolved, and at the same time maintain the discipline which is required to ensure that subsystems will work together when they are brought together in the final systems assembly and/or operation.

CASE STUDY: MISSION PROFILE

The first-cycle mission analysis and profile presented here was prepared for an M.I.T. Ocean Engineering Systems course. The author, LCDR. E. S. McGinley II, prepared the report as a course requirement for the degrees of Naval Engineer and Master of Science in Naval Architecture and Marine Engineering. LCDR. McGinley had extensive previous submarine duty, and served as Duty Officer aboard the U.S.S. RUNNER. The "delivery system" which is described here is purely the creation of the student and does not represent any system which is known to presently exist or has been proposed for existence. It was prepared within the extremely limited time available to a student within the total demands of the academic curriculum, but it is important to note that such time constraint is a characteristic of the management of major ocean systems and that the excess of time available to a system manager over that utilized by the student is not more than, say, 40 to 50 hours per week. To this extent, such a first-cycle analysis becomes highly realistic.

SUBJECTIVE HUMAN PURPOSE

A system for delivering up to 50 people (who may or may not be trained in SCUBA) secretly to a foreign shore anywhere in the Atlantic Ocean-Mediterranean Sea area. The length of time between delivery and retrieval would be no more than five days. Nondetection is of primary importance, particularly when within 1000 miles of the target area. Of secondary importance is cost, held, if possible, to less than $50 million per unit. There are to be four units capable of widely-separated deliveries simultaneously at all times. Maximum time for delivery should not exceed one month for any location within the delivery area.

The system is to be cost-effective, military or otherwise, so that secondary mission capabilities are to be developed in the system.

[*Neither student nor instructor is, by this example, advocating the construction of such a system. It is, however, an excellent case study which involves all of the elements of sea systems at a level which is practicable for the student.*]

MISSION ANALYSIS

I. Phases of Mission

A. Four major means of delivery

1. Spacecraft
2. Aircraft
3. Surface vehicle
4. Subsurface vehicle

Characteristics

1. Is impractical at present for up to 50 people and $50 million/ unit.
2. Has been used extensively with paratroops in the past. However, the element of covertness is seriously lacking, and its one advantage (speed) is not a requirement of this system.
3. Is lacking in covertness to a high degree, at least if it is used for the entire run from port to dropoff point.
4. Is probably the optimum in covertness to varying degrees according to its design. It would probably be possible to meet time and cost requirements of the system.

B. Delivery vehicle selection

1. From (A) the subsurface vehicle is selected.
2. This means that at some point close to dropoff, there must be a secondary means of delivery, due to draft limitations of a large submersible capable of carrying 50 passengers, and also due to its "overexposure" at this point, which requires surface or near-surface operations.

C. Mission phase selection

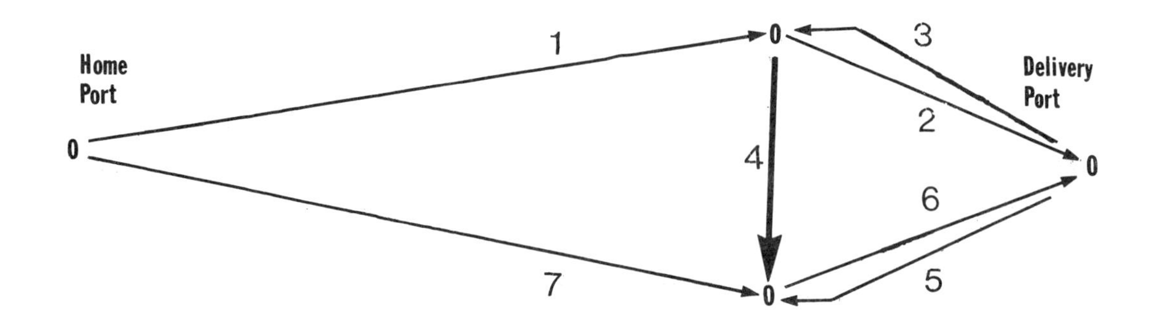

Figure 1-4.

Sub-Phases	Phase of Mission	Description
1	A	Transit Out
2,3	B	Secondary Vehicle(s) Delivery of Men and Return of Ship
4	C	≤ 5-Day Waiting Period
5,6	D	Secondary Vehicle(s) Sortie and Retrieval of Men
7	E	Transit Back

II. Quantization of Mission Analysis by Phases

A. Transit Out (Phase A)

1. Objects to be collected

a. Up to 50 men + submarine crew.

b. Food, fuel and other supplies for round trip (80 days maximum time, including roughly one month each way plus 5 days' mission time plus safety factor).

c. Secondary (piggy-back) vehicle(s). NOTE: secondary delivery system will hereafter be referred to in mission analysis as "piggy-back vehicle," although this does not rule out possibility of putting swimmers themselves in the water to swim to the target area.

2. Location of objects

   a. At designated "home" ports of vehicles.

3. Constraints on transit

   a. Subsurface transit when possible.

   b. In-transit repairs, etc.

4. Obstacles to transit

   a. Storms/sea state (may be serious constraint if any at-sea transfer is used).

   b. Submerged transit:

      1) bottom sonar, aircraft magnetic detection, mines, underwater natural or man-made (when close in to shore) obstacles, other ships.

   c. Surfaced (or snorkel) transit:

      1) same as submerged transit, plus radar and visual sighting from shore.

5. Time limits

   a. ≈ 30-day maximum to delivery point (including transit).

   b. As short loading time as possible in port (12 hours should be within reason).

6. Processing

   a. Food for 50 + crew for maximum of 80 days.

   b. Atmosphere for maximum of 80 days (if submersible has no need of surfacing).

c. Adequate space for crew and passengers.

d. Servicing of piggy-back vehicles on way to phase B, if necessary.

7. Distance for transit (see Fig. 1-5)

a. Home bases: assuming the four units are based on the east coast of the United States, at New London, Norfolk, Charleston, and Key West, respectively. This spreads them out as far north-south as possible and still keeps them within the Continental U.S., and at submarine supply bases.

b. Maximum one-way distance: 8400 miles, to Cape Horn.

c. Average one-way distance for 80% of area covered: 3600 miles. This includes everything except northern Norway, Africa south of 10°N, South America south of 10°S, and the Mediterranean east of Majorca.

d. NOTE: Distance traveled by piggy-back vehicle would be short in comparison with these distances, and therefore is ignored.

8. Routing intermediate stops

a. Fueling stops: only constraint is that, to fulfill original goal, no fueling will take place when mother sub is within 1000 miles of target area.

9. Objects to be delivered

a. Up to 50 men plus light personal equipment.

b. Piggy-back vehicle(s) for carrying passengers from dropoff point in to shore.

TRANSIT DISTANCES

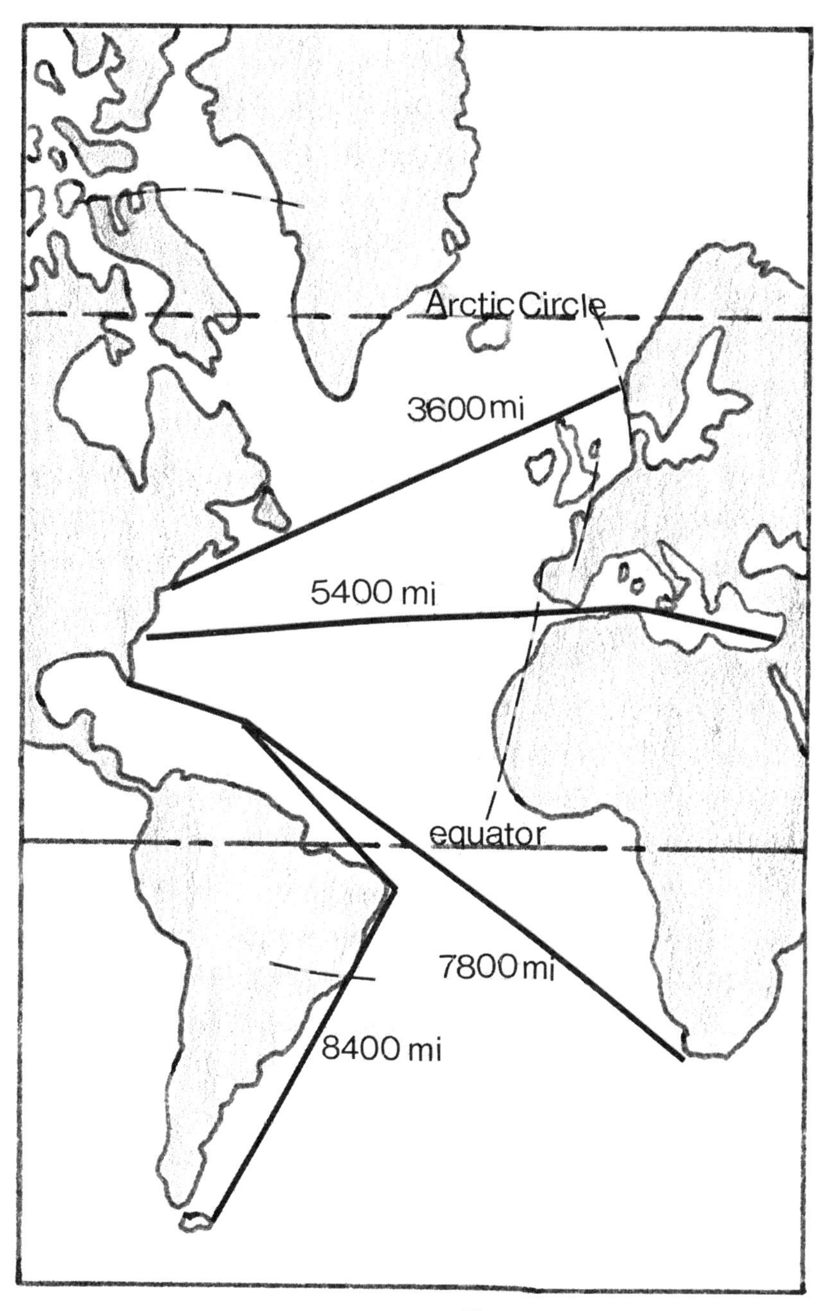

– – – – : Indicates 3600 mile ranges

Figure 1-5

10. Addresses and precision of delivery

    a. Addresses: dropoff point for piggy-back vehicle would be a short distance offshore (say, around 10 miles or less, depending upon needs and circumstances), as close as possible as the mother sub could get and still have a reasonable safety margin for detection.

    b. Precision: accuracy of at least 1/2 mile at dropoff of piggy-back vehicle(s) to ensure reasonably good placement of men on shore target.

11. Obstacles to delivery

    a. Sea state: depends upon piggy-bank vehicle/swimmer selections as secondary delivery systems. Sea state 5 preferred if possible (<30% of year), but problems here in this subsystem may be difficult.

    b. Wind: if piggy-back system operates on surface. Winds accompanying sea state 5, or about 18-20 knots.

    c. Others: same as obstacles to the transit itself.

12. Constraints on delivery

    a. Subsurface when possible.

    b. Personnel injury and vehicle damage.

13. Time limits on delivery

    a. dropoff (piggy-back vehicle) to be as fast as possible, depending upon vehicle selected. As separation time lengthens, so do chances for system detection.

14. Probability desired for system success/confidence limits

    a. As high as possible, using constraint of $50 million per unit.

[NOTE: This is a perfectly satisfactory way of specifying such limits; it does, however, virtually assure a cost of at least $\$50 \times 10^6$ per unit.]

B. Secondary Vehicle Delivery of Men (Phase B)

1. Objects to be delivered

   a. Up to 50 men plus small amounts of personal equipment.

2. Object location

   a. Piggy-back vehicle(s) leave(s) mother sub after having taken aboard the personnel.

3. Constraints: same as delivery constraints in phase A.

4. Obstacles: same as delivery obstacles in phase A.

5. Time limits: as fast as possible to minimize detection.

6. Processing

   a. Underwater delivery: supply of air to passengers until debarkation.

   b. Surface delivery: none.

7. Distance for piggy-back delivery

   a. Distance will be limited by available water depth (i.e., how close in to shore the mother sub may approach). A rough estimate may be obtained using data as follows:

      1) average continental shelf slope = 2 fathoms/mile.

      2) considering the mother sub should always work in water at least 100 feet deep (for submerged delivery or for ability to submerge quickly if delivery would be surfaced).

      3) from 1) and 2), then, average distance to shore from piggy-back dropoff point would be $\frac{100}{(2)(6)} = 8.3$ miles.

4) rounding off, and adding same, then, it can be said in general that the piggy-back vehicle(s) would have a one-way trip of about 10 miles, and counting the trip back, after dropping off the passengers, total distance would be 20 miles.

8. Delivery precision

   a. Error should be no greater than 1/2 mile from dropoff point.

9. Success probability: same as phase A.

C. Five-Day Waiting Period (Phase C)

1. Constraints, obstacles: same as phase A, parts 3 and 4.

2. Time limits: up to five days.

3. Processing: food, air for crew only.

4. Distance

   a. Such that sub would return to rendezvous in time.

   b. Sub has option to bottom if outer configuration permits.

5. Rendezvous precision: 1/2 mile, so that piggy-back vehicle(s) may be vectored in properly for pick-up of personnel.

6. Success probability: same as phase A.

D. Secondary Vehicle Retrieval of Men (Phase D)

1. Objects to be retrieved: up to 50 men plus personal equipment.

2. Object location/precision

   a. Surface vehicle: on beach, 1/2 mile.

   b. Submerged vehicle: off beach, beyond surf, and deep enough for submerged recovery, 1/2 mile (this based on swimmers and piggy-back vehicle(s) having crude portable sonar homing devices).

3. Remaining considerations: same as phase B.

E. Transit Back (Phase E)

This is merely the reverse process of phase A, and thus no additional requirements are added.

III. Possible Auxiliary Missions

A. Coastal defense sub: using troop spaces for torpedo stowage.

B. Oceanographic research sub

1. Empty troop spaces for scientists/equipment.

2. Possible escape trunk lockout chamber.

C. Target sub for naval exercises.

[A common error among system designers is the presumption that there will be or may be substantial periods of time in which the system may be used for auxiliary missions. If a system does indeed meet some human need, this is rarely, if ever, the case.]

MISSION PROFILE

ROUGH TIME FRAME	EVENT
-	1. Decision to send troops: notification of ship through respective commands.
1/2 day - 2 days	2. Arrival at pier of ship and troops (ship from operational area, troops from base):
1/2 day - 1 day	a. Loading of troops and equipment. b. Loading of ship's patrol supplies (food, spare parts, weapons, fuel). c. Briefing of key personnel.
	3. Departure:
4 hours	a. Exit from port to diving point. b. Submerge to transit depth as soon as possible.
	4. Submerged transit to target area:
Up to about 30 days	a. Fueling: must be arranged to allow no replenishment during time sub is within 1000 miles of target. b. Transit to point of piggy-back vehicle(s) dropoff. Can be to within water depth not less than 100 feet to allow complete submergence.
	5. Piggy-back vehicle(s) delivery of personnel:
3 hours maximum	a. All passengers into vehicle(s), or into lockout chamber(s) if final design utilizes swimmers alone. b. Final navigational posit and vectoring instructions to piggy-back vehicle(s). c. Release. d. Transit to beach and dropoff.
3 hours maximum	e. Return to mother sub; location visually, by infrared, or by sonar

ROUGH TIME FRAME	EVENT
	6. Waiting period:
5 days maximum	a. Sub retires some distance from area. b. Circles/bottoms/generally avoiding detection. c. Returns at time and place of prearranged rendezvous (up to 5-day period).
	7. Piggy-back vehicles' retrieval of personnel:
6 hours maximum	a. Departure, with latest navigational fix and vectoring instructions. b. Location of personnel (infrared or sonar) and pickup. c. Return to ship after similar locating procedures.
30 days maximum	8. Submerged transit back to home port: a. Same as transit out.
	9. Arrival in port: a. Offloading/debriefing.

[Although these times appear reasonable, it is a general rule that initial estimates which are based on the best and most detailed analyses are nevertheless much too low. Such analyses <u>should</u> be made and the results increased by factors which may vary from 20% to several hundred percent depending upon the novelty of the operation.]

LIFE CYCLE

Using the overhaul cycle cost graph below (Fig. 1-6), it can be seen that an overhaul cycle of about four years gives very little increase in submarine life cost over the three-year cycle, and is even less in cost than the two-year cycle.

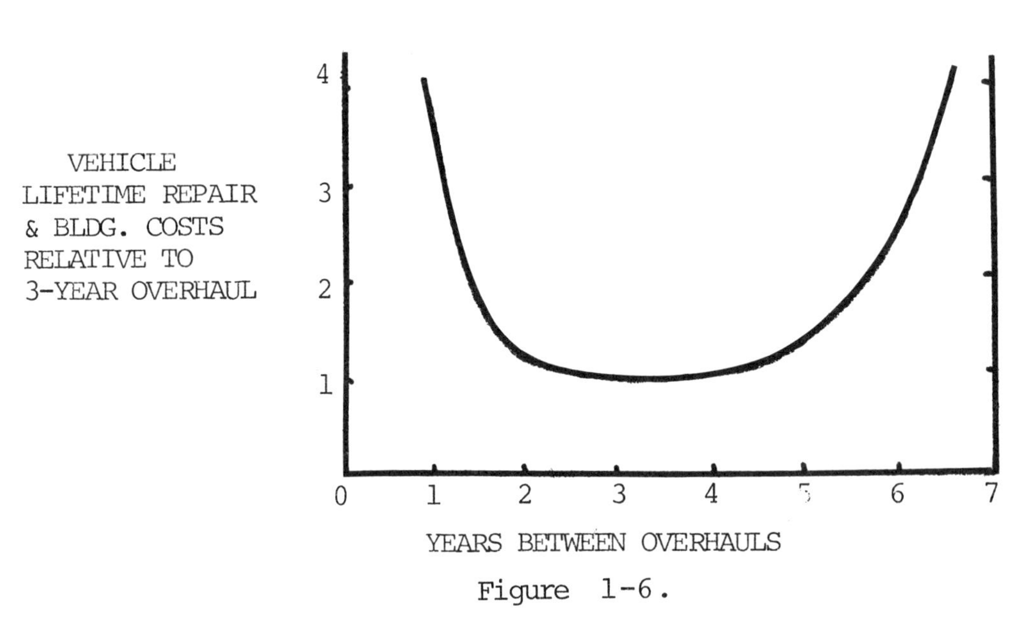

Figure 1-6.

Taking the four-year cycle, and iterating a few times, the following optimum life cycle chart was obtained, using a total of five mother submarine vehicles (the least number possible in order to have four on continuous alert);

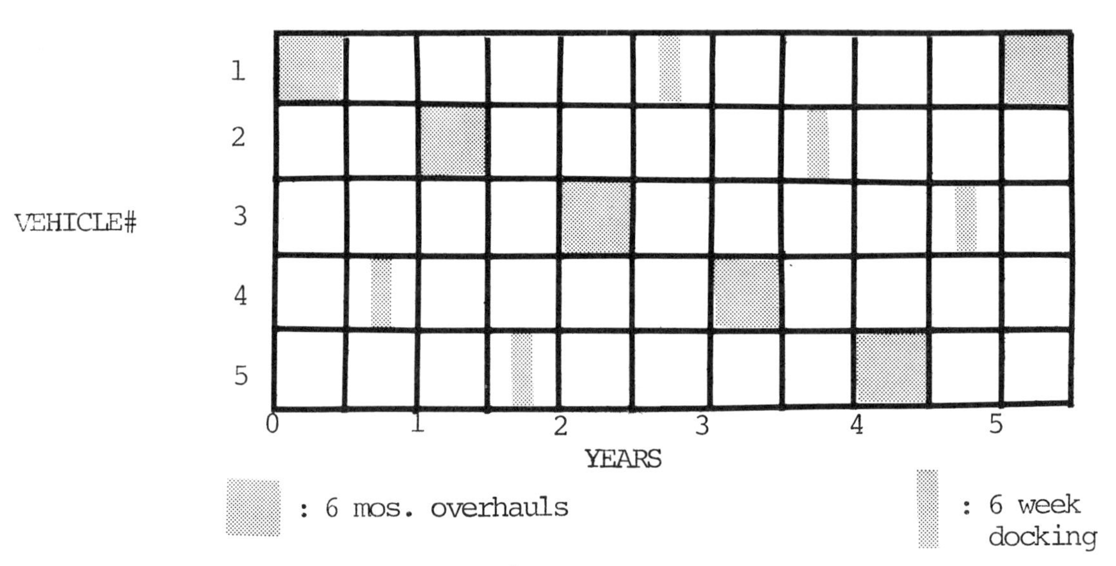

Figure 1-7

From this, typical yearly and overhaul cycles were worked out (see Fig. 1-8). Expected schedule between overhaul periods would be roughly 60% at-sea operations (utilizing secondary missions developed for vessel) and 40% in port for minor maintenance, on call.

Conclusions:

1) Five submarine "mother" vehicles.

2) Four-and-a-half-year staggered overhaul and short docking cycles.

3) With five overhauls per submarine, life expectancy would be 27 years each. This is not unrealistic with present-day state-of-the-art design and construction.

TYPICAL YEARLY CYCLE

HOLIDAYS (2)	OPS (7)	PORT (3)	OPS (6)	PORT (4)	OPS (6)	PORT (3)	OPS (6)	PORT (4)	OPS (6)	PORT HOLIDAYS (5)

JAN FEB MAR APR MAY JUN JUL AUG SEP OCT NOV DEC

Number of weeks indicated in parentheses.

Figure 1-8

ORIGINAL SUBSYSTEM DEFINITION

A. Hull and Structure (of mother sub)

1. Superstructure, tanks, and sail (if utilized).

2. Hull, hull accesses (including escape trunk(s)).

3. Internal compartmentation (i.e., bulkhead arrangements).

4. Piggy-back vehicle stowage facility (if utilized).

B. Power and Propulsion (of mother sub)

1. Main power plant (less shafting).

2. Associated auxiliaries.

3. Associated spaces design.

C. Stability and Control (of mother sub)

1. Main propulsor and shafting; main shaft seals.

2. Hovering system.

3. Control surfaces (up to but not including hydraulic rams).

D. Environmental Sensing, Communications, and Navigation (of entire system)

1. All equipment associated with above functions.

2. Associated spaces design, excluding control space.

E. Object Delivery and Retrieval (piggy-back vehicle(s))

1. Design of piggy-back vehicle(s), including only the following: hull and structure, power and propulsion, stability and control, and life support.

F. Command and Control (of entire system)

1. Sensor display arrangements.

2. Control systems operation and systems (up to and including hydraulic rams).

3. Associated control and auxiliary machinery spaces.

G. Life Support (of mother sub)

1. Crew and passenger berthing/messing.

2. Food and medical stowage spaces.

3. Air revitalization.

4. Air conditioning.

5. Associated equipment and spaces.

H. Test and Training (of entire system)

1. Coordination and detailed planning of mission profile.

2. Development of crew/passenger training plans and requirements. Coordination of this system's personnel requirements with various commands involved as a liaison function.

I. Terminals, Construction, Maintenance, and Logistic Support (of entire system)

1. Selection of pier facilities and determination of alterations required.

2. Tender alterations required.

3. Shipyards: special requirements.

4. Fleet replenishment ships: alterations required.

CHAPTER II

PLATFORMS

EPIGRAMS

System requirements should dictate platform requirements and not vice versa.

A ship is not the system.

Ship design is not necessarily platform design.

Any platform coupled to the sea surface is sea-state-limited.

Aircraft are not designed for occasional contact with the interface--submarine design should be like aircraft design.

General Purpose

The general purpose of a sea platform is to provide a survivable locale from which or on which the general function of the sea system can be accomplished. Where the mission requires spatial movement, then propulsion may also be a platform requirement and, where the power required for propulsion is the largest power requirement of the system, then it has been common to assign major power system responsibility to the platform designer. For this reason, the platform as a system is generally divisible into subsystems of hull and structure, and power and propulsion.

The nature of the platform is heavily dependent on the environment in which it will operate. Such environments include space, high altitude atmosphere, normal atmosphere, the air just above the free surface, the air/water interface, just below the free surface, the ocean, and the sea bed. From a design standpoint, therefore, we may categorize platforms (including a subset called vehicles) in the following categories:

a. Spacecraft and satellites

b. Missiles and rockets

c. Aircraft

d. Seaplanes and amphibious helicopters

e. Hover craft and hydrofoil planing craft

f. Displacement forms

g. Semisubmersibles

h. Submersibles and submarines

i. Near-bottom vehicles

    1. Near-bottom vehicles in shallow water

    2. Near-bottom vehicles which are semisubmersible

    3. Near-bottom vehicles which are fully submerged

j. Bottom crawlers

k. Tunneling or tunnel vehicles

It is immediately apparent that this quite large range of platform vehicle types is such that each has its own major technologically-limiting factor, and the choice of the particular environment in which the platform will operate will have a major effect on the entire sea system. It is important for the system designer, therefore, to be aware of the limiting problem or conditions for each of these classes of vehicles and to be aware of the effect on interfaces which the choice of one of these classes of vehicles will entail.

Atmospheric Vehicles (Missiles, Aircraft, Hydrofoils, Hover Craft)

Vehicles which operate in the atmosphere or outside the atmosphere and which are launched or retrieved from the land fall outside the scope of "sea systems." Vehicles which are launched and recovered either from the water or from other floating platforms are, however, greatly affected in their design by the characteristics of the launch and retrieval process.

The most obvious characteristic of aerospace vehicles is the efficiency of the airframe or missile structure. The very ability to utilize the environment requires a continuing design effort to increase payload and reduce basic vehicle weight. Three basic design trends result from these characteristics, as follows:

a. Optimization of structure to match maximum operating loads;
b. Optimization of control to ensure operation within design limits;
c. Meticulous care to eliminate resonant vibration modes within the operating regime.

A few reminders of specifics will reinforce these points. The trend in aerospace industry for increasing use of titanium, fiber-reinforced plastics (boron and carbon filaments), for sandwich structures, and for composites emphasizes the importance of strength to weight ratio in structural design. The use of altitude- and acceleration-sensitive automatic control devices in missiles, the use of radar for turbulence avoidance and the extensive use of operation limit warning devices are examples of optimized control to ensure operation within design limits. Hydroelastic problems which have been encountered in operating designs include divergence, flutter, vortex-induced vibration, and fatigue in aircraft. The failure of the early Comet aircraft and the flutter failures on the initial versions of the Lockheed Electra are classic cases in point. In both instances, the failure modes were subtle, requiring in the case of the Electra an initial springing of the engine nacelle due to hard landing before the flutter mode could be induced.

Missiles have also experienced destructive resonant excitations in operating design. These have included "sloshing" and Helmholz-type resonances in liquid chambers and rather spectacular combustion instabilities in solid propellant missiles. Resonant modes of these types are greatly aggravated by the inherent flexibility of the light airborne structures.

The attention which must be paid to the fundamental problems of aircraft and missile design is so great that the uninitiated are apt to overlook the fact that, for systems which land or are launched from the sea or sea-based platforms, the limiting design constraint is most probably during the launching and retrieval stage. Indeed, a number of otherwise operational craft have come to grief or have been canceled partly as a result of this constraint. These include the Seamaster, the Sea Dart and the Navy version of the F-111 (i.e., TFX or F111-B). Because of the previously-mentioned efficiency of structure, it is vital that the launch or retrieval pulse be appropriately distributed to the main structure and that its maximum value be controlled. Two major design features must be carefully developed in order to meet these objectives:

a. The force generator in the catapult or retrieval mechanism;
b. The prediction and control of relative motions between platform and aircraft.

Catapults and launchers have in general employed gas eject to achieve the appropriate pulse. Traditionally, the use of compressed air flasks which vent through a controlled valve into a plenum chamber has been employed. A major design hazard has been explosive dieselization resulting from the atomization and heating of hydrocarbon lubricants or other materials which have been inadvertently introduced into the design. A systems manager ought not approve the design of any novel launcher or retrieval mechanism without demonstrated assurances that this possibility has been fully investigated.

More recently solid propellant gas generators have been employed. The simplicity, reliability and low volume associated with them have on occasion outweighed the additional operating cost.

Retrieval mechanisms include the trip wire or other catching device which is in turn attached to a piston which compresses a gas, fluid, spring or combination thereof to achieve the deceleration. The required deceleration characteristics are achieved through the geometry of the piston and orifices and the elastic characteristics of gases, fluids and springs.

The estimation of relative motion between platform and vehicle is a problem which is sufficiently generic that it will be treated later under the category of object delivery and retrieval. It should be pointed out, however, that carrier landing systems have constituted and do constitute a major engineering design undertaking, and at present even the most effective are limited by some sea state beyond which it is unsafe to operate.

When the aerospace vehicle is intended to land in the water, then for such a vehicle the critical design factor is generally slamming or impact. The large forces of impact result from the fact that, as a body enters the water, not only is the body decelerated but the mass of water surrounding the body must be accelerated to body velocity at the boundary and elsewhere, in accordance with the requirements of fluid flow. To a first order this flow field may be determined on the assumption that the fluid is incompressible and nonviscous and is represented by ideal potential flow. It has been convenient for hydrodynamicists to equate the integrated effect of the mass of fluid accelerated with a hypothetical additional mass of the body which is known as the added mass. When such substitution is made, the force on a body entering the water may be expressed as follows:

$$\int_{to}^{t} \frac{\partial F}{\partial t} dt = mV_o - (m+m_a)V_t$$

Here m is the mass of the entering or reentering vehicle, m_a is the added mass, V_o is the velocity at the vehicle entry, and V_t is the velocity of the vehicle in the water. Differentiation of this equation (see Szebehely and Todd) yields

$$F = (m+m_a) \frac{\partial^2 Z}{\partial t^2} + \frac{\partial m_a}{\partial t} \frac{\partial Z}{\partial t}$$

The change of added mass with time is thus of vital importance in the total load. The added mass at any instant of time is, in turn, a function of the instantaneous configuration of the body geometry on the directional motion of the body. It can be determined if the potential flow field about the body is known or can be estimated. Techniques for estimating the added mass of a body in the vicinity of the free surface have been given by Landweber and Macagno (Journal of Ship Research, June 1967).

Potential flow analysis is not applicable for blunt bodies at the moment of impact. It should be apparent that ideal flow analysis will result in infinite forces at the moment of impact. In actuality, the compressibility of the water and the yielding and/or the deflection of the structure will place an upper limit on the actual force experienced. Pontoons or other reentry devices must therefore be carefully designed with respect to the coupling with the hull proper, the configuration of the pontoon on water reentry and, in particular, the configuration with respect to the local surface of the sea (i.e., the surface as disturbed by waves).

The most obvious example of a vehicle which must enter the air/sea interface is that of the seaplane. To date, nearly all of the seaplanes which have been designed have been constrained by the nature of the sea with respect to areas and times with which they can land. Thus, the majority of seaplanes will choose to land in lagoons or sheltered harbors, and on the open sea only in calm weather or under emergency conditions. Experimental craft, such as the Sea Dart and the Sea Master, never reached the production stage, partly due to technical difficulties associated with the landing process and the potential for catastrophe associated therewith. Other craft which are faced with the necessity for emergency entry on the air/water interface or occasional inadvertent entry on the air/water interface include such vehicles as hover craft. This is another type of vehicle for which the design problems of open-ocean operation have not as yet been fully resolved. Some craft must avoid inadvertent reentry, such as high-speed planing craft where the reentry conditions must be precisely oriented or disaster will ensue. Hydrofoil craft will also be required to reenter the free surface during emergency conditions when there is a power failure or when some failure involving the foils occurs.

A large number of system constraints will be introduced as a result of slamming and impact for systems which utilize platforms which reenter the water. Among these are:

a. The high probability that the maximum g-loading will be determined by the slamming requirement.

b. The high probability that the maximum vibration energies and the spectral distribution of vibration energies will be determined by the slamming and impact situation.

c. The orientation constraints on the platform and its appendages during reentry. These constraints include:

   1) Restraints on orientation with respect to direction of seaway and wind, and constraints on the attitude at reentry of the platform and its ability to perform maneuvers during the reentry phase;

   2) The nature and maximum value of burst power requirement. This may be a constraint if the decision to reenter is based on a go/no-go basis just prior to entry through the air/water interface or if launch runways are limited in size;

   3) The requirement for appropriate distribution of spray and splash during the reentry phase. Failure to provide for this can have vital systems effects as, for example, its effect on visibility in the cockpit, or on the ingestion of fluid in the propulsion system;

   4) The heavy configuration constraints on payload volume which will generally result if the shape of the hull (V-form) is constrained by reentry considerations. This

will be particularly true if there is a subdivision of space as a result of requirements of the main structural girder.

Vehicles which are fully coupled to the free surface, such as displacement craft, may or may not find that slamming and impact are the limiting conditions. If the vehicle is of sufficiently deep draft and size, then there will be very few, if any, conditions in which the forefoot of the vessel will be raised out of the water and require reentry conditions. When slamming and impact are not the major design limitations for vehicles fully coupled to free surface, then the major systems problem is associated with the maximum motions and orientations which the vehicle will experience. It is therefore important to stipulate the maximums of sea state and swell which will be encountered at various points in the mission profile and to determine for the vehicle the spectrum of motions which will result therefrom. Surprisingly enough, this has not been the general practice in the design of sea systems with which the author is familiar, but, in those instances where it has been accomplished, major system deficiencies were uncovered and major design changes were made, resulting in highly successful system performance. The major design barrier to the approach recommended here has been the difficulty in adequately determining the appropriate spectrum of motions and orientations. If it is recognized that overpreciseness in determining the spectrum is not required, then more approximate but satisfactory results can be obtained. Numerous theoretical spectra exist, including those by Darbyshire, Neuman and Pierson. These spectra are generally related to sea state and the systems manager ought to assure that an appropriate spectrum is chosen which is based on empirical evidence for the areas in the ocean in which he expects the system to operate. If then the transfer function for the platform in regular sinusoidal waves is known, then the spectrum of motions may be obtained by simple linear superposition. While the complete validity of such linear superposition may be questioned, enough empirical evidence exists (see studies by the Stevens Institute of Technology) to indicate that a reasonable design spectrum results.

When the motion spectrum is known and the speed and volume requirements have been established, then a first cut may be made on the platform configuration. This approximation will then determine the following interface characteristics:

a. The maximum orientations and motions which subsystems located aboard the platform will experience.

b. The number of cyclic variations of motion and the spectral distribution of the motions, which will be experienced by subsystems.

c. Configuration constraints on the volume, such as those imposed by structural configuration requirements of the main structural plates and frames. Foundation limitations imposed by the maximum deflections of hull and deck plating and structure and cyclic variations in deflection resulting from deflections of hull and deck plating and structure.

d. Frequency of occurrence of wetting and drying, of subsystems located in various portions of the platform.

e. Temperature regimes which will be experienced by subsystems exposed to the hull, noting that this temperature regime will be bimodal, depending upon whether the platform is continuously exposed to the water or is exposed to the atmosphere where a very markedly different range of temperature characteristics is to be experienced.

f. The major hydrodynamically-induced effects, such as cavitation, free-surface collapse, air entrainment, and vortex-induced phenomena will be determined by the choice of platform.

g. The availability of atmosphere and atmospheric oxygen to various locales and places of the platform.

h. The availability of over-the-side, over-the-fantail, or over-the-bow access for the transfer of goods and material.

i. The distribution of stack gases, noxious gases and other pollutants which are produced by the platform and its propulsion system and the configuration with respect to the platform.

If a conventional displacement hull has been chosen, the system manager need not have major concern with respect to the choice of hull form, optimized power for a specific form, structural configuration, stability, seakeeping characteristics, etc. These aspects of ship design have been so fully explored by the Naval Architect profession that many differences in design between differing naval architects are only nuances insofar as the performance of the total system is concerned. Indeed, a major problem of sea system designers is to force the rejection or modification of the conventional hull form to meet system requirements at the expense of conventional ship performance. For example, it is the author's view that most oceanographic ships should have a centerwell located near the position of minimum ship motions through which test and towed gear can be lowered or raised. The existence of such a centerwell can cause structural and speed and power inefficiencies as compared with a standard hull. These inefficiencies are trivial as compared with the hazard, the time and the inconvenience of raising and lowering gear over the fantail or over the side. Despite the excellent experience of those few oceanographic ships that have employed the centerwell (USNS Mizar, for example), very few oceanographic ships have been so designed.

Another example is the catamaran which pays very heavy structural penalties through the loads imposed on the bridge which connects the hulls together. Such vehicles are, however, highly mission-effective for drill rigs, where stability of the surface platform is important, or for small submersible tenders where the at-sea recovery of the submersible is a system requirement.

The system manager will therefore face his most difficult decisions when dealing with displacement forms which are highly nonconventional. Nearly all displacement forms are sea-state-limited; that is, there will be some combination of sea and swell in which the damage which will accrue to the structure is unacceptable or, in the alternative, safety of life and structure are imperiled. At the present time, it is hard to conceive of any class of floating structures or fixed platforms which have been designed to operate in the open ocean, which have not experienced complete catastrophe at high sea states. Each year the total number of conventional ships in the merchant marine which are complete losses as a result of stranding, foundering, collision, or other casualty of the sea is in excess of 200 ships per year. The loss rate on drill rigs and floating offshore platforms remains high despite improved design. For example, over a ten-year period, 1955-1965, at least twenty-three major accidents involving a loss in excess of one million dollars took place with respect to oil rigs of all types and varieties. The loss rate on unmanned buoys in the open sea is very high indeed and, in point of fact, there have been only a few instances in which buoys deployed in the open sea have had any appreciable life span. These buoys have included the Nomad buoy and the Monster buoy designed by Convair, the Isaacs buoy of the Scripps Institution of Oceanography, and one or two other unique designs. It is true that large displacement platforms, such as aircraft carriers or supertankers, can be shown to have acceptable motions even in the highest sea states of record. Nevertheless, such platforms have sustained severe damage in extremely high seas because of the breaking of waves over the superstructure. The slamming impact forces associated with such wavebreaking are extremely difficult, if not uneconomic, to design and as a consequence, although survivability is not at stake, effective operation, even for such large ships, is precluded in extremely high sea states. Table 2-1 is a table of the Pierson-Moskowitz Sea Spectrum which is characteristic of the type of seas which are experienced in the mid- and north-Atlantic oceans. In the north Atlantic, sea state 5 can be expected to be encountered about 10% of the time, whereas sea states 8 and 9 are very rarely, if ever, encountered. The systems designer must determine the sea state in which he will desire his platform to operate, the sea state in which he desires his platform to survive and, if the survival sea state is less than the maximum experienced sea state, he must provide some form of alerting or warning system together with some form of transportation system to ensure that the platform is removed to more sheltered waters

SEA STATE	SIG. WAVE HEIGHT IN FEET	SIG. RANGE OF PERIODS IN SEC.	PER. OF MAX. ENERGY IN SEC.	AVERAGE PERIOD IN SEC.	AVERAGE WAVE LENGTH IN FT.
0	0.10	0.34 - 1.09	0.87	0.62	1.31
0	0.15	0.42 - 1.33	1.07	0.76	1.97
1	0.50	0.77 - 2.43	1.95	1.39	6.57
1	1.00	1.09 - 3.43	2.76	1.96	13.14
1	1.20	1.19 - 3.76	3.02	2.15	13.76
2	1.50	1.34 - 4.21	3.38	2.40	19.70
2	2.00	1.54 - 4.86	3.90	2.77	26.27
2	2.50	1.72 - 5.43	4.36	3.10	32.84
2	3.00	1.89 - 5.95	4.78	3.40	39.41
3	3.50	2.04 - 6.43	5.16	3.67	45.98
3	4.00	2.18 - 6.87	5.52	3.92	52.54
3	4.50	2.31 - 7.29	5.86	4.16	59.11
3	5.00	2.44 - 7.68	6.17	4.38	65.68
4	6.00	2.67 - 8.41	6.76	4.80	78.82
4	7.00	2.89 - 9.09	7.30	5.19	91.95
4	7.50	2.99 - 9.41	7.56	5.37	98.92
5	8.00	3.08 - 9.71	7.81	5.55	105.09
5	9.00	3.27 - 10.30	8.28	5.88	118.22
5	10.00	3.45 - 10.86	8.73	6.20	131.36
5	12.00	3.78 - 11.90	9.56	6.79	157.63
6	14.00	4.08 - 12.85	10.33	7.34	183.90
6	16.00	4.36 - 13.74	11.04	7.84	210.17
6	18.00	4.63 - 14.57	11.71	8.32	235.45
6	20.00	4.88 - 15.36	12.34	8.77	262.72
7	25.00	5.45 - 17.17	13.80	9.80	328.40
7	30.00	5.97 - 18.81	15.12	10.74	394.08
7	35.00	6.45 - 20.32	16.33	11.60	459.76
7	40.00	6.90 - 21.72	17.46	12.40	525.43
8	45.00	7.32 - 23.04	18.52	13.15	591.11
8	50.00	7.71 - 24.28	19.52	13.87	656.79
8	55.00	8.09 - 25.47	20.47	14.54	722.47
8	60.00	8.45 - 26.60	21.38	15.19	788.15
9	70.00	9.12 - 28.73	23.09	16.41	919.51
9	80.00	9.75 - 30.72	24.69	17.54	1050.87
9	90.00	10.35 - 32.58	26.19	18.60	1182.23
9	100.00	10.91 - 34.34	27.60	19.61	1313.59

Table 2-1. The Pierson-Moskowitz Sea Spectrum

in time to avoid the destructive sea state. It cannot be overemphasized in this text that failure to so provide invites catastrophe and that catastrophe at sea is not an uncommon occurrence, and has been experienced with the most competent of designs throughout history, even to the present generation of supertankers.

The systems designer should by no means presume that his free surface problems are resolved through use of the semisubmersible or the submersible which is moderately close to the free surface. While it is indeed true that the motions of the semisubmersible will be greatly attenuated over that of the surface displacement form, nevertheless, substantial interactions with the free surface can occur. One of the most commonly overlooked of these interactions is that of shoaling. When the velocity of water which passes over the submersible is approximately equal to the $\sqrt{gh}$ where h is the depth of submergence, then the flow field must be critically examined. If, for example, the flow field passes from a value of $\frac{V}{\sqrt{gh}}$ of greater than 1 to a value which is less than 1, then the deceleration of the fluid will not take place without the formation of a hydraulic jump, a consequent loss of energy, and a change of the flow patterns associated with the submersed object will occur. For example, if a hydrofoil or pitching fin is exposed to this regime, then heavy vibrational patterns can be set up which result from the formation and decay of such hydraulic jumps or rollers. Another frequently-overlooked phenomenon is the net vertical lift force, which results from the passage of waves over a body. This is a Bernoulli-type effect which results from the increase in velocity induced by the wave. The effect is quite independent of the phase of the wave, since some velocity component will be present at all phases of the wave and therefore a net lifting force will be present. The magnitude of the lift force is proportional to the square of the velocity and varies with depth in proportion to e^{-2kh}. This variation in depth results in a very unstable situation. If, for example, a train of waves passes over the object in which the amplitude of the train of waves varies by a factor of 10, the force on the object which tends to lift the object up will vary by a factor of 100. In addition, if the object is controlled in depth by a ballast, then the object will be drawn to the surface unless extremely high rates of ballast movement are provided for. Many other subtle systems effects associated with such things as cavitation, air entrainment, surface biologics, will be introduced by the use of a near-surface semisubmersible and the systems designer should be very wary of such problems, and should require an accounting of these problems and a method of dealing therewith on the part of the subsystems manager.

The true submersible which operates continuously at depths of water in excess of approximately 300 feet and which operates in water whose depth is 1,000 feet or greater has the great advantage of being free of the many difficult problems of the free surface. Not many systems designers take advantage of this freedom, since it is all too easy to impose requirements which will force the submersible to operate or to be designed to operate on the free surface in heavy seas. This emergency design condition has not been considered a constraint by many designers, but in fact makes critical differences in the weight and type of structure which must be employed in the submersible. In an overly-crude analogy, this is somewhat analogous to designing an aircraft to make occasional contact with the land between airfields. Such a requirement would preclude efficient airplane design just as a free surface requirement now precludes and would preclude efficient submarine design. If the system designer, by strength of will, is able to ensure design conditions which will preclude operation of the submersible in all but the calmest sea states, then the entire nature of his design constraint is markedly changed. The chief problem which he will now face is one of hull and structure and, more importantly, one of optimization of payload with respect to displacement. In general, most equipments for submersibles have been placed within the pressure hull and the hydrodynamic and pressure-hull boundaries have been identical for the significant portion of the vehicles. As the trend toward deep submarines increases, and as the pressure-hull design becomes increasingly critical with respect to materials, fabrication and construction, there can now be observed design trends which a) maximize the amount of machinery and equipment that are free-flooded, pressure-compensated, or contained in a separate compression compartment outside the pressure hull, and b) minimize the volume and equipment which are enclosed in the hull. In both shallow and deep water, submersibles which move in the direction of either end of the spectrum can be identified; for example, in shallow water the wet diver transport vehicle is an example of a submersible which carries humans and has no pressure hull, and in deep water the cable-controlled underwater recovery vehicle is a vehicle, consisting almost entirely of free-flooded machinery and components, which is utilized for unmanned work on the ocean floor. On the other hand, the conventional military submersible has almost all of its equipment inside the pressure hull and in deep water such submersibles as the Aluminaut and the Alvin are examples of vehicles which have a high ratio of pressure-hull volume to vehicle volume. There is not evident, therefore, a simple principle for determining the envelope of a pressure boundary for a submersible, but there are evident some basic approaches and tradeoffs that would guide the designer to reasonable if not optimum design.

Component Location

Initially, it is necessary to identify those components which must be exterior to the pressure hull, free-flooded or pressure-compensated, and those which must be interior to the hull. Such classification itself is a function of technology and current states of the art; and, for numerous components, it will be many decades before there exists any change in the indicated categorization.

Exterior Components. In the category of components which must be exterior, the following can be identified:

Propellers

Control surfaces

Emergency ballast

Sonar transducers

Electromagnetic antennae

Running lights

Anchors

Manipulators and mating attachments

Towing, mooring, and docking attachments

Interior Components. Among components which must be interior, the following can be identified:

Man (at depths greater than 1000 ft or in vehicles having depth excursions greater than 30 ft)

Life-support and atmospheric control

Hand and foot controls, seats, and other physical connections to man

Voice communication transducers

Control signal generators

Vacuum tubes and other non-pressure-compensated electronics

Emergency power

Emergency ballast release

If the glass-hulled submersible is excluded from the categorization, the list can be expanded to include visual displays and lights.

Exceptions. The sparseness of this list is somewhat surprising until one recognizes that, whereas the majority of submersible components can be classified as "usually exterior" or "usually interior" to the hull, the applications are not mandatory, and exceptions can be found in current submersibles or submersibles being designed. For these components, individual analysis of each is needed to determine whether it belongs inside or outside the pressure boundary.

Weight and Volume Components

The weight and volume relationships on which choice of component location is based are relatively simple. If the component is located inside the hull, then the following relationship obtains:

$$\frac{\text{Weight of the component} + \text{Incremental weight of hull} + \text{Weight of ballast}}{\text{Incremental displacement of hull}} = 1.0$$

Here the incremental weight of the hull is the tonnage that results from the increase in hull size and the increased weight which results from the penetrations required to accommodate the component. In the case of shaft and shaft seals, for example, incremental weight may be decisive in the placement of propulsion motors.

The weight of ballast becomes an item where the geometric configuration of the component (a) is such that, when the component is enclosed in a pressure capsule, the volume it occupies results in a net buoyancy, or (b) requires the addition of stability lead (although this may be accommodated outside the hull).

Incremental displacement of the hull is the hull-displacement increase resulting from the larger hull size required to accommodate the component plus the displacement of compensatory or stability lead if located outside the pressure hull.

If the component is located outside the hull, then the relationship is as follows (W = weight, D = displacement):

$$\frac{\text{W of the component} + \text{Pressure compensator W} + \text{Incremental W of hull} + \text{Buoyancy material W}}{\text{D of the component} + \text{Pressure compensator D} + \text{Buoyancy material D} + \text{Incremental D of hull}} = 1.0$$

Here the weight of the component may be quite different from its weight in air as a result of the design changes required to adapt the component to a pressurized fluid-filled environment.

The incremental weight of pressure hull results from the penetration required by location outside the hull and is, in many cases of electrical penetration, quite trivial. Also associated with the penetration, the incremental displacement of the pressure hull can be neglected in most cases.

For a weight-limited system, comparison of the resulting total weight for each alternative can be the decisive factor in the choice; and, for a volume-limited system, comparison of displacements can be equally definitive. If the weight, volume, reliability, maintainability, configuration penalty, and cost of the component in its free-flooded or pressure-compensated configuration are the same as for configuration inside the hull, then the choice is determined by comparing the pressure-hull weight to displacement ratio (W/D) to the buoyancy-material weight to displacement ratio (W/D). If the W/D of the buoyancy is smaller than that of the hull, then the component should be located in the free-flooded spaces and vice versa.

The basic relationship tends to override other considerations to the point of either forcing the design to be a maximum pressure hull with minimum exterior components or a minimum pressure hull with maximum exterior components. As a consequence, the shallow-water vehicle for which low W/D can be obtained in the pressure hull with conventional materials has, generally, a large pressure hull, whereas the deep-ocean vehicle with a pressure hull of unconventional material tends to have a minimal pressure hull.

It is evident from the foregoing that the W/D of the hull and the W/D of the buoyancy material which are utilized in a submersible are critical in choosing the major design configuration of the submersible. The W/D values available to the designer are a function of the depth requirement, materials available, fabrication techniques, and ability to analyze the structure. To a first order, the choice of materials determines the ratio (W/D)of the hull and buoyancy elements for any given depth. It is an immediate oversimplification to base the first rough iteration of hull dimensions on the characteristics of the parent material under consideration. On the contrary, any material must be considered within the context of its fabricability, the type of structure with which it is compatible, the ability to be welded, joined, or penetrated, its degradation by flaws and imperfections, and uncertainties resulting from limitations in nondestructive methods of test (see Pellini). In a broad categorization, one can differentiate between the metallics and the nonmetallics as shown in Figure 2-1. It can be seen that a very substantial difference exists between the utilization of either high-strength steels or titaniums, and this includes the possibility of aluminum and the nonmetallics. The nonmetallics include fiber-reinforced plastics, such as fiber glass with epoxy resin, carbon-reinforced plastics, boron-reinforced plastics. It also includes ceramics and solid glass. At the present time very few, if any, nonmetallic hulls have been constructed for manned operation other than on an experimental basis. The systems manager should therefore be very wary of any proposal to provide a platform made with such material. Nevertheless, the developing state of the art for such materials is moving quite rapidly, and one can anticipate the increasing use of nonmetallics for unmanned vehicles at depths up to 1000 feet in the next

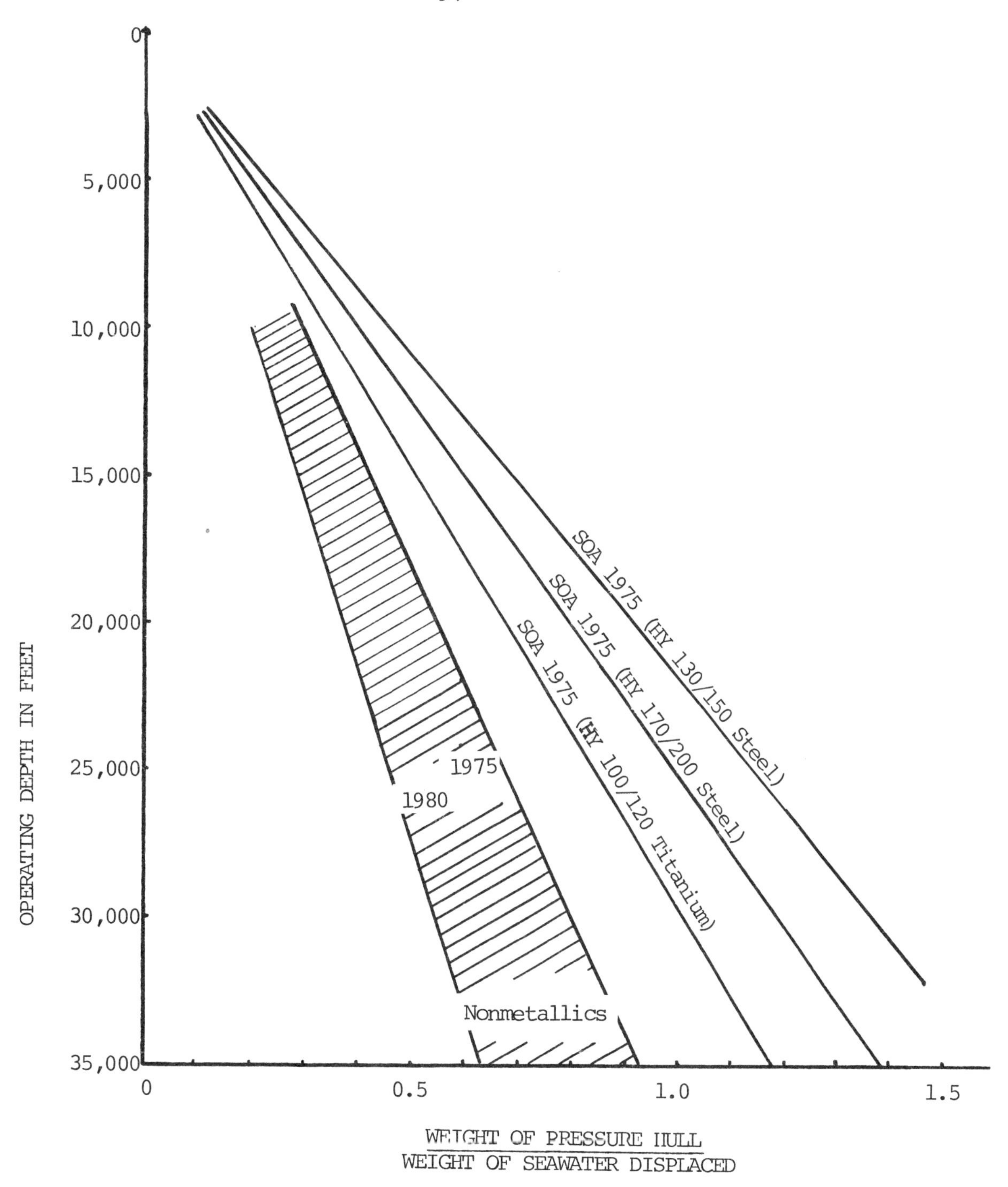

Fig. 2-1. Projection of maximum strength-weight characteristics of pressure hulls for Navy small submersibles.

few years and for depths up to 20,000 feet in the next decade.

For most submersible design, the limiting structural stress is that related to one of the modes of buckling, rather than the principle of compressive stress that derives from the normally-deflected structure. The ability to compute the depth at which a structure will buckle is extremely difficult, particularly in the case of optimally-designed metal structures, since such structures will be designed so that elastic yielding will occur at the same time that buckling becomes critical. Such structures must therefore be analyzed, utilizing theories of inelastic buckling which, in turn, limits the analysis to that of simple spheres or ring-stiffened cylindrical structures. Even here, many modes of buckling must be differentiated: the mode of generalized instability in which the structures buckle as an entity; the mode in which the buckling occurs between frames in an accordion style; and other higher harmonic types of buckling which can occur in the plates with many modes either around the periphery of the hull or between the stiffeners. If an optimal structure is required, and if operating conditions will result in effective utilization of the structure at or near the critical buckling point, then it is absolutely mandatory that large-scale (approximately one-half scale) structures be built and tested before a final design is determined. It is even highly desirable to test one prototype of a given class of submersible to destruction in order to determine the depth and the mode of failure which will actually be experienced. Even with the benefit of computation and test, the depth at which a structure will in fact fail is critically dependent upon the manufacturing tolerances. The systems designer will find that his costs are highly dependent upon the degree with which tolerances must be kept, and that rather large penalties can be paid if extreme efficiency is required of the structure. In addition, highly efficient structures must be protected against subsequent deformation, heat treatment, chemical action, erosion, and other extraneous effects which would have the potential for greatly reducing the depth at which collapse will occur.

A projected estimate of the W/D ratio which can be effectively realized for hulls of different materials and in different time scales is shown in Figure 2-1. Dates have been changed from earlier estimates in this chart to reflect some deceleration in the national support of research and development. The subsystem designer's choice in submersibles is thus limited at present by the state of the art in materials fabrication and design. While rapid progress can be made in these areas, the systems manager should assure himself of "alternatives" which, though less optimal, will meet system schedules and time.

The safety alternatives in submersible design are chiefly in provision of buoyancy materials which can compensate for the additional weight of hull resulting from misestimate of hull weight or from the need to shift in selection of hull material or structure. Two approaches to buoyancy have been taken: use of a) materials whose net density as compressed is less than sea water and b) buoyancy structures.

In the case of buoyancy structures, the problems of buoyancy are the same as those of the pressure hull except that the structure can be a simple sphere without hatches or penetration, and the size can be considerably smaller. As a result, materials for buoyancy structures will be available several years ahead of materials for manned pressure structures. The disadvantages of buoyancy structures arises from the requirement that the material and structure has a reliability as great as the pressure hull or, in the alternative, that the vehicle be safe against mutually induced implosion. In addition, the spherical configuration adds another inconvenient shape for packaging within the overall vehicle envelope.

The alternative to buoyancy structures is buoyancy materials. Liquids, solids, and gases are suitable for consideration. The number of fluids with a density significantly lower than that of water is very limited. Indeed, the only fluids which have been employed are the higher fractions of petroleum, ranging from kerosene with a W/D of 0.72 to gasoline with a W/D of 0.60. When structure is added for containment of these fluids, the effective W/D of gasoline rises to about 0.80 - 0.85, and fluids having a W/D greater than this value are practically ineffective.

Solid materials which can both resist compression and remain neutrally buoyant also are very limited. In fact, except for solid liquids such as paraffin, solid materials actually are multiple-pressure structures in which the failure mechanisms are so random and statistical that only an approximate collapse depth can be ascertained.

The most effective and useful materials for buoyancy have proved to be the syntactic foams, which actually are composites of glass spheres in an epoxy matrix. The spheres are sufficiently small (400μ) that the failure of one or more will not induce mutual implosion or seriously affect buoyancy. Syntactic foams having a depth capability of at least

20,000 ft are available in W/D of 0.7, and they should be available shortly as proved material at 20,000 ft with W/D ranging as low as 0.65. The great advantage of the syntactic foams is that they can be cast in place and can fill voids of arbitrary and unusual shape. In design, therefore, a syntactic foam can be employed as a highly flexible "negative lead."

Subsystem Interactions

The selection of a submersible of long endurance relieves the systems manager of a great many problems associated with the free surface. These include freedom from interference by ice. As a result, large submersible tankers for Northwest Passage operation are now under consideration. These not inconsiderable advantages are coupled with a number of constraints on other subsystems. These include:

a. The nonavailability of atmospheric oxygen;

b. The nonavailability of celestial or electromagnetic navigation references;

c. The loss of the atmosphere to guarantee a fixed ambient pressure within the vehicle;

d. The limitation on use of hull or piping for thermal reduction and heat rejection;

e. Limitations on numbers and types of penetrations through the hull. For example, main propulsion shaft seals at depths in excess of a few thousand feet will be difficult if not impossible to accommodate.

f. Limitation on access and stowage imposed by hull shape and configuration. For example, the need for all nonpermanent equipments to be removable through a limited size hatch (e.g., 25 inches). Such a limitation has proven critical; for example, in the design of swimmer transfer vehicles for which the hatch diameter compatible for entry by swimmer fully equipped with SCUBA (approximately 40 inches) is not compatible with structural efficiency of the transfer vehicle pressure hull.

The shape of the hull will also prove troublesome, the ideal structural sphere being in general incompatible with space arrangement for electronic cabinets or instrument modules, or orientation of the human body, etc. On the other hand a pressure hull which would be optimized for volume considerations is not yet amenable to calculations for buckling depth or mode.

g. The structural requirements of the hull will also limit the ability to locate and place structural mounts and hard points. In some instances this has necessitated the use of an internal "bird cage" which permits the spreading and distribution of normal loads to the hull and minimizes localized shear.

Near Bottom Vehicles and Platforms

The selection of a platform or vehicle which operates at or near the bottom quite obviously introduces additional design problems. The system designer should again differentiate between platforms which operate in depths of less than 200 feet (approximately), those which operate in deeper water but have a connection to the free surface region, and those which operate in deep water with no effective interaction with the free surface.

The most difficult design problems are associated with platforms and vehicles which must operate for protracted or continuous periods of time in unsheltered waters whose depth is 200 feet or less. The system designer will be well advised to avoid such platforms if there is any other mechanism for accomplishing his mission (i.e., the use of tunnels, helicopters, the construction of breakwaters, operation in deeper water, providing means for rapid dispersal of platform and associated equipments, etc.). If the designer is so unwise as to elect to construct a permanent or semipermanent installation, then a most rigorous and searching review of the following design factors must be made before the subsystem manager is authorized to proceed with that alternative.

a. Annual variation in local spectrum of sea and swell with detailed accounting for reflection, refraction and amplification factors due to topography of the bottom, the shoreline configuration and other local hydrographic anomalies. Estimates of magnitude and frequency of worst-case conditions (i.e., 10-year, 50-year or 100-year storms) must be included.
b. Annual variation in local distribution of wind velocity and direction together with pattern and types of storms which may be experienced. Estimate of magnitude and frequency of worst-case estimates are also required.
c. Local distribution of current velocity and direction and estimates of maximum values and vertical distribution.
d. Local patterns of sediment movement under prevailing winds and currents and patterns of sediment movement during characteristic storms.

e. Estimates of the interaction of the proposed platform with each of these environmental phenomena. Particular attention should be paid to shoaling phenomena.

f. Local estimates of ice and surface debris and estimate of the effects of environmentally-induced interactions between the platform and the ice and/or surface debris.

g. Estimate of hazard to navigation and probability of collision under adverse environmental conditions.

h. Full investigation of properties of sea bed, including bearing strength of sediment and subsoil potential for scour and probable scour patterns.

In coping with this environment the designer may have the choice of a vehicle or platform which is neutrally buoyant and proud on the bottom, one which is barely resting on the bottom or one which derives substantial support from the bottom. In the first instance, the major problem is stationkeeping, particularly in heavy weather. This is compensated by the ability to move from station during adverse conditions. In the latter instance, fixed position is maintained at the expense of design for heaviest sea state. Unfortunately, the soil mechanics problems of the sea bed are not as yet on a fully scientific basis. A number of major platform failures have occurred because of fluidized movement of the soil under one or more supporting legs of the platform with consequent tilting and/or structural failure. If it is recognized that the upper layers of sediment on the sea floor are, by and large, porous media, then the designer will assure that the foundations are installed below the lowest level of anticipated scour and below the lowest level of significant hydrodynamic movement on the basis that the sediment is a porous media exposed to current and wave velocity at the sea bed.

The intermediate case includes such vehicles as wheeled submersibles, tractors and sleds. It is the author's experience that, where currents and wave effects are small, the use of wheels or traction is a decided disadvantage. This is true since the control forces and resistance for a hovering craft are small and wheels and tractors are subjected to sharp impact forces in uncertain terrain. Where there is a significant current, however, wheels and tractors may be the only effective means to cope with drag-induced forces and moments due to cross-currents. The initial commercial use of manned submerged bulldozers in Japan is a forerunner of a substantial class of such vehicles which will have the requisite mobility to leave the area in the presence of storms and, at the same time, to operate in the face of tidal currents and waves.

When operating near the bottom in deep water, vehicle and platform problems are much more simplified unless an umbilical cord or other attachment to the surface is required. Three configurations of significance to the systems designer can be identified.

a. The submerged vehicle is by itself negatively buoyant and is held in position with respect to the bottom by a float (ship, buoy, platform) at the surface.

b. The submerged vehicle is by itself positively buoyant and is held in position with respect to the bottom by a clamp affixed to the bottom. The clamp is in turn connected with a ship or buoy at the free surface.

c. The submerged vehicle is by itself neutrally buoyant and is coupled to the free surface with minimal forces necessary to sustain the umbilical.

The first configuration poses the most difficult mooring-line and umbilical problems. The analysis of the cable tension is well nigh intractable because of the dynamic behavior of the cable under end-point loadings which, on the one hand, are coupled to the free surface and, on the other, to a dynamically free body. The avoidance of resonant modes is extremely difficult. Where the operation is confined to a relatively small area, mode b is far superior and, where the operation is moving in a continuous track, operation in mode c is preferred.

Subsystem Interactions

The major subsystems interactions for vehicles operating near the bottom results from the character of the bottom and include:

a. The introduction of silt into machinery and condenser systems;

b. The reduction in visibility due to disturbance of sediment;

c. Reflection, refraction and masking of sonar signals due to the nature of the terrain;

d. The introduction of benthic biologics and biologic fouling agents into subsystems;

e. The exposure of externally-mounted subsystems to possible impact.

PLATFORM

(POWER AND PROPULSION)

EPIGRAMS

The most effective sustained fusion power plant is the sun.

Atmospheric oxygen makes a winner out of fossil fuel.

Society tabus are the limiting factor on the use of nuclear power systems.

The fuel cell would be competitive if someone could build one. Closed-cycle submersible systems would be competitive if someone would build one.

Power and Propulsion

The clear advantage of submerged platforms and vehicles from the standpoint of forces and moments would dictate their more extensive use, were it not for the problems of power and propulsion. The availability of solar energy and atmospheric oxygen have to date made surface-operated power systems the least costly and therefore the most extensively employed. Thus, in conventional discussions of power and propulsion the greatest concentration will be on comparison of such atmospheric systems as the gas turbine, the diesel engine and the steam generator. The extensive literature on these systems belies the fact that from a systems standpoint the tradeoffs between these atmospheric oxygen competitors are often only marginal. The systems designer ought to recognize therefore that his major choice dilemma will arise when, for other systems reasons, he desires to use a power source that is not air-breathing. He must therefore take a more generic view and recognize that the choice of a total power plant begins with the selection of an energy source, and then the conversion of the energy source to desirable transmitted power.

Table 2-2 lists the relative ranking of energy sources in terms of the Btu's per pound of fuel and oxidizer (or energy storage material). Environmental constraints have inhibited or advanced the use of most of the elements in this table, and they should be understood by the systems designer.

Only one fusion power source now exists, i.e., the sun. Its use is greatly constrained by the diffusion of solar radiation and the large arrays which are needed to absorb adequate quantities of solar energy. Its diurnal and seasonal availabilities are also significant factors in its effective use. When combined with the rotative energy of the earth and moon, the sun can be conceived as the prime source of energy for conversion into the mechanical energy of wind, wave, tide and current. Despite the sole reliance on these power sources of only one hundred and fifty years ago, many system and component designers fail to employ them today. A notable exception in modern ocean system design is the SKAMP buoy which utilizes a computer-controlled "sail" to maintain the buoy in an approximate position in the ocean. The author anticipates that the effective design of offshore floating islands will similarly require sailing appurtenances to maintain approximate position in the ocean to avoid large, expensive and power-consuming propulsion devices.

The most prominent new source of power is fission, either with nuclear reactor or from isotopic sources. The primary reduction from the idealized efficiency results from the reactor shield. For example, the weight of primary and secondary shielding for 16 kilograms of U235 is approximately 1500 tons (see SNAME TRANS, Vol. 65, 1957, Rickover et al.).

Table 2-2

Energy Storage per Pound - Fuel Oxidizer Systems

Fuel Oxidizer	Energy Stored Btu/lb
Fusion (ideal)	$\sim 10^{11}$
Fission (ideal)	$\sim 10^{10}$
Fission (weight of shield included)	$\sim 5 \cdot 10^{4}$
Hydrocarbons in atmosphere	$\sim 2\text{-}3 \cdot 10^{4}$
Hydrogen + Oxygen (ideal)	$\sim 6 \cdot 10^{3}$
Hydrazene + Hydrogen Peroxide (66%)	$\sim 2 \cdot 10^{3}$
Hydrogen + Oxygen (cryogenic storage)	$\sim 1 \cdot 10^{3}$
Hydrogen + Oxygen (compressed gas)	$\sim 1 \cdot 10^{3}$
Storage Battery, Silver Zinc	$\sim 10^{2}$
Storage Battery, Lead Acid	$\sim 5 \cdot 10^{1}$
Flywheel (200,000 psi)	$\sim 2 \cdot 10^{1}$
Elasticity of Spring	$\sim 5 \cdot 10^{-1}$
Capacitor Storage	$\sim 10^{-3}$

The resulting energy storage in Btu per lb is thus only about ten to twenty times as effective as the use of hydrocarbons where atmospheric oxygen is available. The net result is that tradeoffs which would ordinarily be in favor of nuclear power for surface ships can be quickly dissipated if special premiums for crew, special arrangements for harbors and other artificial constraints are introduced into the cost of nuclear-powered ships. The U.S. Maritime Commission merchant ship Savannah has not been an economic success due in large measure to these artificial constraints. More recently, Japan and West Germany have constructed nuclear-powered merchant ships in the hope that they will not be so inhibited and the system advantages of nuclear power can be realized.

When considering nuclear power for submersibles, an additional order of magnitude advantage over hydrocarbons is introduced because of the nonavailability of atmospheric oxygen. Figure 2-2 is a comparison submarine power plant system specific weight in lbs/Kw-hr as a function of mission time in hours. As would be expected from the foregoing figures, the nuclear-power system becomes superior to the idealized fuel-call system (for which the major part of the weight is fuel and oxidizer) at somewhat between 100 and 200 hours.

This time superiority is such that the system designer ought to look quite closely at nuclear power in the next decade for both surface and subsurface applications. If the pressurized water reactor is employed, the nuclear reactor has the additional potential of high safety in the ocean environment. The excellent shielding properties of water are such that for all catastrophes except explosion (prevented by the use of

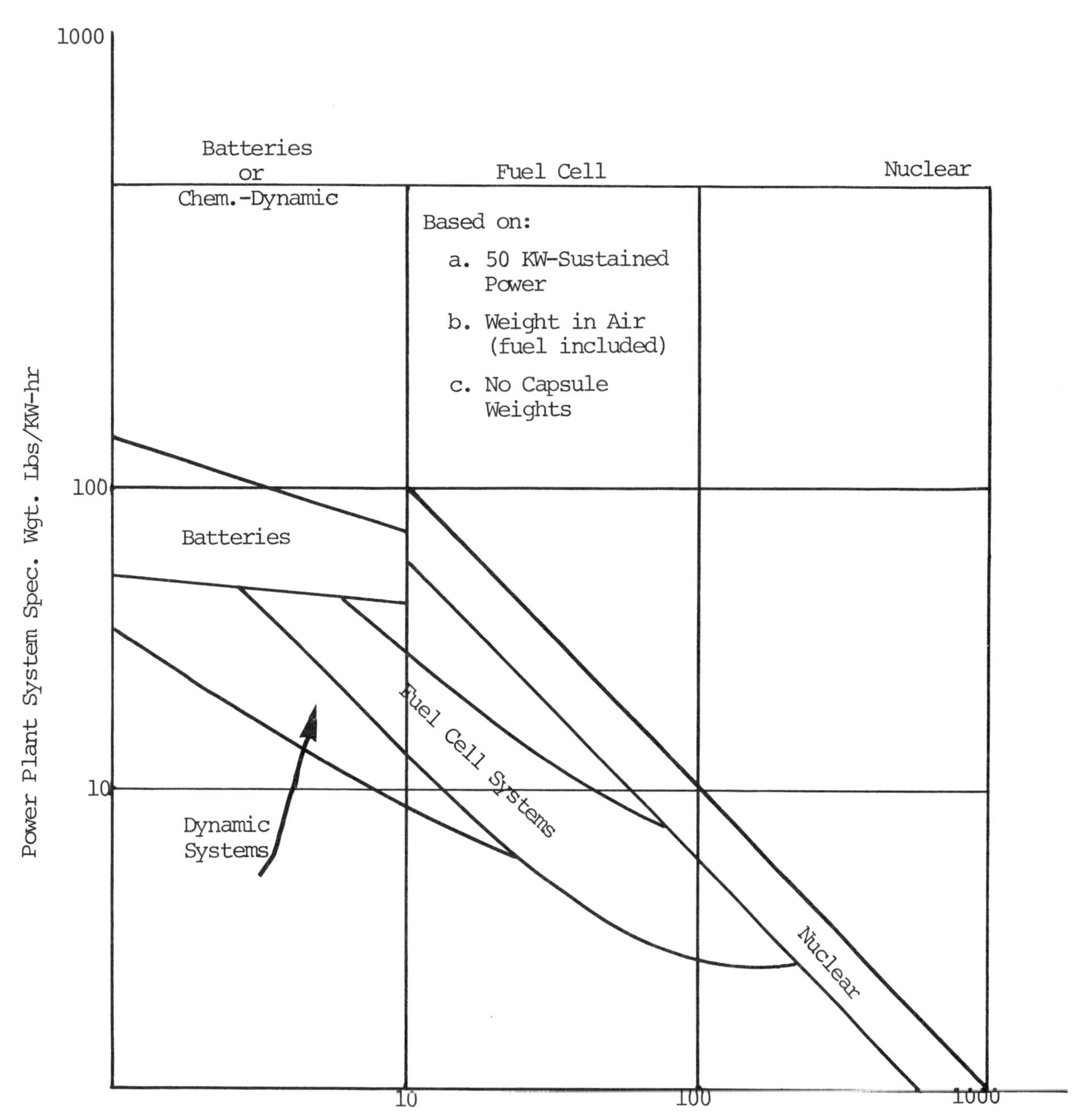

Fig. 2.2. Power plant system weight versus fuel endurance

pressurized water) the radioactive effects will be localized even in the event of a breech of the shield.

When atmospheric oxygen is available and when nuclear power is denied by social constraints, then the power plant choice devolves quickly to the burning of hydrocarbons and becomes very much a function of platform characteristics. In general, the large displacement ships have employed the boiler-fired steam turbine or the slow-speed diesel engine. At the high-speed small-craft end of the spectrum the maximized version of the gas turbine has come to dominate the power plants. In the intermediate range of high-speed patrol craft a healthy competition is developing between high-speed diesel, steam and gas turbine and hybrid combinations thereof. The system manager should be cautioned that to date the prime design consideration for the gas turbine has been for aircraft. The marine environment is markedly different, particularly from the standpoint of corrosive environment, splash, spray and water entrainment. In addition, the characteristic mission times for aircraft are short as compared for most marine missions.

When atmospheric oxygen is not available as in the case of high-altitude missiles as well as for submersibles, then fuel-burning systems lose much of their competitive advantage. In stochiometric ratio eight times as much oxygen by weight is required as is fuel if hydrogen and oxygen are the ingredients. Unfortunately, both hydrogen and oxygen are volatile and hypergolic and it is necessary that both be contained in some manner, either cryogenically in isolated Dewars, compressed in bottles, or chemically combined in a stable or quasi-stable configuration.

Once a decision is made to eschew the atmosphere, other oxidants such as fluorine or chlorine are available and many fuels besides hydrogen can be considered. Figures 2-3 and 2-4 show the heat of combustion of the various elements with oxygen and fluorine as a function of atomic number. From a combustion standpoint, the most attractive are beryllium and oxygen and lithium and fluorine. Hydrogen and oxygen and hydrogen and fluorine are only 60% as effective. However, the extreme difficulty of packaging and the toxicity of fluorine have mitigated against its use and the high cost and toxicity of beryllium and lithium have similarly prevented their use as fuels except in exotic application. Quite some development has taken place with respect to the use of boron as a fuel and it is often found as an additive to hydrocarbons. In rocket applications, where both volume specific impulse and density specific impulse may be important, many of the "exotic" fuels and oxidizers can be found. In submarine applications, however, the efficiency of the fuel is quite secondary to the efficiency of the oxidizer and, above all, the safety and compactness of the container. As a general rule, the system designer ought to insist that the hydrogen be safely locked in a hydrocarbon, such as kerosene or fuel oil or other stable hydrocarbon. For extreme applications, where weight is important, hydrazine (N_2H_4) or ammonia compounds might be considered. Far more crucial to system performance is the manner in which the oxidizer is carried. One of the most efficient chemical containers is hydrogen peroxide (H_2O_2). Unfortunately, this is quite unstable and is always diluted with water. Mixtures of H_2O_2 and H_2O, where the hydrogen peroxide is as high as 80%,

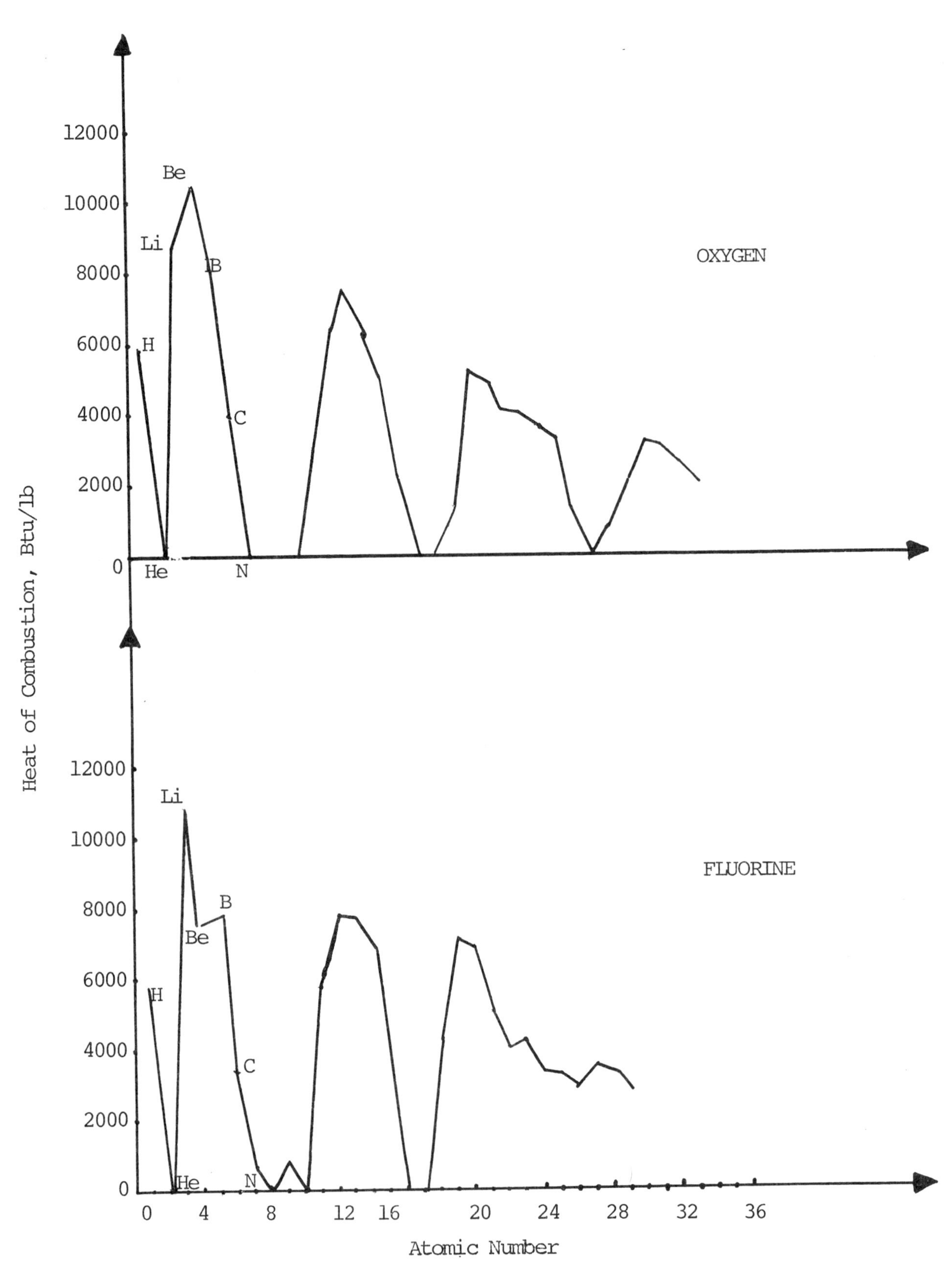

Figures 2-3 and 2-4. Heat of combustion of element plus oxidant versus atomic number.

have been employed in rocket propulsion, but would be unsafe for submarine application. As shown in Table 2-2, the combination of N_2H_4 and 66% H_2O_2 would yield an efficiency of 2×10^3 Btu/#. This is further reduced by the weight of hydrazine and hydrogen peroxide tankage. In the author's view this represents the clear upper limit for submarine energy storage and the designer should be wary of systems that purport to approach this limit.

The storage of oxygen in tanks under high pressure results in a weight of about 0.5# per cubic foot of atmospheric oxygen including the weight of the bottle. This would result in an efficiency of approximately 1000 Btu/# if hydrazine were employed as fuel. This is roughly competitive with utilization of the heat of fusion of molten salts such as lithium hydride. Proposals for submarine energy sources based on thermal storage of this type have been made and ought to be feasible.

It is noted that all of these storage systems are superior in energy storage to the silver zinc and lead acid storage battery. At the present time, however, the silver zinc or lead acid battery dominates the small submersible power sourve. This dominance results from the form in which the energy is extractable, i.e., low-voltage direct-current electrical energy. The use of combustion or heat storage requires the additional inefficiency of conversion from thermal energy to either mechanical or electrical energy. The possibility of development of the fuel cell at modest to moderate power levels (25-500 kw) has been mesmerizingly attractive. Such developments are in process, but fuel cells for underwater application (except at low-power levels) are not available today. The alternative technique for converting heat is through one of the thermodynamic cycles (Rankine, Stirling, Brayton). The high ambient pressures associated with submergence preclude open-cycle systems because of the high back pressure and consequent thermodynamic inefficiency. Closed-cycle systems are feasible, but have seen limited development. The author has a personal preference for the hot-air Stirling cycle as feasible for underwater applications capable of meeting environmental constraints. The Stirling cycle, requiring as it does only the application of heat to a closed-cycle process fluid and a heat-rejection sink, comes close to the ideal Carnot cycle. Heat can be obtained by direct burning of oxidizer and fuel or through use of thermal storage. As indicated in Figure 2-2, such a system ought to be competitive with other power sources now in use. For longer endurance, the acquisition of additional oxygen by occasional visits to the surface is a feasible solution for commercial application.

In summary, one of the most crucial decisions to be made when selecting a submersible for a platform or vehicle is the form of energy storage, since it will determine the time interval at which the submersible must return to the surface and consequently the need for and the type of surface support.

Power Requirements and Propulsion

The selection of energy source and its conversion to some form of mechanical or electrical energy is but the first step in the power train. The final stages of power and propulsion must also be ascertained and the energy source matched. This, like so many aspects of system design, is also an iterative process. Independently of propulsion, most platforms will have a substantial power requirement. These include the hotel load for life support, power for electronic equipments and sensors, power for tools and manipulators, and power for control. The first two will inevitably generate a requirement for emergency power and for highly-controlled and regulated voltage. It can be expected, therefore, that a number of battery systems which float on the main power supply will be required, together with a number of static inverters (dc to ac connection) or motor generator sets. It is an unfortunate tendency on the part of systems managers to attempt to reduce the number of such subsystems and to combine power requirements. One of the major design problems in any system is the interference, power drain and loss of regulation that occurs when many subsystems operate from the same power supply. It is therefore vital that appropriate reservoirs of power be provided that are highly decoupled from the prime source of power and whose utilization is the sole province of a single component or subsystems manager. Similarly, the reduction of power transmission requirements will avoid difficult and costly interference problems.

A companion systems problem is that of cooling and heat rejection. At first analysis, the immediate availability of ocean water for cooling and the generally high thermal conductivity of the hull suggest that the problem is relatively easy. This advantage is often more than compensated for by the compactness of equipments within the hull and the generation of heat by electronics equipment and other energy dissipators throughout the platform. It is again an unfortunate tendency to assign the total cooling responsibility to a single subsystem. In such an assignment, the misestimate of cooling requirements by a single subsystem designer can have repercussions throughout the system. It is vital, therefore, that the assignments of the hull for cooling requirements, assignments of inlet areas for condenser scoops, the assignment of space for cooling ducts and fluids, etc., be an early part of interface identification. It is also a paradox of system design that each subsystem manager should have the option of responsibility for his own cooling requirements.

Most prominent of platform power requirements is generally that for propulsion and control. Here three speed regimes can be identified: a) above 5 knots, b) 2-5 knots and c) below 2 knots. These differentiations are based on hydrodynamic regimes, the first being one in which the mean forward velocity is great enough to permit use of foils and other lifting surfaces to generate forces; the third regime is one in which the generation of force by means of purely viscous means (jets, wakes, shear flows) is not prohibitive and in which the generation of flow by lifting surfaces is difficult. The second regime is the intermediate one for which some hybrid approach may be required.

The system designer should approach his selection generically recognizing that within each generic choice there will be many tradeoffs and requirements. From the generic standpoint, however, the systems designer should note two fundamentally different mechanisms for force and moment generation in a fluid. The first of these is by generation of flow fields which are to a first order describable by the equations of ideal potential flow, i.e., by sources, sinks, doublets, and vortices. Whenever possible, this is the most desirable from a power standpoint, since the only energy losses result from the induced drag associated with the generation of circulation and the skin drag associated with the wetted area of the hull.

The second of these is by the generation of dissipative flow fields which are not describable by the equations of ideal potential flow. Jets, wakes, secondary flows and sheer flows are in this category. Drag chutes are a prime example of force generators which are highly dissipative of energy.

In the moderate high-speed regime above five knots, the use of jets, wakes and secondary flows are highly inefficient, if not prohibitive, and the ability to exert force in almost any direction through the appropriate use of hydrofoils precludes the need for this type of mechanism. The most effective propulsor which has been employed to date is the screw propeller. This device, since it employs a lifting foil appropriately oriented to the flow for optimum design conditions, results in extremely high hydrodynamic efficiency (n=0). At or near the free surface, the primary problems associated with the propeller are those of cavitation and of wake-induced vibrations. The cavitation problems can be resolved by utilizing large wheels with slow rotative speed, by the use of multiple propellers, or by the use of counter-rotating propellers. Some attempts have been made to resolve cavitation by the use of ducting which decelerates the flow entering into the propeller, but such ducts generally introduce a drag penalty and have been extremely difficult to design both efficiently and to achieve the desired purpose. When the vehicle on which the propeller is mounted is towing another object or has some system configuration which results in an extremely high drag acting on the vehicle, then the efficiency of the propeller will be greatly decreased unless there is a duct around the propeller which would normally be associated with the ideal potential flow about the otherwise unimpeded vehicle. The most common form is a type of nozzle known as the Kort nozzle generally employed on tug- and towboats. Another mechanism for resolving problems of towing or artificially-increased drag is through the use of propellers of variable pitch in which the pitch of the propeller can be adjusted to take care of the appropriate load which is required without change in speed of the towing vessel. For vehicles above five knots, control forces for maneuvers can usually be easily achieved through the use of lifting surfaces such as rudders or planes mounted at appropriate points along the hull.

Below two knots of forward velocity, appendages such as rudders and planes are ineffective, and of even greater concern is the fact that many of them will operate with force reversal as a result of the greater contribution of viscosity to the flow pattern around the appendage and the effect of flow around the hull proper. This force reversal phenomenon, which is shown schematically in Figure 2-5, is extremely difficult to handle in automatic control systems. At these low speeds, then, the alternative is the use of jets to generate force in the appropriate direction or the use of revolving wheels of hydrofoils which move through the water rapidly enough to generate the appropriate flow field and thus to produce adequate force by circulation. The simplicity of jets recommends their use in the extremely low-speed situations and, in particular, for such applications as maneuvering or mooring or berthing a large ship in a confined harbor. The alternative approach is embodied in two basic configurations: a) the Voigt-Schneider propeller and b) the Haselton tandem propeller. In the first instance, a flat rotating disc is located

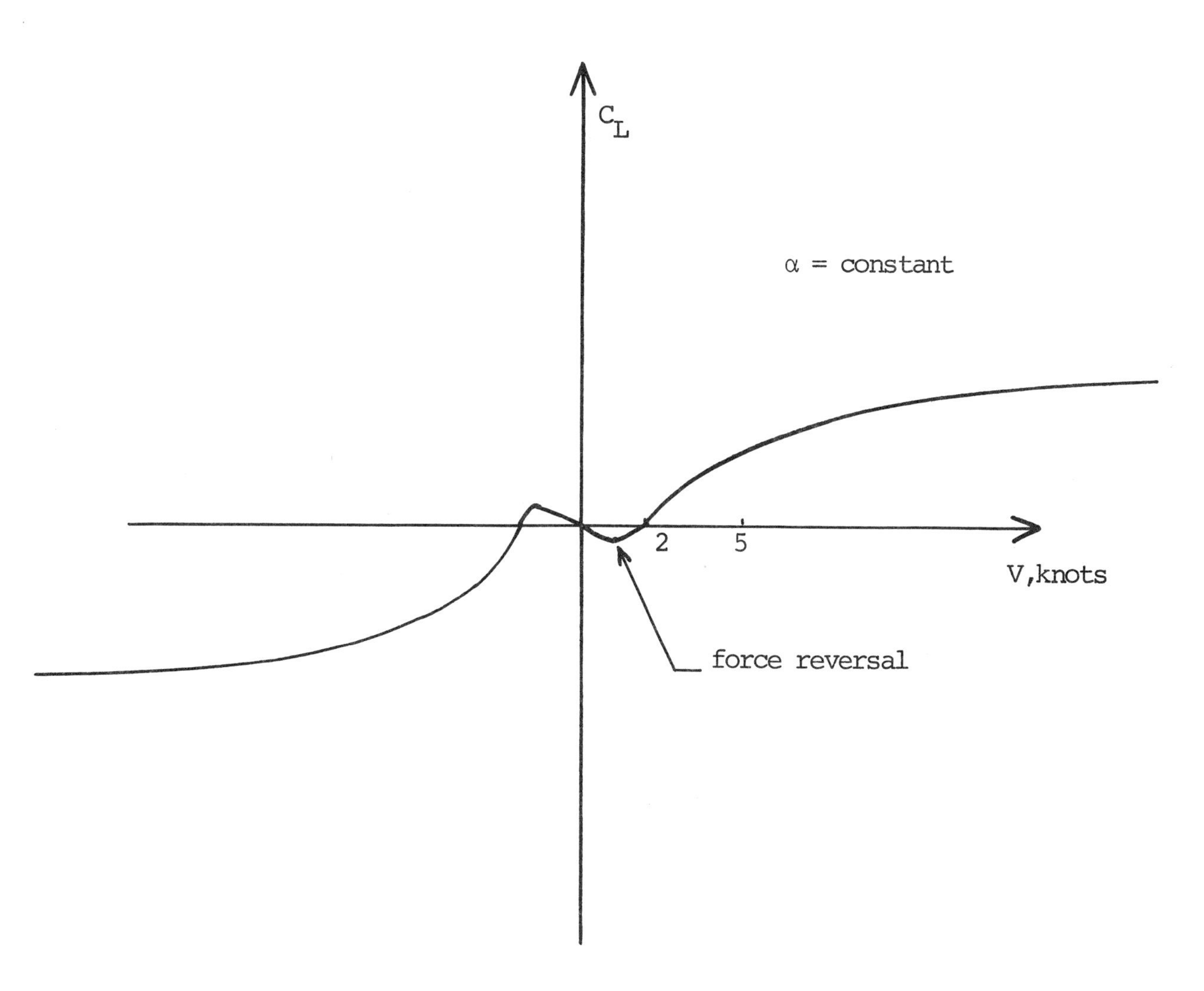

Figure 2-5. Lift Velocity Curve for Force Reversal

flush with the hull on which are mounted a series of projecting adjustable pitch hydrofoils. The rotation of the disc moves the hydrofoil through the water and the adjustment of the pitch in a cyclic manner permits the generation of force in any direction in the plane of the disc. The Haselton tandem propeller consists of a ring which encircles the hull both fore and aft of a submarine-type vehicle. Small foils are projected from this ring around the periphery of the hub. These foils are adjustable with respect to angle of attack cyclically in essentially arbitrary phase. This again permits the generation of force in any direction and, with the two hubs, permits the generation of moment about any axis. The result, then, is a full six degree of freedom. The mechanical complexities of these vis-à-vis the more usual configuration of jets has at present precluded their use in all but experimental installations. This may indeed be more a cultural phenomenon than an engineering phenomenon, since most comparative analyses of six-degree-of-freedom vehicles would indicate that the parts count for a Haselton tandem propeller or of a Voigt-Schneider propeller will be not greater than and indeed is often less than the parts count for the equivalent six-degree-of-freedom jet system. At speeds in excess of two knots and less than four knots, the system designer is faced with a great deal of difficulty since both the jet system and the idealized lift system are quite inappropriate. With a forward velocity in excess of two knots, the issuing jet tends to be accelerated by the flow and by virtue of that acceleration to behave much more like idealized flow and therefore to generate less force than it would as the fully-separated jet. In this intermediate regime, such schemes as the Voigt-Schneider propeller, the Haselton tandem propeller, or the use of rotatable propulsion pods, seem to be the only acceptable solution. The system designer should note in this instance that sensitivity of choice depends on very small differences in the velocity of operation and, indeed, on very small differences in the estimate of the currents which the system will face in the environment during its normal operation. The mission profile should therefore be prepared in meticulous detail before final propulsion and control system choices are made.

CASE STUDY

A DISASTER-RELIEF SYSTEM
(A First Iteration)

The case study which has been chosen for the platform selection and for power and propulsion is that which was done by Lieutenant Charles N. Calvano and Lieutenant Bruce Luxford. This system is a disaster-relief system whose purpose is to provide forces for long-term relief of disaster areas, such as has been experienced recently in Peru. A critical aspect of their system was deemed to be power and propulsion, since it would be anticipated in a disaster that power sources would not be available in the disaster area and that the provision of such power sources would be an important part of the total mission. In analyzing their system, Lieutenants Calvano and Luxford found that the power constraints very quickly led them to a single unique system whereas the alternatives on ship and platform were quite large and required more narrowing down. Their case study exemplifies, therefore, the problems that the systems manager has, and which he initially anticipates, that his major problem area will be one having multiple tradeoffs and his minor problem area being one with the least number of tradeoffs. Quite the contrary is true. As a general experience, the systems manager will find that where he has a major problem, at best two or three unique solutions will be available and at worst only one solution will be available.

I. SYSTEM SUMMARY

A. Subjective Human Purpose

Provide for the deployment of task forces for temporary or long-term relief of disaster areas accessible from the seacoasts and Great Lakes of the continental United States.

B. Mission Description and Résumé of Mission Analysis

System is intended for relief of "nonpolitical" disasters--thus excluding war-caused damage (e.g., nuclear attack).

Types of Disaster Expected

1) Earthquake (disaster area airports unusable)
2) Hurricane/Typhoon
3) Tidal Wave
4) Floods (airports unusable)
5) Earth or mud slides (e.g., Los Angeles area)
6) Large fires
7) Other severe weather damage (heavy snow, ice, etc.) (airports unusable)

Disaster Magnitude to Be Designed for

1) 1000 seriously-injured persons
2) 50,000 homeless (may include less serious injuries not requiring hospitalization or doctor care)

Other

Alaska excluded

From the mission analysis, the supplies and other facilities needed at the disaster site will require a minimum of 280,000 ft^3 of volume.

At least some medical aid must be on scene in four hours or less, with all aid arriving within 48 hours.

System Success Probability must be:

1) 70% with 75% confidence for quick-arrival contingent

2) 90% with 75% confidence for 48-hour contingent

IIA. MISSION ANALYSIS--COLLECTION, TRANSPORTATION, DELIVERY REQUIREMENTS AND CONSTRAINTS

I. Objects to Be Collected

A. Quantity and Types

1) 1000 pints whole blood of various groups

   a) Up to 250 pints may be replaced by packed red cell units

2) 20 Doctors

3) 50 Nurses

4) Crew of Disaster-Relief Vehicles (DRV), if not already aboard

5) 30 Technicians/Recorders (T/R's)

6) 100 Red Cross Volunteers (RCV's)

   a) Up to 75% may be drawn from volunteers at disaster scene

B. Location

1) Blood at Red Cross centers and blood banks in "homeport" of DRV

2) Doctors, nurses, T/R's and at least 25 RCV's in "homeport" of DRV

3) DRV crew from Navy or Coast Guard in DRV "homeport"

   a) Unless permanently assigned

   b) Unless other manning agency seems more appropriate source

C. Constraints on Object Collection

1) Blood availability

a) Air shipment to homeport of DRV, plus local supplies of a fair-sized city (75-100,000 population) should ensure sufficient quantity

2) Doctors, Nurses, Technicians/Recorders, Red Cross Volunteers available in sufficient numbers

a) Precompiled Red Cross rolls should ensure sufficient numbers

b) Considered "located" when reached by telephone by Red Cross

3) Crew availability (unless permanently assigned)

a) Previously-arranged-for procedures by Navy, Coast Guard or other agency

D. Obstacles to Object Collection

1) Transportation to DRV

E. Time Limits

1) All objects aboard in $\leq$ four hours

II. Objects Previously Located on DRV or Prepacked (Containerized) for Rapid Loading

A. Quantity and Types

1) Medical supplies for 1000 injured (seriously)--88,000 ft^3

2) Medical supplies for 50,000 homeless--220 ft^3

3) Medical supplies (general) for injured and homeless--426 ft^3

4) Medical facilities--7200 ft^3

5) Other relief supplies (portable shelters, sleeping bags, clothing)--130,560 ft^3

6) Other required facilities

a) Water distillation plant

b) Kitchens

c) Refrigeration facilities

i) Food
ii) Medical supplies

d) 4-1000 kw electrical power sources (mobile)

III. Transportation

A. Distance

1) Fixed by DRV homeport locations relative to disaster areas

2) Must be such as to conform to transportation time limits

B. Time Constraints

1) Doctors and Nurses: at least 10 of each on scene in $\leq$ 12 hours from notification; remainder in $\leq$ 48 hours

C. Obstacles

1) Early arrival medical contingent: sea state 6 ($\leq$ 20 foot waves); wind $\leq$ 26 knots; plus 3-foot swell

2) Rest: sea state 7 ($\leq$ 33-foot waves); wind $\leq$ 32 knots; plus 5-foot swell

IV. Delivery of Objects

A. Quantity and Types

1) Same as Sections I and II

B. Address

1) Disaster scene

C. Precision

1) Pier

2) Other suitable landing area consistent with delivery capabilities of DRV

D. Constraints

1) Sea state 6 for early contingent, 7 for rest

E. Type Delivery

1) Any of the following types satisfactory, within constraint and time limit requirements and assuming facility availability

a) Pier or wharf

i) Self-unloading
ii) Unloading by shoreside facilities

b) Beaching

c) Boats or barges or lighters or amphibious vehicles

d) Helicopters

e) Combinations

f) Amphibious DRV

F. Delivery Time Limits

1) All personnel ashore $\leq$ 1.5 hours

2) Medical supplies: commence arrival at scene $\leq$ 1.5 hours

3) First food available for consumption $\leq$ 12 hours

4) Food for infants $\leq$ 1.5 hours

5) At least 100 gallons water available $\leq$ 1.5 hours and every 3 hours thereafter

6) All shelter, clothing and food preparation equipment $\leq$ 12 hours

7) Water in quantities for prolonged subsistence needs must be available in $\leq$ 12 hours

8) 4-1000 kw electrical power sources available $\leq$ 6 hours

G. Obstacles to Delivery

1) Weather, sea state

2) Damaged or destroyed harbor facilities

V. Probability of System Success, Reliability, Confidence Limits

A. System success $\equiv$ provision at disaster scene of above

within time constraints

1) Required probability of meeting 48-hour transportation time limit, 4-hour collection limit and 12-hour delivery limit of second contingent must be 90% with 75% confidence limit

2) Required probability of meeting first contingent constraints is 70% with 75% confidence limit

B. System reliability   probability that all system components will function as designed

1) Required reliability: 85%

IIB. MISSION PROFILE (FOR A SINGLE DISASTER-RELIEF VEHICLE OR UNIT)

TIME (hrs)

0 — 1) Notification of disaster occurrence (or imminence) received at DRV homeport

Immediately

A. Red Cross informed

i) Blood dispatched to DRV by Red Cross transportation

ii) Red Cross Volunteers notified and proceed to DRV individually

B. Doctors, Nurses, Technicians/Recorders notified and proceed to DRV individually (may be through Red Cross)

C. Navy, Coast Guard or other agency assigned DRV manning responsibilities informed

2) Blood, personnel arriving

3) Vessel activation (if appropriate) commences

+4 — 4) Blood, personnel all at DRV and aboard

A. High priority team of Doctors, Nurses and first aid supplies dispatched

5) DRV with main shipment leaves homeport for area

+12 — 6) High priority medical team arrives at disaster scene

+48 — 7) DRV with main shipment arrives at disaster area

TIME (hrs)		
+49.5	8)	All disaster-relief personnel, food for infants and limited water available ashore at disaster scene (from DRV)
	9)	Medical supplies commence arriving ashore
+60	10)	First meal ready for consumption by victims; large quantities of water available
	11)	Limited emergency electrical generation provided for key surviving hospitals or other critical needs (4-2500 kw units)
+120 (or less)	12)	Resupply unit arrives to replenish stocks of first DRV or to relieve it
DISASTER TERMINA-TION	13)	DRV LEAVES and returns to homeport where Red Cross and manning agency prepare it for reuse.

IIC. DISCUSSION OF TRANSPORTATION DISTANCES AND DRV HOMEPORT LOCATIONS

Transportation time limits are governing. To get high-priority team on scene by notification + 12 hours, and main shipment by N + 48 hours, after four hours allowed for collection, the high-priority team has eight hours transportation time and the main shipment has 44.

In view of fivefold difference in allowed transit time and vast differences in sizes of shipments (20 people plus first-aid supplies vs. 105 people plus more than 200,000 ft^3 of supplies), two different platforms appear almost axiomatic.

The high-priority segment seems to point to air transport. Since several of the foreseen disaster types can be expected to make runways unusable at the scene, helicopters or C5-A type A/C seem to be indicated. The large volume of needed supplies for the main shipment indicates a ship-type platform (not excluding GEM's, hydrofoils or prelocated, stationary structures or floating vehicles).

Considering the main shipment transportation first:

Transit Speed	Range (44 hours)	Maximum DRV Homeport Separation
15 kts	660 n.m.	1320 n.m.
20 kts	880 n.m.	1760 n.m.
25 kts	1100 n.m.	2200 n.m.

Distances (Sea) Between Major Cities Accessible from Seacoasts and Great Lakes (To Cover Entire Seacoasts and Great Lakes)

Boston to Canadian Border	240 n.m.
Boston to New York	320 n.m.
New York to Norfolk	280 n.m.
Norfolk to Charleston, S.C.	460 n.m.
Charleston to Jacksonville, Fla.	200 n.m.
Jacksonville to Miami	330 n.m.
Miami to New Orleans	660 n.m.
New Orleans to Mexico	550 n.m.
San Diego to Los Angeles	125 n.m.
Los Angeles to San Francisco	340 n.m.
San Francisco to Portland, Ore.	625 n.m. (inland)
Portland to Seattle	420 n.m. (inland)
Duluth to Chicago	580 n.m. (inland)
Chicago to Detroit	620 n.m. (inland)
Detroit to Buffalo	220 n.m. (inland)
Buffalo to Canada	330 n.m. (inland)

Notation "inland" means a significant part of the trip would be in restricted waters and the real port-to-port distances have been increased by 10 to 30% to allow for the fact that full speed could not be maintained in such areas.

Placement of "homeports" for main-shipment platforms (two-piece ships), assuming a 15-knot sea speed.

Forty-four hours @ 15 knots = 660 n.m. Thus, for full coverage, homeports can be up to 1320 n.m. apart.

Homeport Summary: Two-Piece Ships

1) Norfolk, Va.

2) Jacksonville, Fla.

3) New Orleans, La.

4) San Francisco, Calif.

5) Portland, Ore.

6) Chicago, Ill.

At Least Two Ports on Any Coast--In case one homeport is struck by a disaster which also destroys its own DRV, the nearest homeport with a DRV intact will send aid. No two DRV homeports are more than 1320 miles apart.

IIIA. SUBSYSTEM DESCRIPTIONS AND COMPONENT CONSIDERATIONS--PLATFORM (HULL)

A. Preliminary Determination of Required Platform Capacity

1) Stowed Items:	211,412 ft^3
2) Quarters	
a) Disaster Relief Personnel	6,728 ft^3
b) Crew	2,208 ft^3
3) Other Needed Facilities	
a) Water Distillation Plant	~4,000 ft^3
b) Refrigeration Machinery	~ 300 ft^3
c) Disaster Relief Kitchens (2)	~4,800 ft^3
	9,100 ft^3
Platform Capacity Required for Disaster-Relief Mission (carry objects to scene, provide life support for Disaster Relief personnel and crew)	229,448 ft^3

If the platform is a displacement hull, it appears almost certain that it will be volume - vice weight-limited. Much of the equipment to be stored and facilities required in low-density, high cube in nature (e.g., the hospital and personnel quarter count for x 30% of volume, by themselves--but contain only beds and lockers).

For any other type of platform, the weight, too, would become critical.

IIIB. ALTERNATIVES CONSIDERED FOR PLATFORM CHOICE

For discussion purposes, those items which must arrive on scene at Notification Time (N-time), plus four hours for collection plus 12 hours for transportation (e.g., 10 doctors, 10 nurses and first-aid supplies), are called the "high priority" shipment. All other objects to be delivered are called the "main" shipment.

Alternative 1

Use a platform or platforms which engages in some form of commercial or military operation until called for disaster relief. Then it (they) undergoes a transformation.

Alternative 2

Use platform(s) intended solely for disaster relief, which is idle until called.

Alternative 3

Use platform(s) which has some components or parts which are idle until required and others which perform some commercial or military function until called for disaster relief.

Discussion

Quick response, when called, is the crux of the problem. A platform which undergoes a transformation will likely lead to long response times. This, however, might be soluble. The biggest drawback is that the platform(s) may not be immediately available to commence transformation. If the platform(s) are engaged in commercial or military ventures they may be days away from the nearest transformation facility.

In regard to a platform which remains idle until used, many drawbacks arise. The largest is the difficulty of having the platform ready to go when needed. Keeping the machinery (i.e., propulsion machinery) idle for long periods would require a deactivation procedure which would not be conducive to quick activation. And, of course, if no deactivation process is carried out, the machinery will deteriorate.

Alternative 3 may be the best of both worlds. Those items which would not suffer from idle time could be left idle until needed. Those subsystems (e.g., propulsion machinery) which would deteriorate if idle could be used during nondisaster periods for some other purpose.

Various Schemes Considered within above Alternatives

Alternative 1 (Transformation)

Scheme A. A "modularized" ship or GEM or hydrofoil or submarine which engages in a commercial or military venture. Upon notification of Disaster, proceeds to a transformation facility to be fitted for the disaster mission. High-priority shipment by air.

Drawbacks

1) Possible long transformation time.
2) Platform may not be near transformation facilities - hence long reaction time.
3) Gem, hydrofoil and submarine development required for such large-volume loads - or else need many platforms.

Advantages

1) Platform does not remain idle for long periods (efficient utilization of propulsion machinery).

2) Cost offset by commercial revenue or military effectiveness.

Scheme B. Use C-5A's for disaster relief. Divert the aircraft from whatever mission they are performing for Air Force, fly them to nearest stock point for disaster supplies, then to disaster area.

Drawbacks

1) Many plane loads of supplies required (70 to 90), requiring 18 to 22 aircraft, if each can make two trips per day.

2) If disaster in a severe flood or heavy weather damage (e.g., blizzards), C-5A may be unable to land anywhere near the area most in need of help (i.e., flooded runways, fields or pastures are unusable even for C-5A. Same if three feet of snow covers them).

3) Likely that sufficient number of aircraft would be unavailable from Air Force on short notice--unless additional C-5A's authorized for the disaster relief mission.

Advantages

1) Technology already developed.

2) Quick transit time for some cargo.

Alternative 2 (Idle until Needed)

Scheme C. A conventional ship, a GEM or hydrofoil craft or submarine which is located in some port in an idle state, but intended for rapid use in disaster work.

Drawbacks

1) Very poor utilization of equipment.

2) Difficult to ensure equipment will work when wanted, after long idleness.

3) Hydrofoil or GEM or submarine technology not developed for large loads--need much development or many platforms.

Advantages

1) Platform technology for ship is developed.

2) Hydrofoil or GEM would give good transit time.

3) If conventional ship is used, few are needed.

Scheme D. A fixed or floating offshore platform stocked with needed supplies and vehicles for offloading itself.

Drawbacks

1) Would need many platforms--one near each potential disaster site.

2) Interface offloading problems in heavy seas.

3) Having a few floating platforms which proceed to disaster area is ruled out by slow transit time (≤ 5 knots).

Advantages

1) Great size is no drawback.

2) Main platform (in systems sense of the word) needs no propulsion machinery.

Alternative 3 (Partially Idle until Needed--Partially Active between Disaster Missions)

Scheme E. "LYKES" concept ship which carries loaded barges with its own loading, offloading equipment for the barges. Barges loaded with disaster supplies are kept in designated ports. Ships engage in commercial trade, utilizing barge-loaded cargo. Upon disaster notification a ship picks up disaster relief barges, blood supplies and personnel in disaster-relief unit homeport and proceeds to disaster area.

Drawbacks

1) Entire ability for quick reaction revolves about availability of an empty ship in port where disaster relief barges are located at time of notification--an unlikely occurrence.
2) Even if empty ship is available, barge-loading time is not insignificant.

Advantages

1) Only equipment kept idle is that which does not suffer greatly from idle periods--i.e., disaster relief supplies.
2) Shipboard machinery is in use and thus kept in working order.
3) Costs offset by commercial utilization of the ship itself.

Scheme F. Same as above, but with containership and containers in place of barge ship and barges.

Drawbacks. Same as above, plus container offloading at disaster scene unlikely to be feasible.

Advantages. Same as barge ship.

Scheme G. A "two-piece" ship consisting of a forward half that is basically an empty box and an after half with the propulsion subsystem, controls, navigation and communication, etc., on it. The forward half is preloaded with disaster relief supplies and prepositioned in various ports. The after half is designed to fulfill a secondary mission within the harbor area (e.g., buoy tender, tugboat), while operating independently between disaster missions. When disaster occurs, the after half mates with the forward hull (either through a coupling mechanism or a hinge-link mechanism) and proceeds to disaster area.

Drawbacks

1) Some development needed in hinge-link or coupling mechanism area (presently-designed devices used with tugs and large barges have been successful in seas up to 32 feet).

Advantages

1) Propulsion machinery, navigation, communication, etc., are not idle between disasters.
2) Commercial or other uses of after end offset system costs.

Considerations in Choosing among above Alternatives and Schemes for the Major System Platform

Scheme A, a transformed ship, GEM, hydrofoil or submarine, is considered unacceptable because of high probability of excessive reaction time.

Scheme B, employing C-5A or similar aircraft, is considered to have at least two fatal flaws. First, the feasibility of carrying out a large-scale logistics mission over an extended period of time is highly doubtful. Second, conditions in the disaster area make the probability of bringing needed supplies close enough to the scene where they are required, too low for acceptance (i.e., floods or heavy snows make it impossible for C-5A to land in population center where help is most needed).

The problems inherent in activating long-unused machinery, plus the poor utilization of that machinery, caused Scheme C to be ruled out.

The floating platforms of Scheme D would have to be very numerous to provide required system success probability. We are convinced that a higher probability of success, with lower costs, can be achieved through other approaches.

Schemes E and F both suffer from the high probability of excessive delay times, making the mission a failure, and are consequently rejected.

Scheme G remains as the concept offering the highest promise of achieving system objectives without squandering funds (Scheme D, for instance, could be made workable at prohibitive costs).

Expansion of Scheme G for Platform Configuration--Main Platform

Figure 2-6

The Two-Piece Ship Concept: (Lip and Groove Joint)

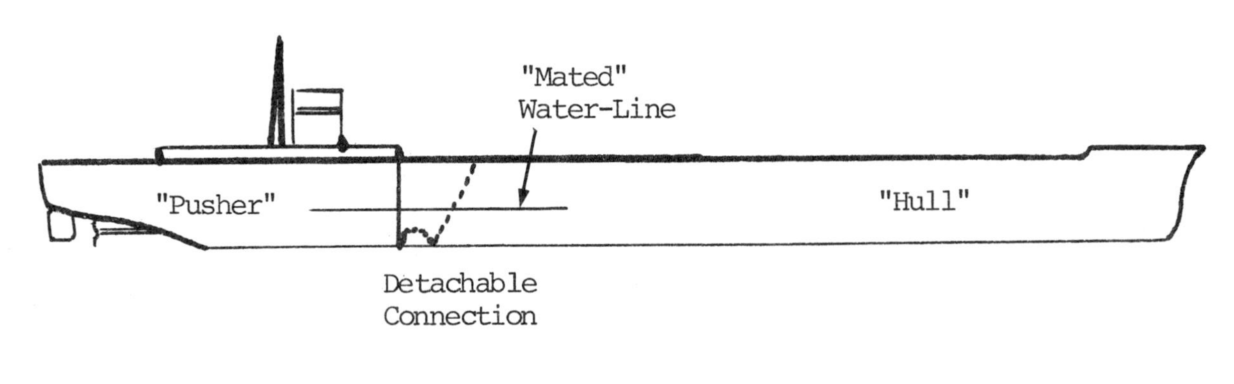

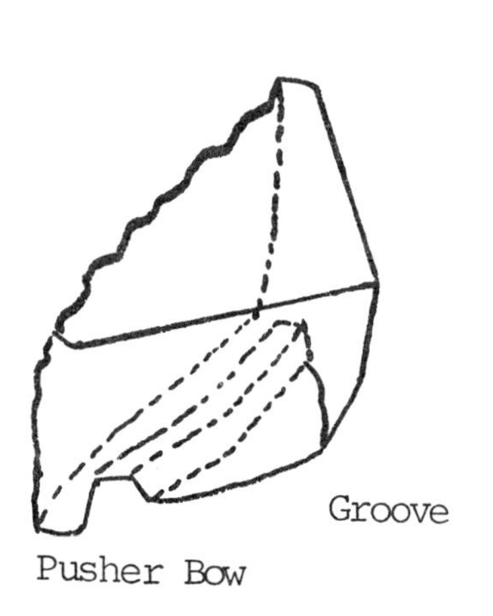

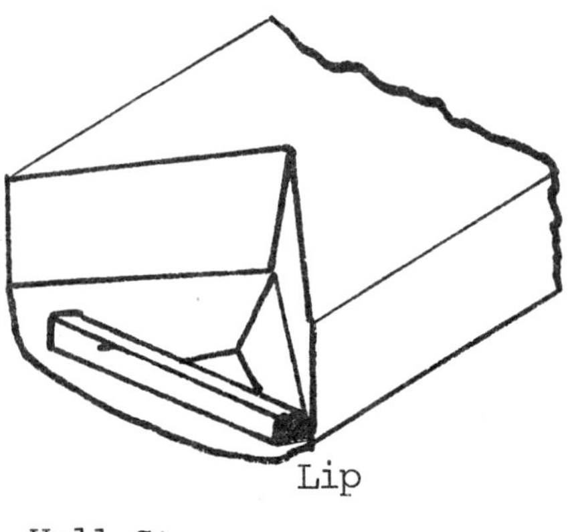

Pusher is fitted with ballast tanks, allowing it to control its draft and longitudinal trim, within limits. The pusher deballasts, moves the bow into recess in the hull stern, then ballasts down to engage the lip of the hull section into the groove in the pusher bow. Additional fittings are provided to prevent excessive "working" of the joint when underway.

The hull sections are previously loaded and left idle in Disaster Relief Unit (DRU) homeports. The pushers are used for some other commercial or government task, e.g., tugs, buoy tenders, etc. The pushers are at all times within three hours of the "hulls" and, on disaster notification, proceed to "hulls," mate, and proceed to disaster area.

The area requiring most careful attention is, at first glance, the lip-groove connection. Preliminary investigation reveals that designs incorporating a similar connecting mechanism have been proposed, but are expected to be very sea-state-limited. At present, a sea state of about three is the highest proposed for such a connection.

The Hinged Joint: (See Reference 3)

Another proposal for a two-piece ship is the utilization of a pair of large pin hinges as the connecting mechanism. This proposal was made with a view toward reducing longitudinal heading moments and stresses to reduce the weight of structural steel required.

Profile:

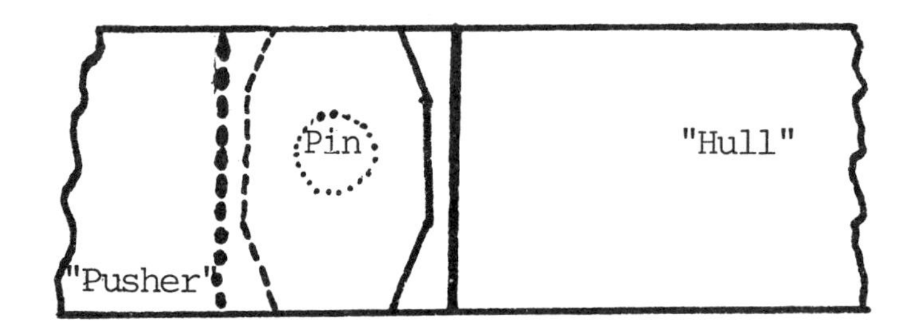

Plan View

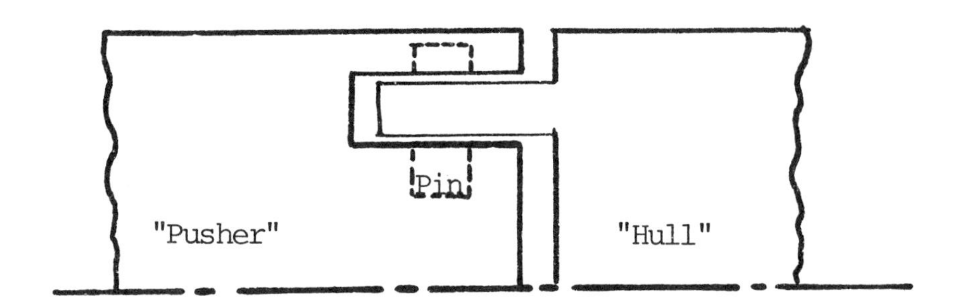

Figure 2-7

If the pin is made of solid steel, a diameter of ~3 feet will be required. This, however, would lead to bearing pressures on the order of 6000 psi, where about 1500 psi is considered to be more in the desired range. It is possible to reduce bearing pressures by making the pin hollow. To achieve 1500 psi bearing pressures on a 2-1/2" thick HY-80 pin, a pin diameter of ~14 feet would be required.

The bearing material might be lignum vitae, lubricated by pumped salt water.

Such a pinned hinge does not appear to be sea-state-limited (see Reference 1). The major drawback, however, is that present designs are not easily connected/disconnected. In fact, a combination of hydraulic rams and overhead cranes is envisioned (see Reference 3) as the tools necessary for coupling/uncoupling. As reaction time is of critical importance, a quicker-acting connection is highly desirable.

The "Sea-Link" (see Reference 2)

A third method for connecting the "pusher" to the "hull" is available and provides the quick-connect/disconnect capability of the lip and groove joint with the seakeeping ability of the pinned hinge. This "Sea-Link" method has been tested in tugboat-barge operations in wave heights up to 32 feet with a tugboat length of only 85 feet.

The "Sea-Link" consists of a rigid pushing frame connected to the barge and hinged about the horizontal axis and connected to the tug by a universal swivel joint. The other side of the tug is connected, via another universal swivel joint, to a steering spar, in turn, swiveled at the barge stern.

In using the "Sea-Link" concept with the two-piece ship, the two-piece ship begins to resemble a barge and pushing tugboat. The major difference is that our two-piece ship will proceed at speeds above 15 knots, and nearing 20 knots. The "hull" will be ship-shaped, rather than barge-shaped, and will be very large for a barge. The "pusher" will be considerably more powerful than any tug to use "Sea-Link" to date.

Plan View:

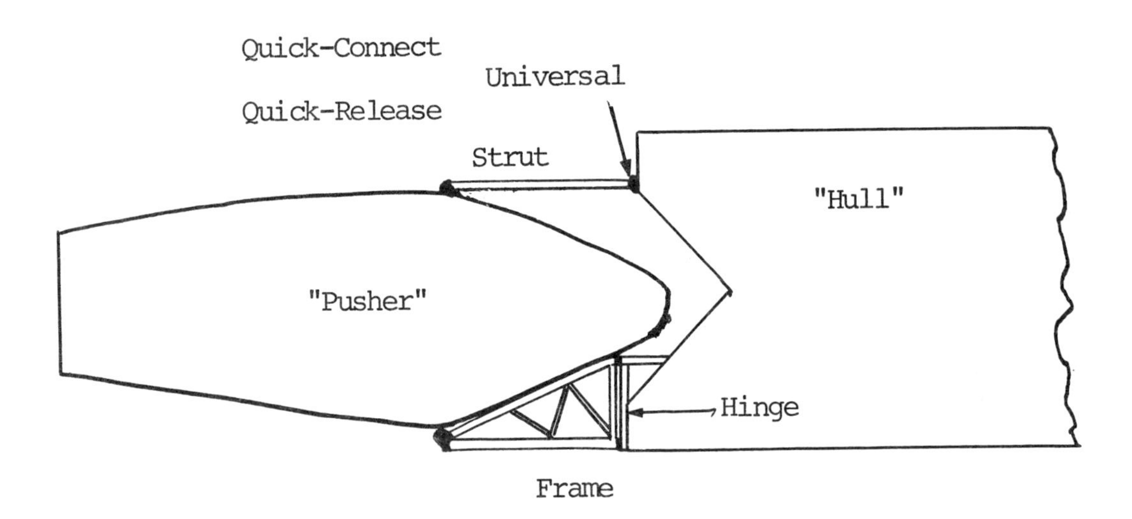

Profile:

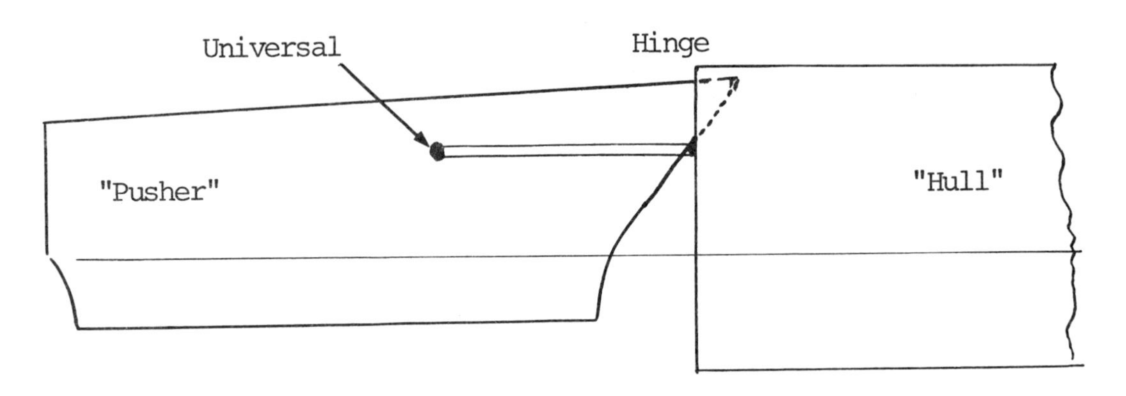

Figure 2-8

IIIC. HULL SELECTION ANALYSIS

For the main shipment vehicle (platform), the two-piece ship concept utilizing the "Sea-Link" is chosen for the following reasons:

1. Of all methods investigated, "Sea-Link" is most independent of sea state and appears to require little, if any, state of the art improvement to meet system constraints (state 7 sea).
2. "Sea-Link" permits very large values of relative roll, pitch and heave between the hull and pusher, thus minimizing connection forces. At the same time, yaw, surge and sway are restrained to give positive steering, pushing and stopping control.
3. Pushing the "hull" has the following advantages, when compared with attempting to tow at high speeds.
   a. Better steering and stopping control.
   b. Saves time by eliminating need to shorten and lengthen tow cable when entering and leaving port.
   c. Towline fouling eliminated.
   d. Overall resistance of tug pushing a barge ahead in less than sum of individual tug and barge resistance, permitting 15-20% speed gain.
4. "Sea-Link" is the only one of the three connecting mechanisms with a quick-release feature. Makeup time of "Sea-Link" is many times less than other methods, requiring only hand-tripping of latches--ballasting and outside-crane assistance are eliminated.

In addition to material listed in the "Preliminary Determination of Required Platform Capacity," pg CAP-1, the two-piece ship will have to carry the mobile generator units, other vehicles and possible boats. It may later prove desirable to make the ship beachable, requiring doors and ramps. To allow space for such added equipment to be carried, the volumetric capacity will be doubled, as a first order guess of required size. Thus the main shipment requires about 560,000 ft^3.

To make "hull" beachable, we will limit overall mean draft to 12'. To keep beam to draft ratio at 4 or below, the beam must be $\leq$48'. Assuming 80% of the product length x beam x draft will be usable space for these items, the length of the "hull" would have to be ~800 feet--much too large a hull for our purpose.

Thus, it is decided to place two "hull" units in each port, with at least two "pushers" assigned. This reduces "hull" size to about 350' length, 50' beam, 12' draft--a more manageable size.

The two hulls will each carry one-half the total required main-shipment goods. Thus, for a small-scale disaster, only one need answer the call and could provide all needed aid without having large quantities of equipment uselessly tied up. For a large disaster, both units in a port would respond.

POWER AND PROPULSION

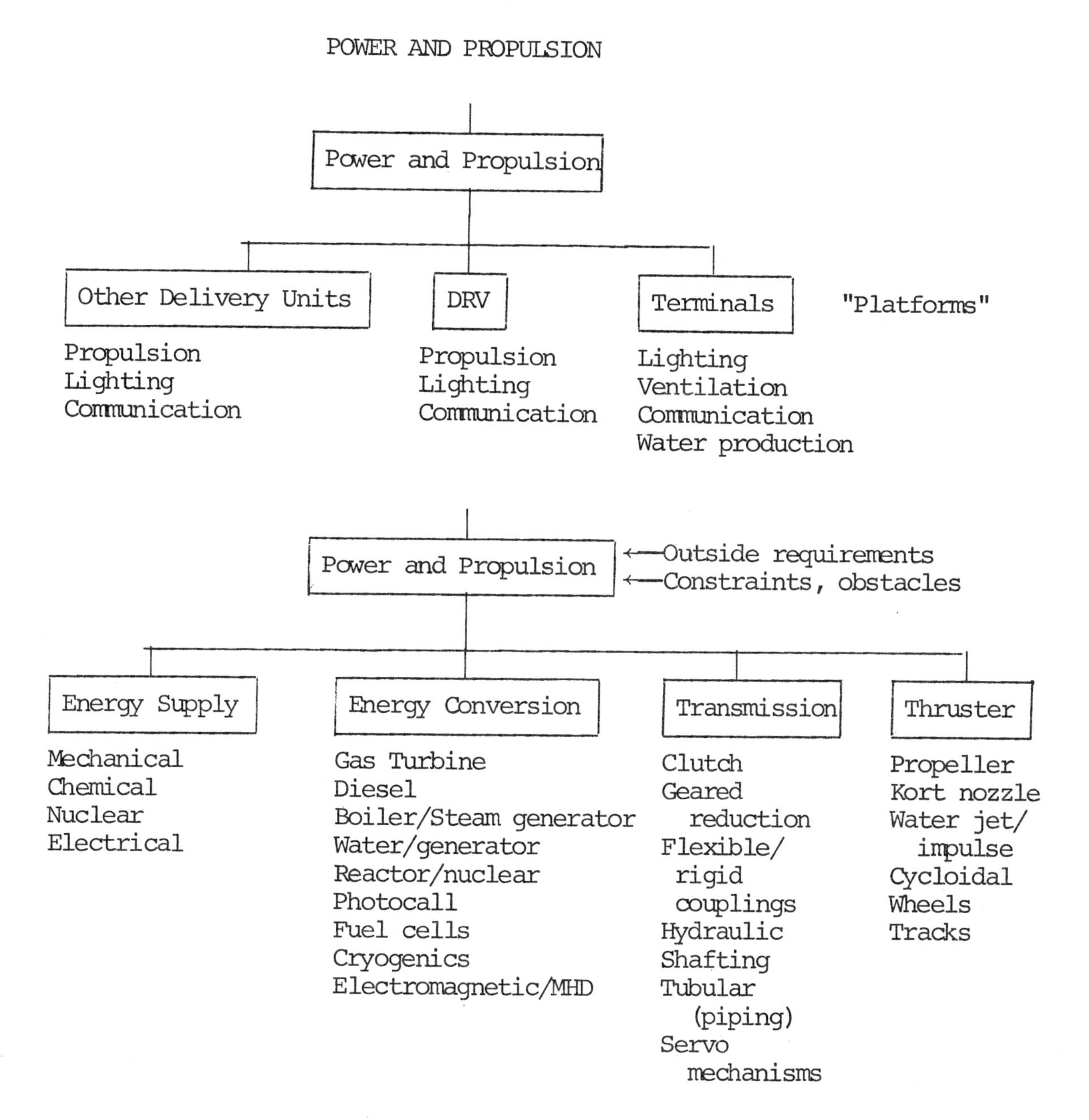

Outside Requirements

Platform mission
Platform lifetime
Reliability
Stability and control
Structure and geometry (shaft down angle and divergence)

Figure 2-9

Obstacles

Meteorological sea state: wind, wave spectrum, currents

Terrain: hills, rivers, rocks

Constraints

Vessel (platform) attitude: roll, pitch, heave

Draft

Terrestrial: (a ship can't go ashore) (pitch a tent in "x" inches of water?)

Meteorological

Self-Generated Obstacles and Constraints

Displacement: weight, volume (include fuel and auxiliaries)

Thrust (SHP)

Vibrations (mounted eng. freq. shall not coincide with natural freq. of vessel)

Noise: (less than 95 db at freq. $\leq$250 Hz, 70 db at freq. >250 Hz)

Heat: (working space not to exceed 90°F)

Response: (maneuverability, reversibility, start up/shut down time)

Reliability:

Standardization/Availability: (sophistication) Off-the-shelf?

Time between Overhaul/Failure: MTBF MTTR

Complexity:

Development/Construction Costs:

Efficiency:

Spare Parts Requirements:

Hazards/Ease of Handling: (toxicity, flash point, corrosiveness)

Disaster Relief Vessel (DRV) Propulsion

Power requirements from Series 60

Container

Δ = 8000 tons   Beam B 53.8'   Depth of side D ~ 30'

Length L 380'   L/B 6.5

Draft T 12'   B/T 4.5

Container (continued)

Letting $V = 18.7$ kt $= \frac{V}{\sqrt{L}} \Rightarrow 1.00 \quad R_r/\Delta = 1.25 \Rightarrow R_r = 6250$

$\nabla = 35\Delta = 0.175 \times 10^6 \Rightarrow \nabla^{2/3} = 3120$

$S/\nabla^{2/3} = 6.5 \Rightarrow S = 20{,}300 \text{ ft}^2 \qquad \nu L = 6550$

$R_F/S = 1.94 \Rightarrow R_F = 39{,}400$

$R_T = R_R + R_F = 42{,}500$

$EHP = VR_T/325.6 = 2440$

$SHP_{nom} = 2440 \{\frac{1}{0.6}^{BH} * 1.05^{app} * 1.25^{max} * \frac{1}{1.1}^{nom}\} = 4880 \text{ SHP} =$ assume

8000 shp/vessel $LBD = (350)(30)(53.8) = 5.65 \times 10^5 \text{ ft}^3$

Assume permeability of 0.5 = $2.8 \times 10^5 \text{ ft}^3$ available for cargo

Pusher

$\Delta = 1500$ tons $\qquad C_B = \frac{35\Delta}{LBT} = 0.48$

$L = 220'$ $\qquad L/B = 4.4$

$T = 10'$ $\qquad B/T = 5.0$

$B = 50'$

$V = 20$kt $\qquad V/\sqrt{L} \approx 1.3$

$R_r/\Delta = 1.25 \Rightarrow R_r = 1875$

$\nabla = 35\Delta = 52500 \Rightarrow \nabla^{2/3} = 1410 \qquad VL = 1.3L\sqrt{L} = 4250$

$S/\Delta^{2/3} = 6.0 \Rightarrow S = 8460 \text{ ft}^2 \qquad R_F/S = 2.03 \Rightarrow R_F = 1.718 \times 10^4$

$R_T = R_r = R_F = 1.91 \times 10^4$

$EHP = \frac{VR}{325.6} = 1172$

$SHP_{nom} = 1172 \{\frac{1}{0.6}^{BH} * 1.03^{app} * 1.25 * \frac{1}{1.1}\} = 2340 \Rightarrow$ assume 2500 SHP/pusher

Only a portion of the total system report of Lieutenants Calvano and Luxford has been presented here. Rather detailed lists of types and varieties of medical supplies and weight, space and volume requirements for the platform were prepared, but were not included since they were not instructive for this chapter. On the other hand, a little more ought to be said about the power and propulsion system. Two basic alternatives confronted the system planners. One was the utilization of all the prime power on board the main DRV with a requirement of setting up power distribution networks or tying into the remains of power distribution networks on the shore, or of setting up special units on the shore itself. Recognizing that the more desirable situation would not be an either/or or a combination of both, it was then decided that common equipments for generating power for propulsion and for power ashore would be employed. This suggested rather compact high-powered machinery for both and very quickly led to the choice of the gas turbine as the prime power generator. In order to transport the power ashore, the planners proposed the use of six APB's (all-purpose vehicles). These are large tired amphibious vehicles which would be employed to deliver objects and personnel from the DRV to the refugee center or to terminals. The APB would be preloaded in the disaster relief hulls. Two of the APB's would be preloaded with 1.000 kw mobile generators which, in turn, are skid-mounted and removable from the APB's. The other four APB's in in each hull will be preloaded with supplies to make the first trip to the refugee center. The power solution, therefore, is one which might not be optimal for the DRV treated as a ship alone, but which meets the requirements of the other subsystems, such as the object delivery system which carries equipments from the DRV to the beach and the onshore disaster-relief system which requires multiple power sources and a high degree of mobility and flexibility in the location and application of the power sources.

REFERENCES

(For Hinged Hull and "Sea-Link")

1. Marbury, Fendall, Jr., "The Motions of Connected Hulls in Regular Seas," M.I.T. Department of Naval Architecture and Marine Engineering, Report No. 67-14, Cambridge, Mass., October 1967.

2. Glosten, L. R., "Sea Link," Marine Technology, Society of Naval Architects and Marine Engineers, New York, July 1968.

3. Boylston, J. W. and Wood, W. A., "The Design of a Hinged Tanker," Marine Technology, SNAME, New York, July 1967.

CHAPTER III

LIFE SUPPORT SUBSYSTEM

EPIGRAMS

Most people are "saturated" at 14 psi.

Oxygen is a poisonous gas.

Major Safety Certification delayed is Systems Operation denied.

There is a dangerous design tendency to permit man to occupy any area where oxygen is available.

Negative ions are happy ions.

Saturation diving is S.O.A. today.

Subjective Human Purpose

Whenever and wherever the operation of a system requires the presence of man, his ability to function within the system requires adequate and appropriate life support. Through evolution man has adapted to unaided survival in a warm open atmosphere with available food and an abundance of fresh water (Tahiti, Big Sur, etc.). All other environments require some level of artificial maintenance. An inadequate life support subsystem will result at least in reduced performance levels, and may become dangerous to health and life. The subjective purpose of the life support subsystem is the provision of a total environment which protects and satisfies the biological, physiological and psychological life functions of the human operators.

The essential elements of a life support subsystem include:

1. Adequate oxygen supply
2. Removal of carbon dioxide
3. Elimination of toxic substances
4. Adequate food and water supply

5. Waste disposal
6. Prevention of accumulation of flammables
7. Thermal protection
8. Pressure and pressure gradient protection
9. Protection from electromagnetic and nuclear radiation
10. Protection from extremes of shock and vibration
11. Methods for use and protection of the aural, vocal, visual functions
12. Physiological monitoring of crew
13. Creation or protection of a psychologically adequate environment
14. Progressive Safety Certification at all design and construction phases

All of the listed elements must be satisfied in any sea system. Nevertheless, a substantial difference in standard of care and complexity occurs between operations at the free surface and those which take place below the free surface or in an enclosed environment. At the free surface the availability of the atmospheric reservoir and the availability of sunlight resolve many of the life support problems. The presumption that this reservoir is infinite or totally available has produced the dangerous and deleterious condition that exists in many cities today and can produce dangerous and deleterious conditions even in open-ocean sea systems. Reliance on nature also introduces other dangerous conditions, such as excess humidity, excess cold, excess wind, diurnal variations in light and dark and the need for visual adaptation, etc.

On the other hand, operations beneath the surface, whether in shallow or deep water, place stringent and critical requirements on the life support system. The provision of an artificial environment and protection against changes in pressure are not the only problems. The small size of the atmospheric reservoir is such that almost every piece of machinery and apparatus will interact in a significant way with the environment, human performance, and human safety so that an item-by-item review of every component from a human factors standpoint is mandatory. With these caveats in mind each element may now be examined.

Oxygen Supply

Providing an adequate safe supply of oxygen is a critical responsibility of the life support subsystem. Although proximity to the free surface and the atmosphere does not automatically fulfill this responsibility, the difficulties encountered there should be minimal. When operating near the free surface, the atmosphere may be obtained by ducting and pumping, i.e., snorkeling. Any snorkel device should be designed to avoid malfunction caused by:

1. Inadvertent wetting, splashing, or submergence
2. Intake of exhaust gases from another system
3. CO_2 buildup in ducts, piping, or tanks.

When access to the atmosphere is denied, as it is in any submerged system, oxygen supply becomes a major design consideration. The subsystem manager must attack this problem with two design criteria clearly in mind: the oxygen source and its backup must be able to provide enough oxygen to fulfill all biological and mechanical needs; the oxygen must be delivered at a partial pressure within critical human limits.

The average oxygen consumption rate for adults is 0.083 lbs/man-hour, on a 24-hour basis, or roughly 2 lbs/man-day. (This is equivalent to about 1 cubic foot/man-hour at 14.7 psia.) It should be noted, however, that this is a daily average, and actual consumption during work periods may rise to .150 lbs/man-hour, while consumption during sleep periods will drop to about 0.050 lbs/man-hour.

The life support subsystem manager should be familiar with all alternative oxygen sources capable of fulfilling the above requirements:

1. Direct Bottle Storage. Compressed air or compressed oxygen may be delivered and stored in pressure containers. Standard oxygen bottles are maintained at a pressure of about 3000 psia, which yields a storage density (of pure oxygen) of about 18 lbs/cu ft. If bottles of compressed air are used, the storage density is about 3.5 lbs/cu ft of oxygen.

Bottles of compressed gas are heavy and awkward to handle. A system manager should realize that any such high-pressure container represents a potential bomb if damaged. A container of compressed air or oxygen also requires extreme protection from fire hazard.

2. Electrolysis of Sea Water. Distilled sea water can be electrolytically dissociated to produce oxygen by passing a D.C. current through a KOH electrolyte solution. These units are presently in use on U.S.N. submarines, and are capable of producing 120 standard cubic feet at 3000 psia.

Power requirements for this method are the limiting design consideration. These units require from 2.6 to 6.4 K.W. per lb of O_2 produced. Therefore, such a source should be considered only when nuclear power is available to the system.

The units require about 1.12 lbs of distilled and demineralized water per pound of oxygen produced. Hydrogen is a by-product of the reaction, and its use or disposal must be considered.

3. Oxygen Candles or Chlorate Candles. A compact means of storage is to be had through the use of chlorate candles. These are composed of a matrix of sodium chlorate $NaClO_3$ and powdered iron. The combustion of this candle produces the following chemical reaction:

$$2NaClO_3 + Fe \rightarrow 2NaCl + FeO_2 + 2O_2$$

The chlorate candle thus provides approximately 585 cu ft of oxygen for each cubic foot of candle.

4. Other Sources. Several chemical cycles (e.g., sodium sulphate cycle) and organic sources (e.g., algae cultures) have been theoretically introduced, but have not been sufficiently developed.

The life support subsystem must regulate the partial pressure at which oxygen is delivered for human consumption. The normal range of O_2 partial pressure is 2.58 to 3.55 psia, with a nominal optimum at 3.06 psia (the normal partial pressure of O_2 at 1 atmosphere).

If the partial pressure of oxygen is allowed to go below 2.50 psia, anoxia (oxygen starvation) occurs. At this level, there is progressive depression of consciousness, inability to concentrate, loss of coordination, increased pulse rate and blood pressure. At 1.50 psia O_2 partial pressure, unconsciousness occurs and the anoxia becomes fatal. Unfortunately, there are very few natural warning signs of this situation, and the incipient stages of anoxia are usually accompanied by mild euphoria.

Care must also be taken to prevent an "oxygen overload." If the partial pressure of oxygen is allowed to reach 1 atmosphere (14.7 psia O_2), extended breathing time may result in lung damage. Research conducted by J. M. Clark and C. J. Lambertson, both of the University of Pennsylvania School of Medicine, established the following rate of symptom development for subjects breathing oxygen at an O_2 pressure of 2 atmospheres.

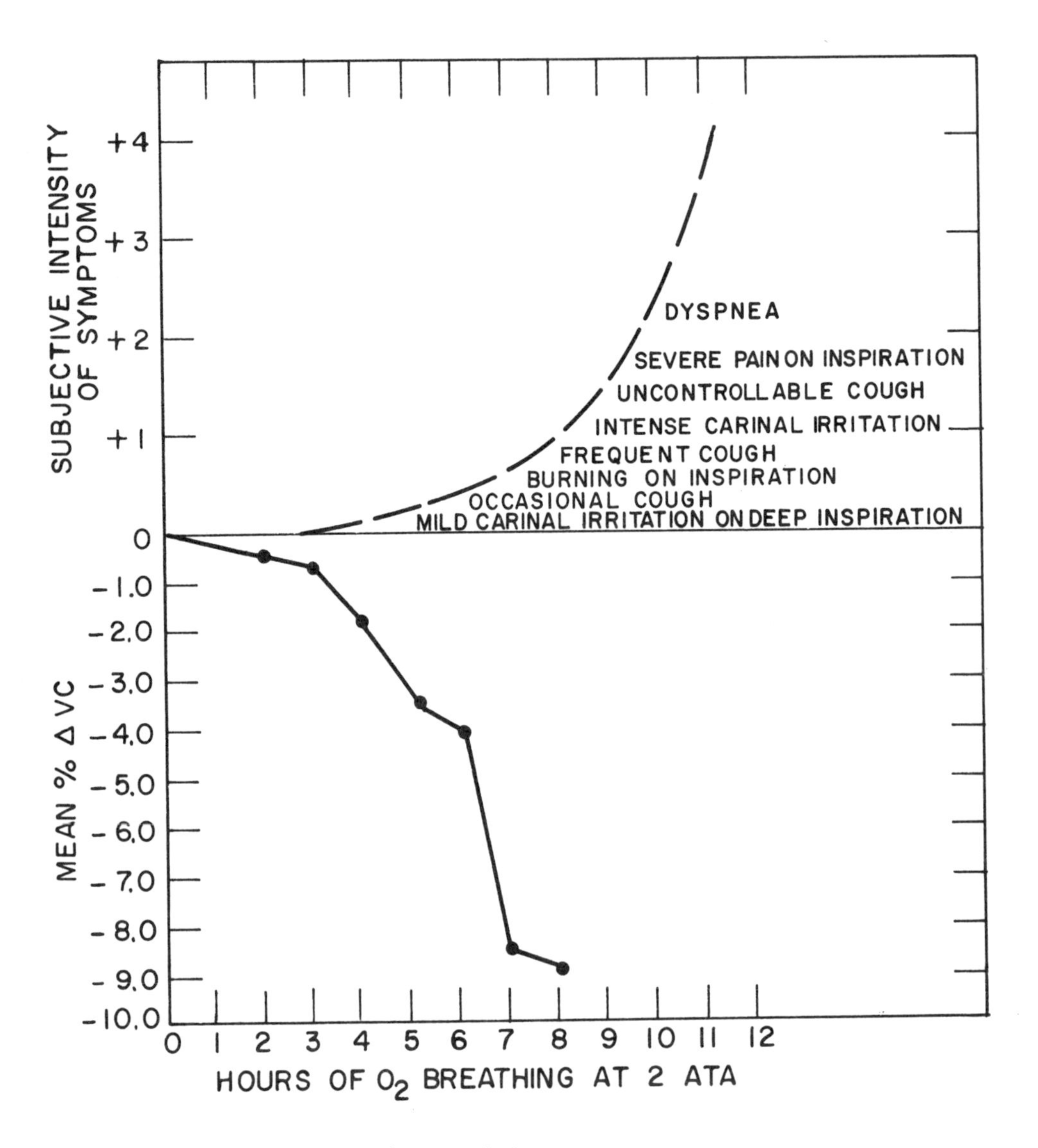

Figure 3-1

From: Clark, J. M. and Lambertson, C. J., "Pulmonary Oxygen Tolerance and the Rate of Development of Pulmonary Oxygen Toxicity in Man at Two Atmospheres Inspired Oxygen Tension." In: Underwater Physiology. Lambertsen, C. J., editor, Baltimore: Williams & Wilkins, 1967, pp. 439-451.

Higher partial pressures may produce dizziness, nausea and convulsions. The breathing of oxygen at an O_2 partial pressure of 4 atmospheres (58.8 psia) can cause epileptic-like convulsions in minutes.

Saturation Diving

It is important for the ocean systems designer to realize that saturation diving is now well within the state of the art of ocean engineering technology. Saturation diving is defined as the use of individuals whose bodies are in equilibrium with an elevated atmospheric pressure and who are supplied with appropriate breathing gas mixtures. While it is beyond the level of detail of this text to fully explain the techniques involved, this information is now readily available (see References). Saturation diving specialist subcontractors are available, and the immensely more efficient use of diving time will more than compensate for increased equipment costs in almost all projects involving even moderate use of free divers.

The subsystem manager should, however, be aware of several potentially troublesome situations which may arise in this area.

1. Saturation diving usually requires a gas mixture (rather than compressed air). Any gas-mixing procedure involves handling of high-pressure gas bottles (fire and explosion hazard), and an extremely accurate metering procedure.

2. Nitrogen cannot be used as a diluent gas in the breathing mixture below a depth of 50 meters, or at pressures greater than 85 psia, because at this pressure it produces a narcotic, euphoric effect (nitrogen narcosis, "rapture of the deep").

3. Helium can be used as a diluent gas in the breathing mixture to a depth of about 300 meters (approximately 440 psia). At this depth tremors may occur. Increasing the depth to 350 meters (approximately 510 psia) has resulted in convulsions which may be due to helium or may be due to trace contaminants in the gas.

4. Hydrogen should be usable as a diluent gas below 60 meters (approximately 100 psia), because at this depth the O_2 content should be less than 4 percent, eliminating the possibility of explosion. Laboratory experiments have indicated that tremors have occurred at depths of less than 550 meters (approximately 896 psia). However, operating experience with this mixture is very limited.

5. Use of a helium or hydrogen breathing mixture impairs vocal communication. At increased pressure, speech becomes unintelligible.

6. Helium gas mixtures have a high thermal conductivity, and divers chill very quickly in water of temperatures below 30°C.

Carbon Dioxide Removal

At sea level (1 atmosphere) the normal partial pressure of CO_2 is approximately 0.0034 to 0.0073 psia (0.023% - 0.05%). If the CO_2 content is allowed to reach 0.73 psia, there is an uncomfortable increase in the breathing rate. At a CO_2 partial pressure of 1.40 psia, there are marked mental effects and eventual (1-2 hours) unconsciousness.

The ratio of the amount of O_2 consumed to the amount of CO_2 produced by human respiration is known as the respiratory exchange ratio, and is a function of diet. It may vary from 0.7 to 1.0, with a mean value of about 0.83. Therefore, the normal range of CO_2 production from human respiratory functions is between 1.1 and 4.8 lbs/man-day.

The life-support subsystem manager should be careful, however, to make sure he has included all CO_2 sources (aerosols, fire extinguishers, combustion processes, etc.) in estimating his CO_2 removal needs.

1. Monoethanolamine Scrubber. This device absorbs CO_2 from a gas flow when cool and releases it when heated. Power requirements are very high.

2. Magnesium Oxide. This unit uses a catalyst to cause MgO to react with CO_2. The reaction can be reversed by using superheated steam to regenerate the removal system.

3. Solid Amine. This unit utilizes silica gel to absorb moisture and CO_2. The system is regenerable, but its reliability has not been clearly demonstrated at this time.

4. Molecular Sieve. A proprietary zeolite is used as a solid absorbent of CO_2. However, the air must be pre-dried.

5. Cryogenic Techniques. The gas stream is cooled to -230°F to -270°F, and the CO_2, water vapor, and trace contaminants are removed in solid form. These units require a large power supply.

6. Electrodialysis. A combination of selectively permeable membranes combined with an electrolysis system can simultaneously generate O_2 and remove CO_2. However, power requirements are high, and reliability has not yet been clearly established.

7. There are many other techniques listed below, which are not yet available for system use, but may merit development for a particular system application:

Photosynthetic gas exchanger

Molten alkali carbonate reactions

Sabatier (methane and water produced)

Bosch (solid carbon and water produced)

Elimination of Toxic Substances

The elimination of toxic substances, or trace contaminants, is the direct responsibility of the life-support subsystem manager, although most of these contaminants are produced by other subsystems. The following sources of trace contaminants are all likely to be found in an ocean engineering system:

1. Incomplete Combustion Processes. Any combustion process will produce CO, carbon monoxide, due to the incomplete combustion of a carbon fuel.

2. Refrigerants. A refrigeration system may allow leakage of freon or other chemicals.

3. Chemical Processing. Various vapors and aerosols are produced in photographic processing, biological specimen preparation, and other processes involving the use of exposed toxic chemicals.

4. Process Fluids and Electrolytes. Power and propulsion systems using electrolytic fluids may be a source of NO and other dangerous substances. Any ballast system utilizing mercury is a potential contamination source.

5. Machinery-Produced Aerosols. Whenever lubricants are applied to bearings, especially in high-speed machinery, an aerosol of the lubricant is produced. This fine mist of droplets must be quickly and effectively removed to prevent both inhalation and fire hazard. (An aerosol consists of particles, either solid or liquid, less than 1 micron in diameter, which are suspended in the atmosphere.)

6. Battery Outgassing. The use of storage batteries (lead-acid, silver zinc, chrome nickel) may produce toxic or hazardous gases such as chlorine or hydrogen.

7. Positive Ion Buildup. Any activity which may produce an excess of positive ions in the breathing gas mixture must be considered. Positive ions depress the normal activity of the cilia of the bronchial tract and lungs.

8. Gauge Fluids. Any gauge which contains a pressure fluid should be checked. Many of these fluids are compounds of mercury and bromine, sources of dangerous contaminants.

9. Other Sources. Other trace contaminant sources include: cleaning solvents, paints, shaving creams, aerosol deodorants, cigarettes, waste disposal systems.

The problem of the elimination of toxic substances should be attacked in three distinct methods, which should begin in the earliest stages of system design.

First, the life-support manager should distribute to all subsystem managers and contractors a list of prohibited substances. This list should cut down the sources of dangerous processing chemicals, gauge fluids, aerosols and lubricants, and refrigerants. After distributing this complete and exact list of prohibited substances, the life-support manager should realize that many of the substances will appear in both his own and other subsystems despite his warnings, and that responsibility to detect and eliminate these substances is essentially undiminished.

Second, a gas sampling and monitoring system must be installed. Units of a monitoring system are needed especially in living areas, food preparation and storage areas, and machinery spaces.

Third, devices for the removal of toxic substances must be provided in the life-support subsystem. Several forms of these mechanisms are now available.

Filters of various types can be used in areas of high contaminant concentration to remove airborne particles and aerosols. An activated charcoal filter unit is a small, cheap device which can be installed easily in a deep submergence system. It is entirely passive, requiring no power supply. However, care must be taken to clean the filter regularly to prevent the buildup of flammable materials.

An "absolute" filter, containing a blower, a prefilter for large particles, and a high-efficiency filter for fine particles, can be used to remove all particulate material and aerosols.

"Purafil," a proprietary chemical, can now be used for the same purposes as an activated charcoal filter.

Electrostatic precipitators are a very effective method for removal of particulate matter and aerosols. They utilize D.C. voltages as high as 15,000 volts to ionize gas molecules which impart their charge to the airborne particles. Charged plates are then used to collect the particles. Electrostatic precipitators do have certain disadvantages, however; they are fairly expensive, they generate electrical interference which may affect instrumentation, and they are sometimes a source of another trace contaminant, ozone.

The chief method of removing non-particulate trace contaminants is the catalytic burner. It can be used to remove CO, H_2, and hydrocarbons (methane, etc.). The catalytic burner consists of an air blower, filter, regenerative heat exchanger, electrical heaters, and a catalyst bed. The burner has a high power requirement, and a high unit cost (around $9,000). The burner must also be carefully monitored because some acids may be produced in the burning process. (HF may also be produced if freon passes through the burner.)

A trace contaminant control system, therefore, must include some filtering processes for particulate matter and aerosols, and a unit like the catalytic burner for hydrocarbon and hydrogen removal. Control of other gases (chlorine, freon, etc.) requires a ducting system which is carefully separated from the living space ventilation system.

Food and Water Supply

Basic dietary requirements for a crew under normal work conditions vary from 2500 to 2900 kilogram calories/man-day. A normal diet consists of 60% carbohydrates, 30% fat and 10% protein. The life-support manager should consider using any or all of the following food forms:

a. Fresh and canned (require elaborate storage and preparation)

b. Frozen (requires a storage freezer)

c. Freeze-dried

d. Dehydrated

Freeze-dried and dehydrated foods are considerably less palatable than fresh or frozen foods, and this consideration should not be taken lightly. Freeze-dried and dehydrated foods are also 30%-40% more expensive than equal portions of fresh or frozen foods. On the other hand, however, freeze-dried and dehydrated foods are 30% lighter than fresh or frozen foods and require 50%-60% less storage space.

Water management can be subdivided into the following categories:

a. Human drinking: 1 gallon/man-day should be provided (0.5 gallon/man-day is absolute minimum)

b. Food preparation: requirements vary from 2-4 gallons/man-day depending on food storage methods

c. Washing and laundry: requirements vary from 10-30 gallons/ man-day.

Therefore, a total of 20-35 gallons/man-day will be necessary to sustain the water requirements for life support.

The method of water supply selected will be dictated by crew size, mission duration, and power availability. The following options are open to the life-support manager:

1. Pure Water Storage (no reuse). A gallon of water weighs approximately 8.3 pounds and occupies approximately 0.13 cubic feet. When the weight and size of storage containers are added, a system capable of providing 30 gallons/man-day will weigh about 250 lbs/man-day. Added to this already high price in weight and space is a large bill for storage facilities and maintenance.

2. Sea Water Still. A sea water still uses electrical power to furnish potable water from sea water. This method is considerably less expensive and space-consuming than storage.

3. Filtration. Used water may be passed through a filtration system consisting of charcoal, an ion exchanger, and bacterial filters to reclaim potable water. These units have low power consumption, but many of the filters are nonregenerable and must be replaced on a semi-weekly basis.

4. Internal Distillation. Used water may be vaporized and

recondensed within a closed system. Filter systems are still required, and care must be taken to keep volatile contaminants out of such a system.

Waste Disposal

The life-support subsystem must provide for the collection and disposal of all solid and liquid waste. Human waste production includes:

Urine (1.3 to 4.4 lb/man-day)

Feces (0.1 to 0.6 lb/man-day)

Perspiration condensate (1.8 to 2.6 lb/man-day)

Food packaging and scraps (depending on food storage methods)

Paper, etc.

Several methods of collection and disposal are currently available:

1. Manual Collection. This requires collection in plastic bags or cans and the addition of chemical disinfectants. While cheap and reliable, this method requires adequate and sanitary storage space. A recent innovation is the freezing of wastes before storage. This method requires a moderate power expenditure, but reduces sanitation problems.

2. Holding Tanks. Holding tanks reduce all the problems of collection at the expense of a large storage space.

3. Discharge Overboard. In this method a small holding tank is discharged overboard by pump or high-pressure air. Such a system will contaminate the surrounding environment, and should not be selected for a system doing any biological sampling. At increased depth (greater than 30 feet) the discharge process becomes difficult, expensive, and (at depths greater than 100 feet) dangerous. Any deep submergence system should avoid a discharge method which requires frequent equalization

of pressure through hull penetrations. Severe accidents have occurred during the discharge process.

Prevention of Accumulation of Flammables

The prevention of the accumulation of flammable substances is particularly difficult in a closed environment. Furthermore, a high-pressure environment increases the likelihood of fire. Therefore, the life-support manager's main task is to prohibit all the subsystems from introducing fire hazards into the environment.

The following hazards have been identified in ocean engineering systems:

1. AVOID USE OF STAINLESS STEEL IN AN OXYGEN SYSTEM. In a pure oxygen atmosphere and at high temperature, stainless steel burns.

2. AVOID USE OF ALUMINUM IN AN OXYGEN SYSTEM for the same reason.

3. The use of WELDED FITTINGS ONLY in an oxygen system should be a system requirement. The oxygen system should be checked in minute detail to assure that all possible leak sources are avoided in design and not by maintenance and operational procedures.

4. AVOID USE OF HYDROCARBON LUBRICANT IN AN OXYGEN OR HIGH-PRESSURE AIR SYSTEM. Preferably use a silicone lubricant (with no hydrocarbon additives). Hydrocarbons in any form should never be present in or near the oxygen supply system.

5. PREVENT DIESELIZATION OF LUBRICANTS. When hydrocarbon lubricants are applied to high-speed bearing surfaces, an aerosol is formed. If this aerosol comes in contact with high-pressure air, self-ignition can occur--no spark is necessary. Any high-speed machinery must be adequately ventilated.

6. AVOID USE OF HYDROCARBONS IN THE VICINITY OF FUSES OR CIRCUIT BREAKERS. Hydrocarbons, whether in the form of lubricants, aerosols, PLASTICS, fuels, etc., will generate free carbon when they come into contact with an electrical discharge at elevated pressure. In a high-pressure, oxygen-rich environment, the life-support subsystem manager should check every piece of plastic within the entire system to prevent the danger of short circuits due to carbon bridges.

7. ALWAYS USE NITROGEN OR OTHER INERT GAS IN AN ACCUMULATOR SYSTEM.

8. ENSURE THE PROVISION OF CO_2 AND DRY CHEMICAL FIRE EXTINGUISHERS IN EVERY COMPARTMENT.

9. ENSURE THAT EVERY SYSTEM MANAGER, SUBSYSTEM MANAGER, CONTRACTOR, SUPPLY DEPOT, AND MAINTENANCE CREW HAS A FIRE-HAZARD CHECKLIST.

Thermal Protection

In order to maintain an adequate life-support subsystem, the temperature, humidity, and ventilation must be controlled. At 50% relative humidity, the comfort zone ranges from 72°F to 78°F. At the higher relative humidities usually present in an ocean system, this range is still a good guideline.

The basal metabolic rate for light to steady work varies from 400 to 800 Btu/man-hour. For sleeping, the rate drops to 250-350 Btu/man-hour. Most of this heat is released to the atmosphere and must be dissipated by the ventilation and temperature control systems.

The two most common temperature control systems are:

1. Forced-Air Convection Cooling, in which cooling air is circulated through the system and then brought into contact with the hull to exchange the heat with the water environment.

2. Vapor Compression Cooling, in which normal refrigeration techniques are used.

The life-support manager must realize that in a mixed gas atmosphere, especially a helium-rich atmosphere, the thermal conductivity of the gas is higher than that of air. Temperatures will have to be maintained at a higher level because the crew will chill much more quickly.

The most serious thermal protection problems occur in the use of divers. The thermal conductivity of the water is very high. The average survival time for an unprotected human in water of 55°F temperature is about two hours. At 33°F, this survival time drops to 15 minutes. The head and extremities of a diver are areas of high heat loss and need special protection. Various forms of thermal garments have been proposed, including:

1. Normal lined-neoprene wet suits
2. Hot water suits
3. Electrically-heated suits
4. Radioactive isotope suits

In evaluating these protection garments, the life-support manager should realize:

1. As depth and pressure increase, the insulating properties of a normal lined-neoprene wet suit decrease. The open gas cells in the rubber are compressed and lose their insulation value.

2. Hot water suits have been tested satisfactorily, with one problem: local scalding and blistering. While the suit is being worn, the diver is fairly insensitive to a nonuniform overheating in a small

area and skin damage may result.

3. Electric and radioactive isotope suits will require further development before they can be considered operational equipment.

Pressure and Pressure-Gradient Protection

The life-support subsystem must maintain the pressure of the artificial environment so that the crew does not encounter a harmful temporal pressure gradient ($\frac{dP}{dt}$), the change in pressure with respect to time).

Most men can withstand a very high positive pressure gradient, that is, the pressure may increase very rapidly with little effect. The main problem is equalization of pressure across the eardrum. A pressure differential of 5 psia may cause damage to, or rupture, the eardrum. All personnel who enter a pressurized environment should have diver training to learn to recognize and alleviate pressure on the eardrum by the pressurizing process.

The negative pressure gradient (a decrease in pressure) is the more dangerous situation. Depending on the rate of pressure decrease, either or both of the following conditions may occur:

1. Embolism. Embolism, or bursting of tissues by gas bubbles, occurs when there is a high-pressure differential between the lung and the atmosphere. The blood vessels and alveoli (air sacs) of the lungs can withstand a pressure difference of only about 2 psia without damage. Therefore, if the ambient pressure should drop by much more than 2 psia instantaneously, without allowing an individual to relieve the pressure in his lungs (i.e., if he holds his breath), damage may result. Air bubbles may enter the bloodstream, and will lodge in the brain causing

severe and permanent paralysis or death.

2. Decompression Sickness (The Bends). Decompression sickness may occur at a much smaller pressure gradient. At increased pressures, the amount of gas which will dissolve into the bloodstream increases. The body tissues act as a membrane containing the gas. The solution and dissolution processes are not instantaneous, and the rate of dissolution varies with different tissues. The pressure difference across a tissue is known as the tissue tension, and affects the dissolution rate according to the following relation:

Ambient Pressure = P_{am}

Tissue Pressure = P_t (pressure of dissolved gas within tissue)

Tissue Tension = $[P_{am} - P_t]$

$$\frac{d}{dt}[P_{am} - P_t] = K[P_{am} - P_t]$$

As previously stated, if the tissue tension is positive (that is, $P_{am} > P_t$), no damage will occur. The situation becomes dangerous when the tissue tension becomes negative ($P_t > P_{am}$). As a basic rule, P_t should never exceed P_{am} by more than 1 atmosphere (14.7 psia). Solution of the differential equation shows that the tissue tension will decay exponentially with a time parameter of K. K is sometimes called the "tissue constant," and each tissue type has its own value of K_1 varying from 5 min^{-1} to 240 min^{-1}.

The rate of pressure decrease $\left(\frac{d}{dt}[P_{am} - P_t]\right)$ must be controlled so that the tissue tension $\left([P_{am} - P_t]\right)$ does not go below -14.7 psia. If it drops below this value, the gas will not have time to evolve from the tissues and bubbles will form within the tissue structure. At a minimum, this will cause pain or mechanical damage in the locally-affected area. At a maximum, severe internal ruptures, internal hemorrhaging, and death can ensue.

Different tissues in the body will have different time constants. The physiological range of tissue time constants is not now known, and it is therefore presumed that a full range of time constants from zero to infinity does, in fact, exist in the body. The centrally critical tissue for decompression is a function of the pressure-time history of exposure. It is computable, but, for purposes of safety and reliability, it is common to use prepared diving tables. The use of such tables will, however, constrain operational flexibility.

Although saturation diving is now well within the state of the art of ocean technology, system and subsystem managers should recognize that the use will require the assemblage of previously-trained and certified individuals, and extensive licensing, certification, and safety procedures before any operations can be initiated.

Whenever a life-support subsystem entails exposing a crew to an elevated pressure environment for any length of time greater than 20 minutes, the manager should contact a reputable saturation diving company.

Shock and Vibration Protection

It is the responsibility of the life-support manager to check the vibration characteristics of all spaces occupied by the crew, and to modify these characteristics if necessary. The human tolerance levels for vibration are highly frequency-dependent, with moderate amplitude-dependence. Performance is adversely affected in the frequency range of 100 cycles/sec (and an amplitude of 10^{-4} inches) to 1 cycle/sec (and an amplitude of 10^{-1} inches). The vibration range of 2 to 6 cycles/sec is especially dangerous because this is the natural resonant frequency range of the human body. All vibrations in this frequency range should be eliminated.

Aural Protection

The life-support manager must maintain noise control in all living and working spaces.

1. In sleeping areas, the sound pressure level should be below 30 db (The sound pressure level is referred to a pressure of .0002 dynes/cm^2.)

2. In general work areas, the sound pressure level should be maintained within 30 to 50 db. In this range, normal speech communication is not impaired.

3. Between 50 to 70 db communication is difficult.

4. The threshold of pain is approximately 120 db (2.9×10^{-3} psia). Sound pressure levels above this limit may result in permanent damage to the eardrum. In this pressure range, the ear is most sensitive to sounds at a frequency of about 8000 cycles/second.

When system operation entails the use of free divers, the life-support manager should also bear in mind:

1. The velocity of sound is different in water, gas mixtures, and air.

2. The ability of a human to determine the direction of a sound source is derived primarily from the pinna (exterior ear structure and lobe), not from binaural reception. When the ears are covered by a hood

or diving mask, this directional ability is severely impaired.

3. The ear is the most frequent spot of bacterial infection for divers. Special care should be taken to rinse the ears after exposure to sea water.

4. Any pain in the ear, or difficulty in clearing the ears and sinuses, is cause for discontinuing all diving activities. If a diver cannot clear his ears quickly and easily, he should not be permitted to enter the water.

Vocal Protection

In an ocean system utilizing a mixed-gas breathing system (especially helium), speech communication is difficult. The essential difficulty is that the sound velocity in the medium has changed. The production of speech involves the generation of fundamentals by the vocal folds and resonances in the oral (first formants) and nasal (second formants) cavity. Fricatives and sibilants are also generated by passage of gas between the lips, teeth and/or various configurations of the tongue. To a first order, helium gas shifts the resonant frequencies of oral and nasal cavity to a much higher value. This produces a "Donald Duck" quality and makes the speech highly unintelligible. Most devices (speech unscramblers) for vesting intelligibility operate on the principle of dividing the produced speech into three parts of the frequency spectrum, frequency-shifting the first and second formant portions down to their more normal levels and then resynthesizing.

Visual Protection

The usual measures of illumination are the lumen or the foot-candle:

1 lumen = 1.5×10^{-3} watt

1 foot-candle = 1 lumen/square foot

In general living areas, 40 foot-candles of illumination is ideal. Reading and examination areas usually require about 100 foot-candles of illumination. Any exterior area to be monitored by television will require very intense illumination.

A system which will operate at the free surface will also require some provision for night-vision adaptation. Standard naval practice is the wearing of dark red glasses prior to standing a night watch.

In designing a system which utilizes free divers, the life-support manager should realize:

1. Bacterial infections of the eyes are a common problem with divers, and medication should be readily available.

2. Any area near the bottom will usually become clouded with silt after moderate diver activity, and visual communication is greatly impaired.

Physiological Monitoring

The life-support subsystem should maintain the capability of monitoring certain physiological functions:

1. Particular attention should be placed on the calcium salts, sugar, and albumin (through urine analysis) to check the dietary and environmental balance.

2. Blood clotting rate should be checked.

3. Psychomotor response, headaches, and drowsiness should be checked as indicators of toxic substances in the breathing mixture.

Psychologically Adequate Environment

The psychological environment in which the operators of a sea system will find themselves is of vital importance in determining the effectiveness of the system. It is not enough that the systems manager merely allow adequate living space for each crew member and, while a value of 1000 cubic feet per man is considered a reasonable living space allocation and while the U.S. Navy currently recommends 100 square feet per man sleeping space, such allocations and requirements are by no means indicative of a successful psychological environment. The provision of many other factors of the environment, such as recreation, environmental stimulation, reproduction of familiar stimuli, regulation of sleeping and working cycles, and personality interactions, are equally vital in this subsystem consideration.

Roughly speaking, a differentiation in psychological environment can be made between systems which have a crew size in excess of 40 people and those which have a crew size less than 40 people. In the larger group, provided that access and mobility is allowable and maintainable in the larger size group, then the ability of individuals to relieve psychological stresses and strains which result from personality conflicts is sufficiently large that less attention need be paid in such environments to personality compatibilities. When the group is less than 40, however, then the ability to "move the oddball around" is greatly constrained and, if there is a series of operations in excess of a week to 10 days, extremely stressful relationships can build up which affect not only the operability of the system, but which can, in fact, result in actual human conflict. It is important for small groups that a psychological evaluation and crew competitional techniques be provided as an essential part of the system. In layman's terminology, one should be particularly careful to identify the following human traits:

a) The dogma level,

b) The need for achievement,

c) The need for social affiliation, and

d) The need to dominate.

It should be fairly obvious that a crew which has substantial numbers of individuals with high need to dominate and high dogma level is one which is prone to conflict. It is thus obvious, but nevertheless true, that conflict will arise when there is a mixture of individuals with high need to achieve and low need to achieve, or a mixture of individuals in which the rank structure and the need to dominate does not have a correspondence.

In addition to personality, the system design must be very careful not to introduce dangers as a result of learned responses. In non-ocean systems, very classic examples of this failure are the high percentage of accidents and fatalities which occur to American and European pedestrians who are first initiated to British traffic systems as the result of their natural tendency to look in the wrong direction when stepping off the street curbing, and the similar high percentage of accidents to residents of Great Britain when traveling in the United States or on the continent. In installations like the Sea Lab, such natural responses as dipping the finger into a fluid to test its temperature, is at atmospheric pressure always accompanied by some visual cue of a gas or a fluid. Such visual cues are not present in saturated high-pressure environments and this natural and normal human reaction becomes a very dangerous one. On sea systems which are located at the free surface, the pitching and heaving and rolling motions of the surface platform will be such that many, many normal human responses with respect to stowage of gear, with respect to placement of articles and utensils of normal living become inappropriate and even dangerous, particularly if items are left adrift during calm periods, and the awareness that they are adrift does not take place until a fully-risen sea is encountered. It is for these reasons that behavioral reviews of all man-operated systems are required to ensure the avoidance of catastrophe during actual operations.

Safety Certification

The safety certification of the overall system is the responsibility of the system manager. However, safety considerations rest most heavily on the life-support subsystem, and there the life-support subsystem manager should be carefully monitoring the safety checkout of his subsystem.

The author highly recommends that safety certification of a system be conducted by an entirely independent (from any subsystem) certification team. The safety inspection and certification should be a continuous process, and should not be delayed to the construction (or even worse, the testing) phases of system evolution. Safety considerations begin to enter the picture as early as the determination of the mission profile, and carry through all the design work.

The life-support manager should also bear in mind that the overall system manager probably will (and definitely should) retain the right to waive safety certification at any time and assume the responsibility himself. Many systems have suffered huge delays due to a snag in the certification process, which could have been corrected without interrupting the system design, construction, or operation.

CASE STUDY: LIFE SUPPORT

The following case study of a "pollution control system," was selected as an excellent example of the proper management of the life-support subsystem. The system contains all the problems associated with a submerged, manned system, with the added constraint that the system cannot contaminate to the slightest degree the surrounding marine ecology. The author, Ronald C. Gularte, submitted the report while participating in the M.I.T.-Woods Hole Oceanographic Institution Joint Graduate Degree Program in Ocean Engineering. Mr. Gularte received the Bachelor of Science and Master of Science degrees in Mechanical Engineering from the University of Southern California, and has had extensive engineering experience with offshore oil platforms.

THE ESTABLISHMENT OF TOLERANCE LEVELS AND FORECASTING PROCEDURES FOR THE INFLUX OF POLLUTANTS ONTO THE CONTINENTAL SHELF AREAS ADJACENT TO THE MAJOR RIVERS OF NORTH AMERICA

by

Ronald C. Gularte

Subjective Human Purpose

The determination of the seasonal as well as long-term effect of the influx of polluted water into the continental shelf areas adjacent to the mouths of the great rivers of North America.

The purpose of the mission is to establish tolerance level for pollutants and develop a forecasting procedure to avert catastrophic losses to plant and animal life on the continental shelves as well as to control the amount of pollution that flows from the mainland.

MISSION ANALYSIS

Information to be Obtained and Objects to be Collected

1) Chemical Composition of Water
   a) Pesticides, fertilizers and industrial wastes
   b) pH
   c) Dissolved gases
   d) Oxidation - reduction potential
   e) Inorganic agents
2) Biological Interrelations
   a) Natural association of organisms
   b) Nutritional relationships
   c) Organic matter
   d) Bacterial distribution
3) Physical Properties
   a) Temperature, salinity and density
   b) Thermal properties
   c) Diffusivity
   d) Absorption of radiation
   e) Viscosity
4) Sedimentation
   a) Suspended organic and inorganic matter
   b) Erosion
   c) Bottom transport phenomena
   d) Deposition rates
   e) Bottom and core samples

Location

The continental shelf adjacent to the following rivers:

1) Mississippi (Gulf of Mexico)*

2) St. Lawrence (Atlantic Ocean)*

3) Colorado (Gulf of California)**

4) Columbia (Pacific Ocean)†

It should be noted that, as ocean pollution becomes of greater concern, other rivers, such as the Ganges (Bay of Bengal), the Indus (Arabian Sea), and the Nile (Mediterranean Sea), also become likely candidates for the mission.

Constraints on Measurements

The data must be taken with as little disturbance to the natural environment as possible.

The specimens that are to be retrieved must be preserved in their natural state until they can be examined conclusively or test-performed. [*Beware Process Fluids*]

The parameters are to be measured within present state-of-the-art techniques with the precision and accuracy defined at the subsystem or component interface.

---

*These rivers are of particular interest in that they contain pesticides, fertilizers, and industrial wastes.

**The Colorado has little industrial waste, but a considerable amount of agricultural waste. The flow is also regulated through a large irrigational system for California and Arizona.

†The pollution in this area is inextricably connected to the lumber industry.

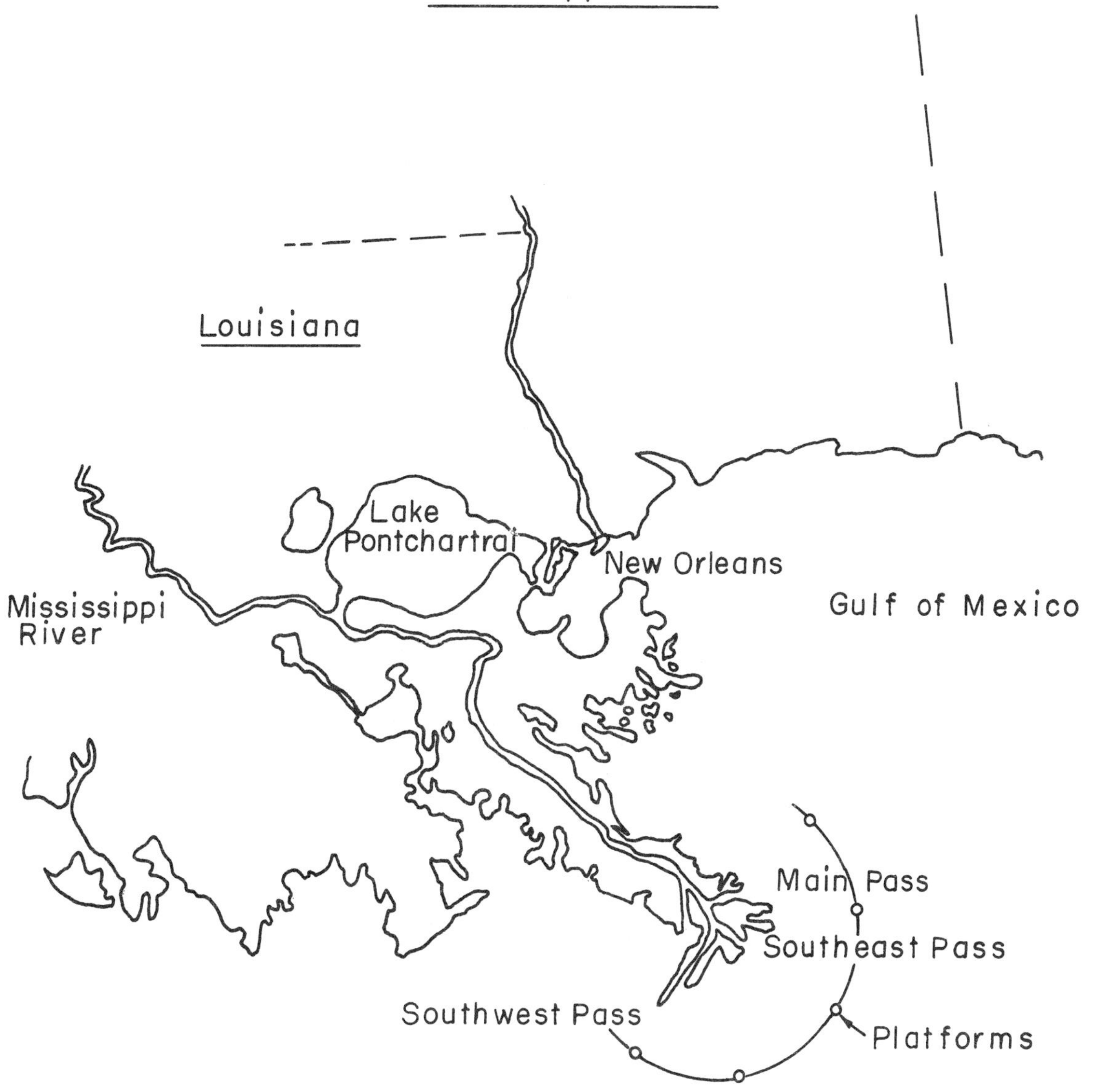
Mississippi River
Louisiana
Lake
Pontchartrai
New Orleans
Mississippi
River
Gulf of Mexico
Main Pass
Southeast Pass
Southwest Pass
Platforms
Continental Shelf

0
25
50

Figure 3-2

St. Lawrence River

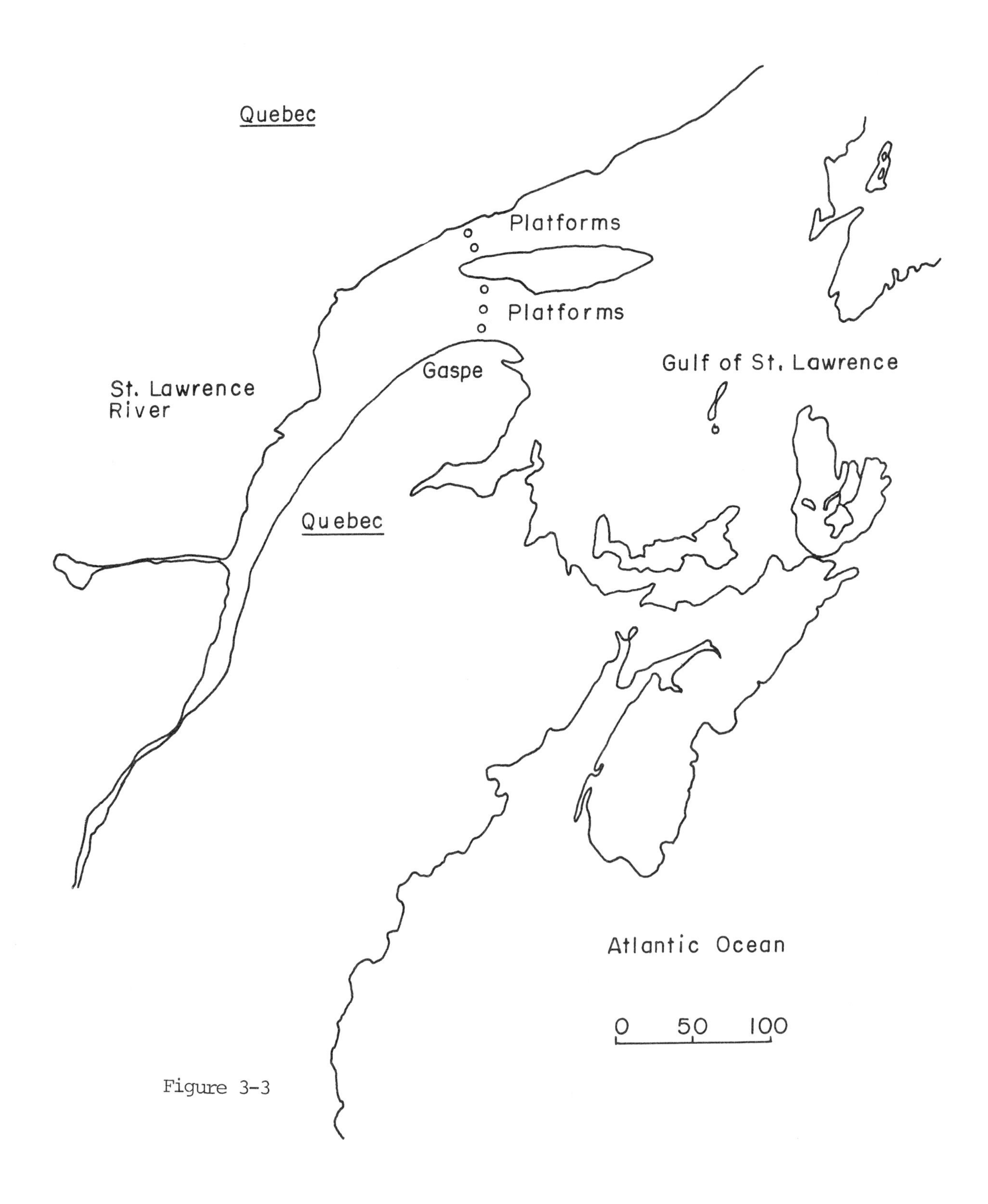

Figure 3-3

Colorado River

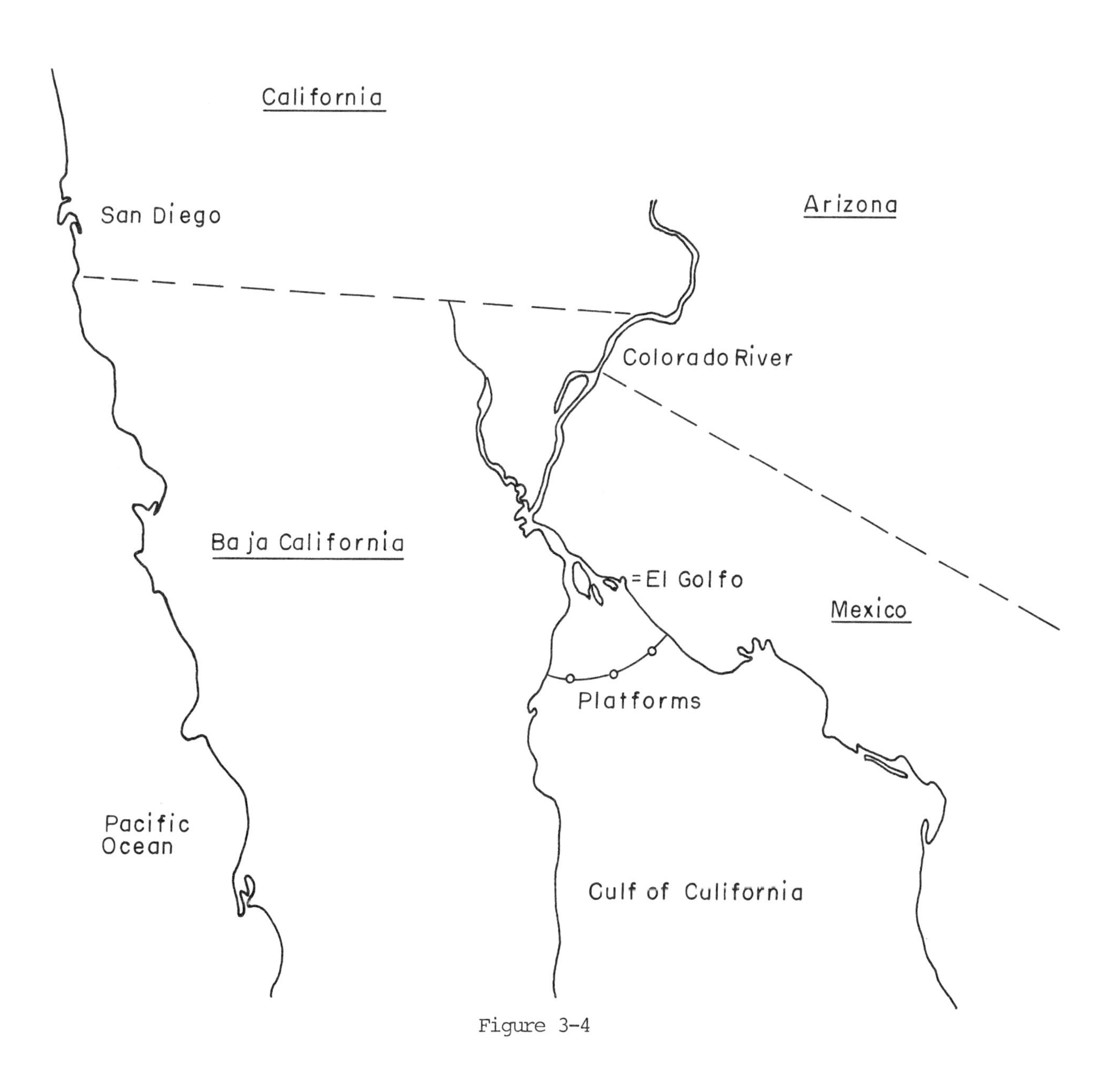

Figure 3-4

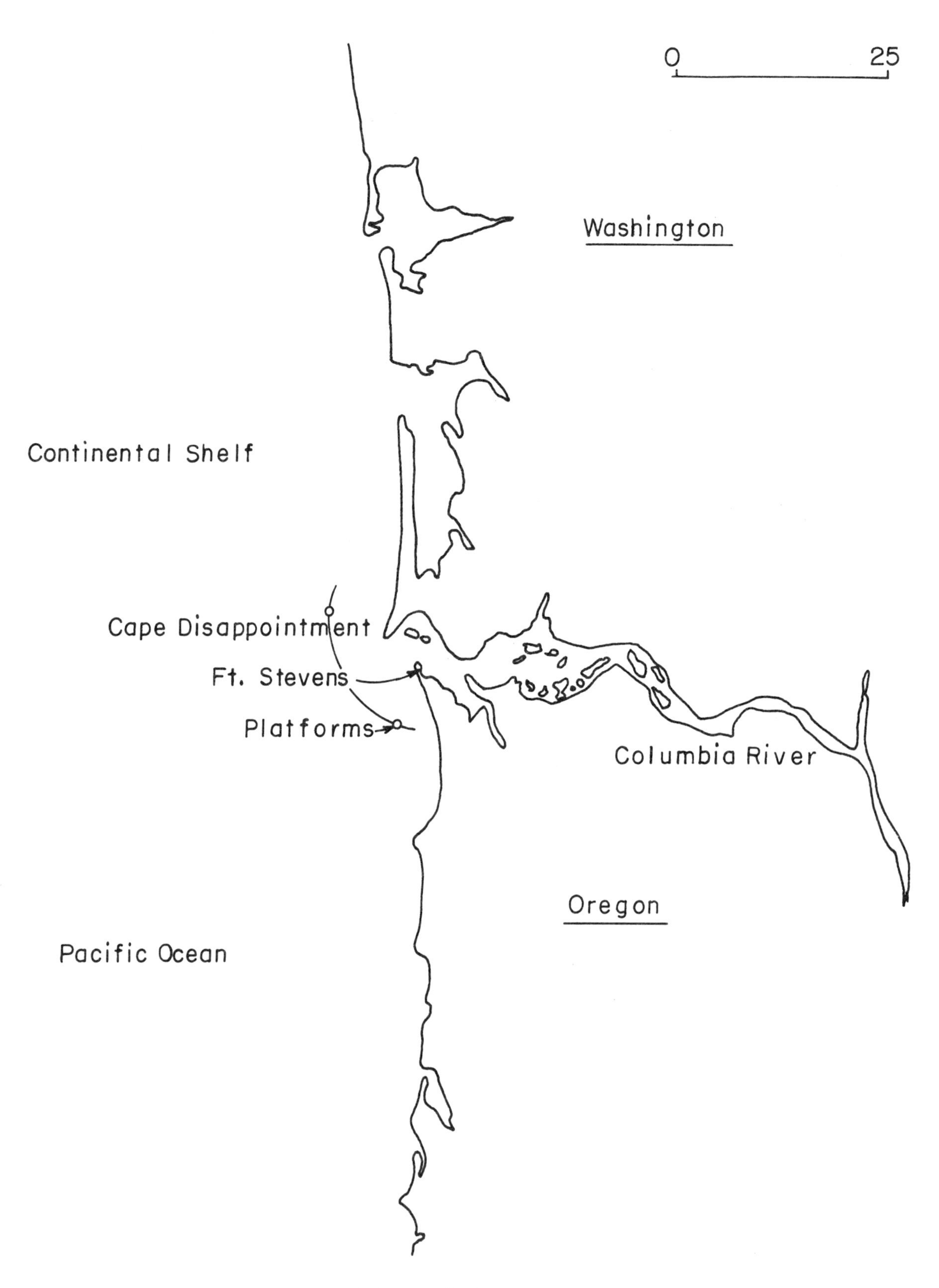
Columbia River
0
25
Washington
Continental Shelf
Cape Disappointment
Ft. Stevens
Platforms
Columbia River
Oregon
Pacific Ocean

Figure 3-5

Obstacles to Measurements

The mission requires that measurements be conducted on a yearly basis and therefore the system must operate in all sea states.

Precautions are to be taken to ensure that as little risk as possible is taken with respect to environment and predators.

Time Limits on Collection of Data

The purpose of the mission is to provide year-round information on pollution and therefore data will be collected continuously.

Quantity of Processing

The information that is gathered on pollution will be of a magnitude which will require computerization.

The quantity (number of monitoring locations) will determine the accuracy of the computerized forecasting procedure.

Time Limit on Processing

The data must be correlated, processed, and predictions made with sufficient lead time so that corrective measures can be taken.

Reliability and Confidence Level

The overall system is to have an 80% reliability with a 90% confidence level.

MISSION PROFILE

Operational Phase

Data Gathering (six months)

(The stations are in place on the continental shelf and data channels are operational.)

- Each crew arrives at its particular site (30 men to each site).
- Final briefing.
- Crew transported to habitat.
- Excursion vehicle made operational (batteries charged, life-support system activated, etc.).
- The first three men of the first crew depart gathering data and performing test in preassigned area, an approximately four-hour excursion.
- First crew returns.
- Specimens and data removed from excursion vehicle.
- Second crew readies excursion vehicle.
- First crew examines, identifies and preserves specimens.
- Data and appropriate specimens transferred to ambient laboratory for further investigation.
- The first shift of the second crew is briefed and departs.
- The first crew has approximately 12 hours to rest and relax.
- The second crew returns, examines, tags and preserves the specimens collected.
- The excursion vehicle is readied for the second shift of the first crew.
- Upon returning the second shift of the second crew departs.
- When they return both crews undergo decompression.

(By this operation a total of 12 hours of immersion time (for collecting data and performing tests) in 48 hours of saturated diving time. Each diver is wet only four hours during this 48-hour period. Initially, 24 hours of testing and data-gathering will be needed.)

Data Processing (two months)

- The data will be preprocessed in the ambient laboratory (observations digitized, etc.).
- The data is then telemetered to a centralized computer center.
- Field data from locations along rivers near large cities.
- The data is analyzed at centralized computer center.
- Pollution forecasting is performed.
- Until desired accuracy is obtained, further tests will be required.

PLATFORM CONFIGURATION

Provision shall be made for living and working space for the following number of men and combinations of laboratories:

1) An atmospheric pressure laboratory for 30 men;
2) A decompression-recompression laboratory with a 15-man capacity, and
3) A dry saturation laboratory for 15 men and an excursion vehicle.

The design depth shall be 1,000 feet of sea water. A minimum free-flooding volume of 1,000 cubic feet per man shall be provided throughout the platform.

Hull penetrations shall be provided for the transfer of men from the atmospheric to the decompression-recompression laboratory. (Hatch diameter a minimum of 48 inches in diameter.) Also transfer chambers will be provided of sufficient capacity to handle a cubic container two feet on each side. This capability shall be provided between the atmospheric laboratory and both the decompression-recompression laboratory and the ambient dry laboratory. A viewing port shall be provided between the atmospheric and the saturated laboratory.

The decompression-recompression laboratory shall have access to men in both the atmospheric and saturated laboratories. It shall be provided with living and working space so the useful work can be performed under lengthy decompression or during recompression.

The dry saturated laboratory shall provide living and working space for the crew as well as a bay for the excursion vehicle. An elevator is to be provided to raise the excursion vehicle into the ambient laboratory.

Provision shall be made for both the atmospheric and the decompression-recompression to mate with the DSRV. Also, a hatch shall be provided on the ambient laboratory for mating with a saturated diving capsule which can also be used to effect a rescue. A docking platform will also be provided to allow the submersible tender to mate with the atmospheric laboratory for the purposes of resupply.

The platform shall be designed to gain buoyancy during descent, providing inherent safety and requiring less operator control during descent. Also, the platform must be hydrostatically, as well as hydrodynamically, stable.

The bottom support structure must be capable of attaching to slope of up to 15 degrees. This structure is to be fastened to the shelf or anchored by some means. The design shall allow for the separation of the support structure from the platform and the raising of the latter to the surface for maintenance. Upon conclusion of maintenance and refitting, some means shall be provided to return the platform to the bottom support structure.

Life Support Requirements and Constraints

Requirements

1. Provide up to 30 men with an essentially self-contained environment for an extended period of time (essentially in perpetuity).
2. Provide an atmospheric, transition, and ambient-pressure life support for depths up to 1000 feet.
3. Emergency life support for five days.
4. Logistic support on 30-day intervals.
5. Will use approximately 70 kw of power continuously.

Constraints

1. Safety is paramount.
2. Hull penetrations allow gas out, water and heat both in and out. Sewage only into sewage system.
3. Smoking allowed.
4. Allow up to five years for equipment development.
5. Provide an ecological system that is physiologically as well as psychologically adequate.

Life-Support Subsystems*

1. Atmospheric control
2. Comfort control
3. Water management
4. Food management
5. Waste control
6. Environmental monitoring and emergency equipment

[*Note that the sewage system is to be treated as a separate subsystem.*]

Three Life-Support Modes

1. Laboratory at atmospheric pressure
2. Decompression-recompression laboratory
3. Lockout-ambient pressure laboratory (includes excursion vehicle)

Environmental Monitoring and Emergency Equipment

The equipment to be monitored depends on the subsystems selected for the particular life-support system.

---

*Nonexpendable, except for food and backup (emergency).

Figure 3-6

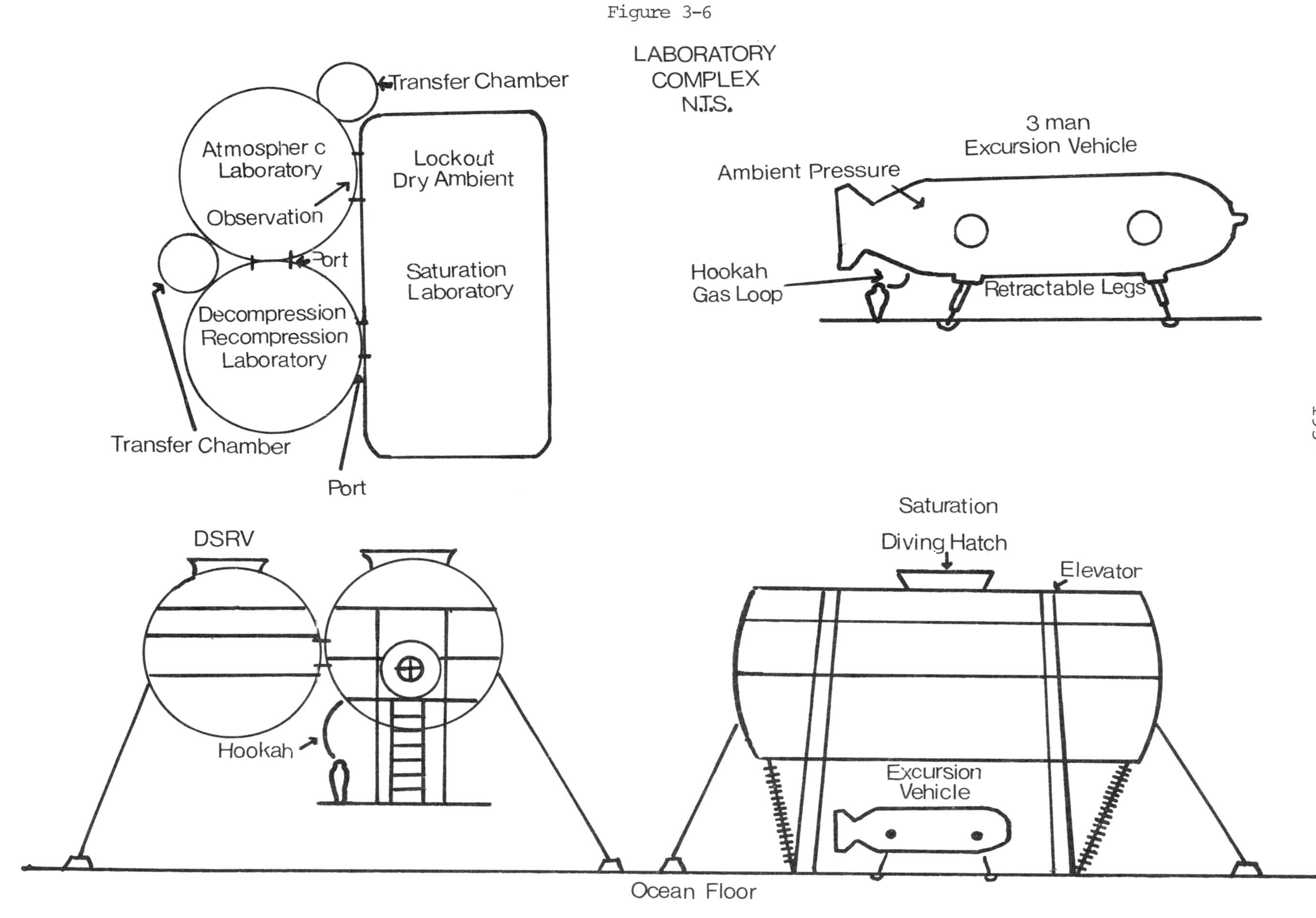
LABORATORY
COMPLEX
N.T.S.
Transfer Chamber
Atmospher c
Laboratory
Observation
Port
Lockout
Dry Ambient
Saturation
Laboratory
Decompression
Recompression
Laboratory
Transfer Chamber
Port
3 man
Excursion Vehicle
Ambient Pressure
Hookah
Gas Loop
Retractable Legs
DSRV
Hookah
Saturation
Diving Hatch
Elevator
Excursion
Vehicle
Ocean Floor

LIFE SUPPORT SYSTEM

System	Subsystem	Method	Weight (lb)	Volume (ft^3)	Power (KW)	Cost ($)
Atmospheric control system	Oxygen supply & carbon dioxide removal	Water electrolysis & molecular sieves	5,600	105	71	150,000
		Activated charcoal	450	25	-	700
	Trace contam-inant control	Absolute filter	450	17	1.7	1,000
		Catalytic burner	750	35	7.0	9,500
Comfort control	Temperature, humidity & ventilation	Forced air convection cooling	500	35	4.5	3,000
	Illumination	Internal incandescent	100	5	4	1,000
Water management	Sea water & recovery stills	Wash water & condensate recovery	3,300	90	15	30,000
Food management	Fresh, canned & frozen	Replenished on 30 day intervals cost 30% less	11,000	450	6	35,000
Waste management	Combination holding tank with discharge capability	Power required essentially only when dumping	2,300	176	10	11,000
Environmental monitoring & emergency equip. (5 days)	Selected for minimum: 1) Maintenance 2) Power 3) Cost 4) Volume	See Emergency System & Environmental Monitoring Breakdowns	(Battery wt.) (8000 lbs) 15,000	220	75 KWH	24,000
		TOTAL LIFE SUPPORT	29,450 lb	1,158 ft^3	194.2 KW	265,200 $

The following list under the heading "Environmental Monitoring" illustrates the parameters to be monitored for the life as well as the emergency support system.

Environmental Monitoring

1. Oxygen analyzer*
2. CO_2 analyzer*
3. Hydrogen analyzer* $< 3\%$
4. Humidity
5. Temperature (internal)
6. Ventilation (circulation blowers)
7. Internal pressure
8. Potable water level
9. Holding tank level
10. Absolute filter blower
11. Catalytic burner blower
12. Activated charcoal blower
13. Water electrolysis monitor
14. Emergency power monitor
15. Sea temperature
16. Sea pressure
17. Clock
18. Noise level (in various areas)
19. Vibration levels
20. Radiation
21. Television monitor
22. Emergency communication (body condition)
23. Fire sensors

The emergency life-support system was selected on the basis of minimum maintenance, power, cost, and volume.

It is a complete entity and is to be completely divorced from the main life-support system.

---

*Continuously monitored.

It should include open and closed-circuit emergency breathing apparatus, also firefighting equipment.

Life-Support Personnel Breakdown

1	Medical doctor
2	Physiologists
2	Engineers specializing in gas analysis and monitoring
2	Specialists in charge of gas mixing
2	Specialists responsible for the decompression program (work 12+-hour shifts)
3	Teams of three divers each (one in excursion vehicle, two gathering specimens of performing tests)
1	Cook and general purpose
11	Scientists and engineers (maintenance, electrical and instrumentation people included here)
30	TOTAL

Selection Criteria in Order of Importance

1.	Minimum cost	*
2.	State of the art	≤2 **
3.	Safety	≤2
4.	Reliability	≤2
5.	Trace contaminants	*
6.	Noise	≤2
7.	Electrical interference	≤2
8.	Minimum volume	*
9.	Minimum weight	*
10.	Minimum power	*

[The following charts illustrate a very thorough analysis of both the quantitative and qualitative characteristics of the alternative designs open to the life-support manager.]

---

*Quantitative criteria
**Qualitative criteria:

acceptable = 1
usable = 2
marginal - not acceptable at this time = 3

QUALITATIVE ANALYSIS

Method	S.O.A.	Safety	Reliability	Trace	Noise	Electrical
Oxygen Supply System (Regenerable)						
Water electrolysis	1	2	2	H_2	1	1
CO_2 Removal System (Regenerable)						
MEA	1	1	2	MEA	2	1
Alkali oxides	3	1	2	None	2	1
Solid amine	3	1	2	Clycol	2	1
Molecular sieves	2	1	2	None	2	1
CO_2 freezing	3	1	2	None	2	1
Combined Single-Step O_2 Supply - CO_2 Removal System						
Sodium sulfate cycle	3	2	3	H_2	2	1
Electrodialysis	3	2	3	H_2	2	2
Photosynthetic gas exchange	3	2	3	None	1	1
Solid electrolyte	3	2	2	CO	1	1
Molten carbonate	3	2	2	None	1	1
CO_2 with H_2 to Produce H_2O						
Bosch	3	2	2	H_2 CH_4 CO	1	1
Sabatier	3	2	2	H_2 CH_4	1	1

1 = acceptable

2 = usable

3 = not acceptable at this time

QUALITATIVE ANALYSIS

Method	S.O.A.	Safety	Reliability	Trace	Noise	Electrical
Atmospheric Control System						
H_2O electrolysis & molecular sieves	2	2	2	H_2	2	1
H_2O electrolysis & MEA	1	2	2	MEA H_2	2	1
H_2O electrolysis & alkali oxides	3	2	2	H_2	2	1
Solid electrolyte	3	2	2	CO	1	1
Bosch	3	2	2	H_2 CH_4 CO	1	1
Trace Contaminant Control						
Activated charcoal	1	1	1	None	2	1
Absolute filter	1	1	1		2	1
Electrostatic precipitator	2	2	1	Ozone	2	3
Catalytic burner	2	2	1	HF from freon	2	2
Temperature, Humidity, and Ventilation Control						
Vapor comp. cooling	1	1	1	Freon 12	2-3	2
Forced air convection cooling	2	1	1	None	2	2
Thermoelectric cooling	3	1	1	None	2	2

1 = acceptable

2 = usable

3 = not acceptable at this time

QUALITATIVE ANALYSIS

Method	S.O.A.	Safety	Reliability	Trace	Noise	Electrical
Water Management						
Stored	1	1	1	None	None	None
Sea water still	1	2	1-2	None	1	2
Wash & condensate water recovered by filtration	2	2	1-2	None	1	2
Food System						
Fresh and canned	1	1	2	Odors, Heat	1	1
Frozen	1	1	2	Odors, Heat	1	1
Freeze-Dried	2	1	1	None	None	None
Dehydrated	2	1	1	None	None	None
Waste Management						
Manual collection, chemical treatment, storage	1	2	1	Odors	None	None
Collection & freezing	2	2	2	Odors	1	1
Holding tank	1	1	1	Odors	None	None
Discharge overboard	2	2	2	Odors	2	2

1 = acceptable

2 = usable

3 = not acceptable at this time

QUANTITATIVE ANALYSIS

Method	Weight (lb)	Volume (ft^3)	Power (KW)	Cost ($)
Oxygen Supply System (Regenerable)				
Water Electrolysis	5,000	50	15	100,000
CO_2 Removal Systems (Regenerable)				
Monoethanolamine scrubber	1,700	60	8	35,000
Molecular sieves	600	55	2	50,000
Magnesium oxide Silver oxide	450	55	2	45,000
Solid amine	1,200	55	2	150,000
CO_2 freezing	350	65	2	85,000
Combined Single-Step O_2 Supply - CO_2 Removal System				
Sodium sulfate cycle	650	150	20	65,000
Electrodialysis	3,000	50	30	300,000
Photosynthetic gas exchanger	6,700	200	120	700,000
Solid electrolyte	900	100	17	160,000
Molten carbonate	1,800	60	12	300,000
Closed Cycle O_2 Supply - CO_2 Removal*				
Bosch	6,000	150	20	325,000
Sabatier	8,000	150	20	525,000

* Method	CO_2 Concentration	O_2 Generation	CO_2 Reduction
Bosch	Molecular Sieve	H_2O Electrolysis	Bosch Reaction
Sabatier	Molecular Sieve	H_2O Electrolysis	Sabatier Reaction

QUANTITATIVE ANALYSIS

Method	Weight (lb)	Volume (ft^3)	Power (KW)	Cost ($)
Atmospheric Control System				
Water electrolysis & molecular sieves*	5,600	105	17	150,000
H_2O electrolysis & MEA	6,700	110	23	135,000
H_2O electrolysis & alkali oxides**	5,450	105	17	145,000
Solid electrolyte	900	100	17	160,000
Bosch**	6,000	150	20	325,000
Trace Contaminant Control				
Activated charcoal***	450	25	-	700
Absolute filter	450	17	1.7	1,000
Electrostatic precipitator	500	25	2.0	7,000
Catalytic burner	750	35	7.0	9,500
Temperature, Humidity and Ventilation Control				
Vapor compression cooling (freon)****	1,300	25	10	20,000
Forced air convection cooling	500	35	4.5	3,000
Thermoelectric cooling****	1,500	20	15	50,000

*Selected because of no trace contaminants since costs are essentially equal.

**Do not meet S.O.A. requirement.

***30-day life.

****Compared on estimated 10-ton refrigeration basis.

QUANTITATIVE ANALYSIS

Method	Weight (lb)	Volume (ft^3)	Power (KW)	Cost
Water Management				
Stored* (30 days)	240,000	4,000 (30,000 gal)	For heating only	230,000
Sea water still 10 gal/man-day	1,100	30	5	10,000
Multi-filter wash water & condensate* recovery	5,000	100	0.15	70,000
Urine & wash & condensate 2 multi-filters & vap. comp.--abs.*	1,000	20	0.65	80,000
Wash water & condensate recovery still 20 gal/man	2,200	60	10	20,000
Food System**				
Fresh & canned	12,000	450	6	35,000***
Frozen	10,000	450	6	35,000***
Freeze-dried	8,500	230	2	50,000†
Dehydrated††	7,500	120	Essentially no power	60,000†

*30-day cycles, nonregenerable.
**To be replenished at 30-day intervals.
***These costs include cost of processing equipment. Replenishment costs approximately 20% less per month (choose a linear combination of these.)
†These costs are fixed and will be replenishment costs also.
††Emergency for five days only.

QUANTITATIVE ANALYSIS

Method	Weight (lb)	Volume (ft^3)	Power (KW)	Cost ($)
Waste Management				
Manual collection, chemical treatment & storage in large cans (feces only)*	300	10	-	3,000
Collection & freezing (feces only)*	400	5	0.4	20,000
Molding tank 5 lb/man-day, assuming condensate & wash water is reclaimed**	1,800	160	-	8,000
Discharge overboard (pump or high-pressure air)	500	16	10	3,000
Atmospheric Laboratory Emergency System, 1 days - 30 men (must be completely divorced from main system)				
150 man/days				
Emergency communication buoy	45	1.5	10 (when charging)	1,500
Emergency breathing KO_2 (potassium superoxide) apparatus 2 hr at work or 8 hr at rest (30 units & 30 refills	420	10	-	1,500
Chlorate candles O_2 supply self-contained oven 150 man-days	800	10	-	1,500
Lithium hydroxide CO_2 removal	500	15	300	1,000
Food dehydrated, precooked	1,000	20	-	8,000
Water stored hot in thermos bottles 1.5 gal/man-day	2,000	30	-	1,000

*Requires 30-day servicing; includes food scraps, kitchen scraps, etc.

**If you use holding tank you still have to be able to discharge overboard, but into sewage system; therefore, will need both holding tank and discharge capability.

QUANTITATIVE ANALYSIS

Method	Weight (lb)	Volume (ft^3)	Power (KW)	Cost ($)
Atmospheric Laboratory Emergency System (Continued)				
150 man-days				
Waste, manual collection & chemical treatment & storage	50	2	-	350
Temperature & humidity, warm clothes & silica gel	600	10	-	1,500
Emergency power (lead acid)	8,000	36	75 @80°F	7,500
Portable fire-fighting equip. 12 CO_2 & 12 dry chemical	720	6	-	500
Medical supply package	200	4	-	600
Decompression Laboratory and Saturated Laboratory Emergency Life Support System (5 days - 5 men in each laboratory - two systems)				
25 man days				
Emergency communication buoy	45	1.5	10 (where charged)	1,500
* Emergency breathing apparatus, potassium superoxide (5 units & 5 refills)	70	15	-	250
Chlorate candles with self-contained over (oxygen)	140	2	-	250

*For immediate emergency until candles and LiOH can be activated
Minimum - maintenance power cost, volume

QUANTITATIVE ANALYSIS

Method	Weight (lb)	Volume (ft^3)	Power (KW)	Cost ($)
Decompression Laboratory and Saturated Laboratory Emergency Life Support System (Continued)				
25 man days (Continued)				
Lithium hydroxide CO_2 removal	80	2.5	50	200
Food dehydrated, precooked	200	3	-	150
Water 1.5 gal/man-day (stored) not in thermos	350	5	-	200
Waste - manually collected & chemical treatment & storage	50	2	-	350
Temperature & humidity, warm clothes & silica gel	100	2	-	250
Emergency power (lead acid)	1,500	6	12	1,200
Portable firefighting equip. 2 CO_2 & 2 dry chemical	120	2	-	100
Medical supply package	35	1	-	100
Inert gas (stored under high pressure)	This would be a function of the internal volume of the laboratory and predicted allowable leak rate			

Final Selection

1. Oxygen Supply: Electrolysis of Distilled Sea Water

Electrolysis was selected because it is the only continuous source of oxygen available to sea systems at this time. Closed-cycle systems will require further development.

Hydrogen is a by-product which can be discharged overboard, used by the gas mixer, or catalytically reacted with CO_2 for water reuse in the electrolysis system.

2. Regenerable CO_2 Removal System: Molecular Sieve

The molecular sieve was the leader in most of the selection criteria except cost, where the MEA scrubber was less expensive. However, the molecular sieve was selected because traces of MEA are produced by the scrubber, and other trace contaminants may react with MEA to produce toxic substances.

In the molecular sieve processes, the solid absorbents are proprietary zeolites. The air must be predried, using silica gel. The sieves are regenerable, using heat, vacuum, and a purge gas.

3. Trace Contaminant Control: Activated Charcoal (Odor Control), Absolute Filters (Airborne Particles), and Catalytic Burner

The activated charcoal was selected for removal of trace contaminants because small charcoal filters can be placed at sources of contamination. The absolute filter was selected (over the electrostatic precipitator) to minimize electrical interference. The catalytic burner was selected to remove hydrocarbons and CO. All three are necessary for "clean" air.

4. Temperature, Humidity, Ventilation: Forced-Air Convection Cooling

The forced-air convection cooling scheme was selected because it was cheap, reliable, simple, and produced no trace contaminants. The condensate can be collected and treated in the water management system. In seas where the water temperature is above 45°F, a dehumidifier will be used to control humidity.

5. Water Management: Sea Water Still

The sea water still was selected for make-up water, wash water, and condensate recovery. No fecal matter or urine is to be recovered. Ample electrical power will be available.

6. Waste Management: Holding Tank

The holding tank was selected for wastes excluding condensate and wash water. The holding must be discharged, using pumps or high-pressure air, into a sewage system (to be developed as a separate entity).

7. Living Area

Activity	ft^2/man
Sleeping	100
Eating	15
Recreation	10
TOTAL	125

For a crew of 30, this necessitates a space of about 3800 sq ft (allowing 7 feet clearance, this results in a volume of 26,600 ft^3).

SATURATION BREATHING APPARATUS

Method	Gas (%) Recovery	Immersion Time	Change in Water Level in Skirt	Umbilical	Miscellaneous
Classical Open Circuit	0	Limited	None	No	
Semi-Closed-Circuit "Mixer"	90	Unlimited	Yes*	Yes	Fed from gas bottles by umbilical. Well accepted
Closed-Circuit "Gas Loop" "Hookah" also electro-lung†	100	Unlimited	None	Yes	Fed from atmosphere inside excursion vehicle, can also be used for forced respiration, saving diver energy and removing CO_2. This system is also lightweight, 4 lbs, including harness. Umbilical is essentially neutrally buoyant

*Level can be controlled manually or with ultrasonic level gauge.

†Not enough known about this unit at this time.

8. Saturation Breathing Apparatus

a.	Classical open circuit	100% loss no umbilical
b.	Semi-closed circuit	90% gas saved mixers - well accepted by French Navy. Fed by an umbilical, gas from bottles. When returns to vessel, will change level in skirt of exit
c.	Closed circuit	4 lbs "gas loop" - "hookah"

Uses a gas load regulator with total gas recovery; fresh gas is supplied to diver with umbilical and polluted gas returned to vessel by same method; no change in water level.

Closed-loop saturation breathing apparatus was selected because immersion time is limited only by diver endurance. Forced respiration can be used conserving diver energy, which also will reduce CO_2 buildup. The unit is light (~4 lbs). Also, since several divers will be working out of the excursion vehicle at the same time and breathing the same gases as within the vehicle, a single monitoring system will suffice.

SATURATED DIVING SUITS

Combination	Energy Form	Umbilical	Immersion Timing	State of the Art
Woolen clothes + electrical resistance heat + neoprene suit*	Electrical - 2 to 400W	Yes	1 hr in water temperatures experienced at 1000 ft	1
Woolen clothes + warm water heat neoprene suit*	Thermal - hot water could be stored in thermos bottles on excursion craft	Yes	Depends on diver endurance and supply of hot water	1
Woolen clothes + radioactive isotope - warm water + neoprene suit*	Radioactive isotope	No	Not known	3

*Either wet or dry.

9. Diving Suits

a. Heated by a web of integrated tubing that carries warm water from an umbilical.

b. Electrically-heated "Piel" heated undergarment (resistance heat 2 to 400 W).

c. Self-contained radioactive isotope (radioactive to warm circulating water).

Thermos hot water was selected for simplicity (SOA). Problems arise in line losses with electrical resistance heating. Radioactive isotope is not acceptable from SOA point of view.

10. Gas Mixer

The precision gas mixer is an integrated breathing gas supply, control and recovery system.

It consists primarily of a gas mixer, a gas analyzer and a gas recovery unit.

The mixer has the advantage of on-site gas mixing, instead of premixing and storage. If the mixture is not as desired for a particular depth and task, immediate adjustment can be made (the mixture can essentially be optimized at the site). The gas mixture is homogeneous and cylinders do not have to be rolled as with premixing. Reanalysis is not necessary as it is analyzed as used. Furthermore, it allows strict adherence to preprogrammed recompression. While decompression is taking place, the gas that is removed is dried, the CO_2 removed, and then recompressed. Next, in the recovery unit, the gas is separated into O_2, H_2, He, etc., for reuse.

Figure 3-7

GAS MIXER SCHEMATIC

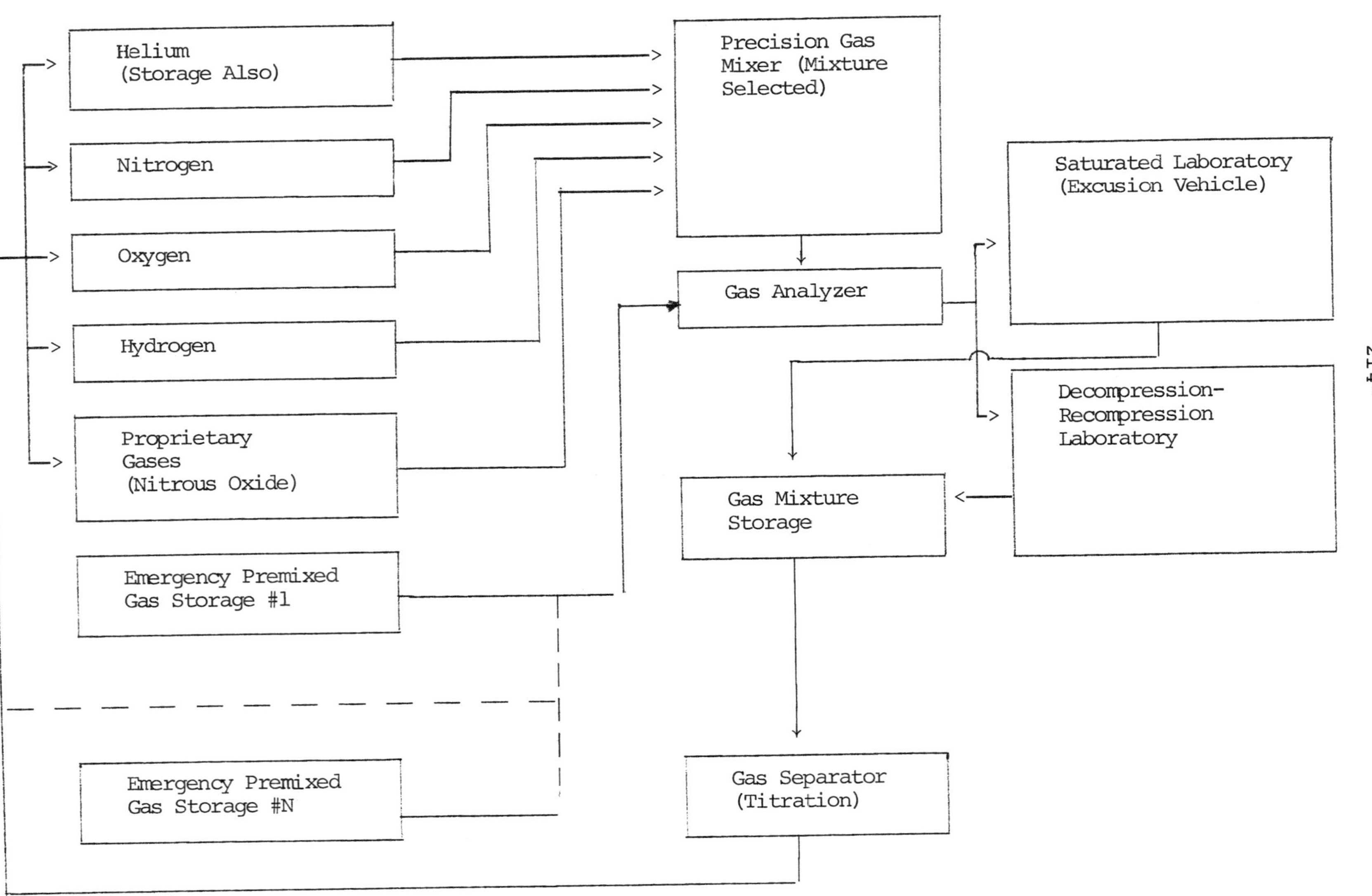

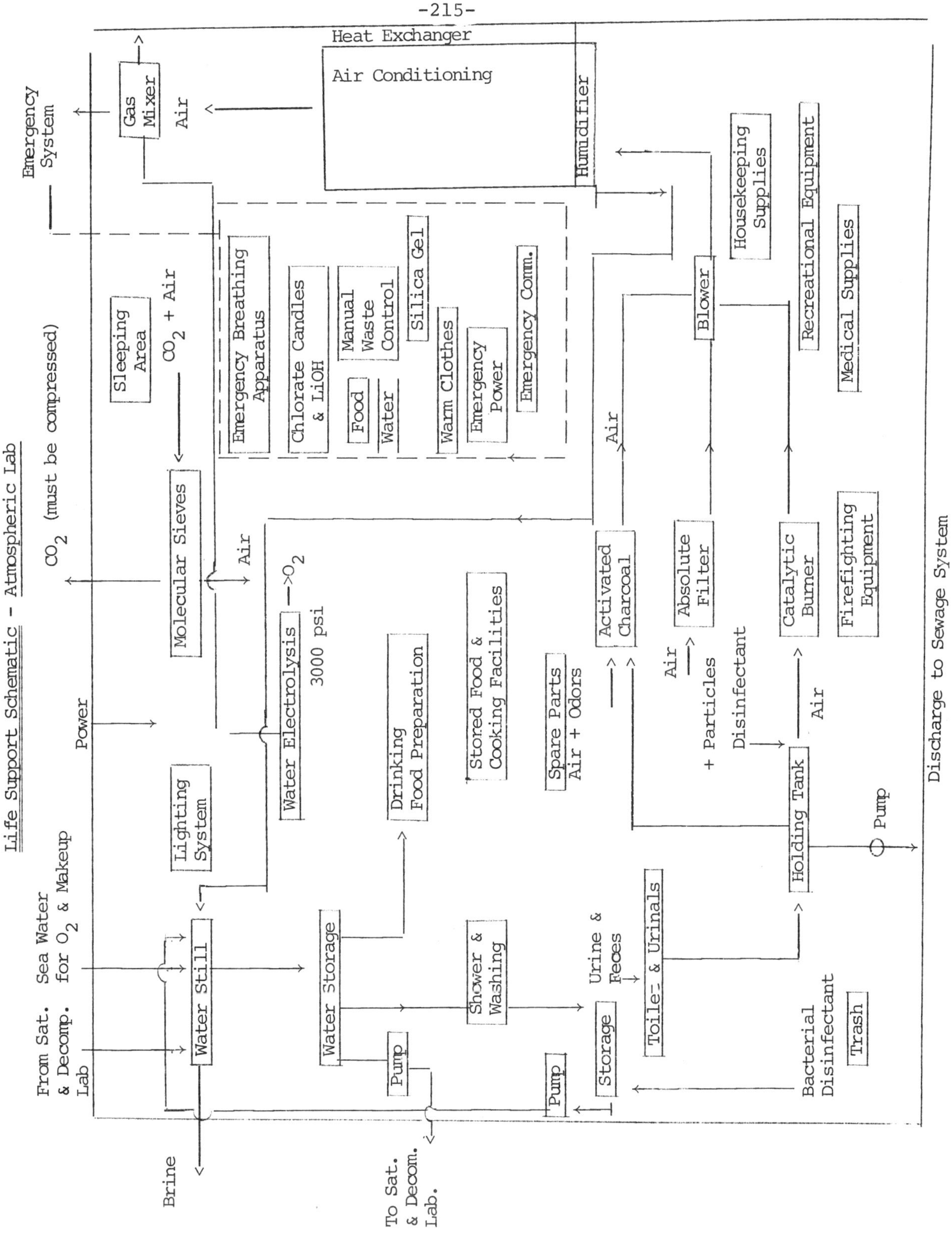

Life Support Schematic - Atmospheric Lab

From Sat. & Decomp. Lab
Sea Water for O_2 & Makeup
Power
CO_2 (must be compressed)
Emergency System
Brine
Water Still
Lighting System
Molecular Sieves
Sleeping Area
CO_2 + Air
Gas Mixer
Air
Air
Water Electrolysis
$\rightarrow O_2$
3000 psi
Emergency Breathing Apparatus
Chlorate Candles & LiOH
Food
Water
Manual Waste Control
Silica Gel
Warm Clothes
Emergency Power
Emergency Comm.
Air Conditioning
Heat Exchanger
Humidifier
Water Storage
Pump
To Sat. & Decom. Lab.
Drinking Food Preparation
Shower & Washing
Stored Food & Cooking Facilities
Spare Parts
Air + Odors
Pump
Storage
Urine & Feces
Toilet & Urinals
Activated Charcoal
Air
Air
Absolute Filter
+ Particles
Blower
Disinfectant
Housekeeping Supplies
Holding Tank
Air
Catalytic Burner
Recreational Equipment
Bacterial Disinfectant
Trash
Firefighting Equipment
Medical Supplies
Pump
Discharge to Sewage System

Figure 3-8

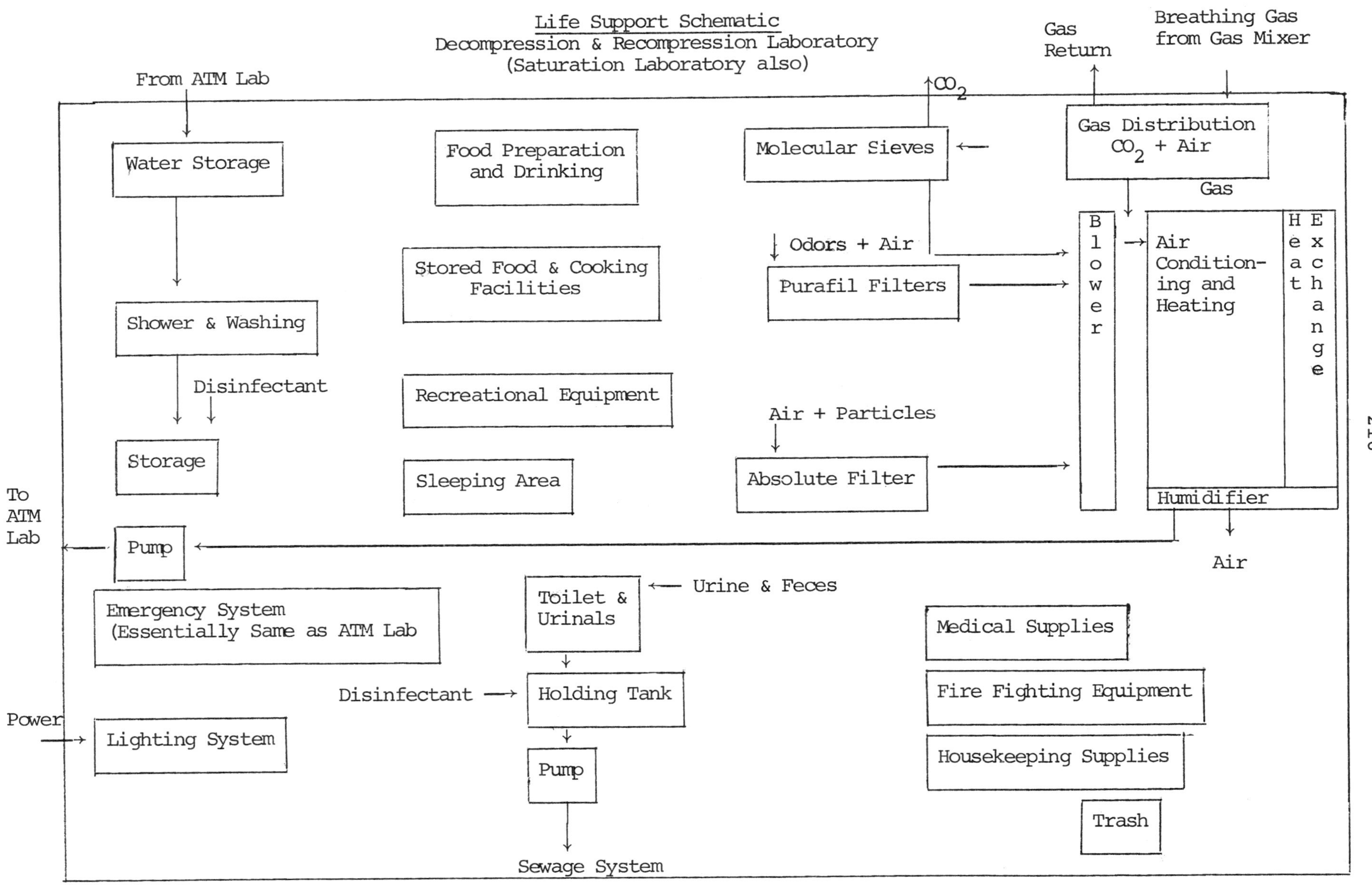
Life Support Schematic
Decompression & Recompression Laboratory
(Saturation Laboratory also)
From ATM Lab
Water Storage
Shower & Washing
Disinfectant
Storage
To ATM Lab
Pump
Emergency System
(Essentially Same as ATM Lab
Power
Lighting System
Food Preparation and Drinking
Stored Food & Cooking Facilities
Recreational Equipment
Sleeping Area
Toilet & Urinals
Urine & Feces
Disinfectant
Holding Tank
Pump
Sewage System
CO_2
Molecular Sieves
Odors + Air
Purafil Filters
Air + Particles
Absolute Filter
Gas Return
Breathing Gas from Gas Mixer
Gas Distribution
CO_2 + Air
Gas
Blower
Air Condition-ing and Heating
Heat Exchange
Humidifier
Air
Medical Supplies
Fire Fighting Equipment
Housekeeping Supplies
Trash

Figure 3-9

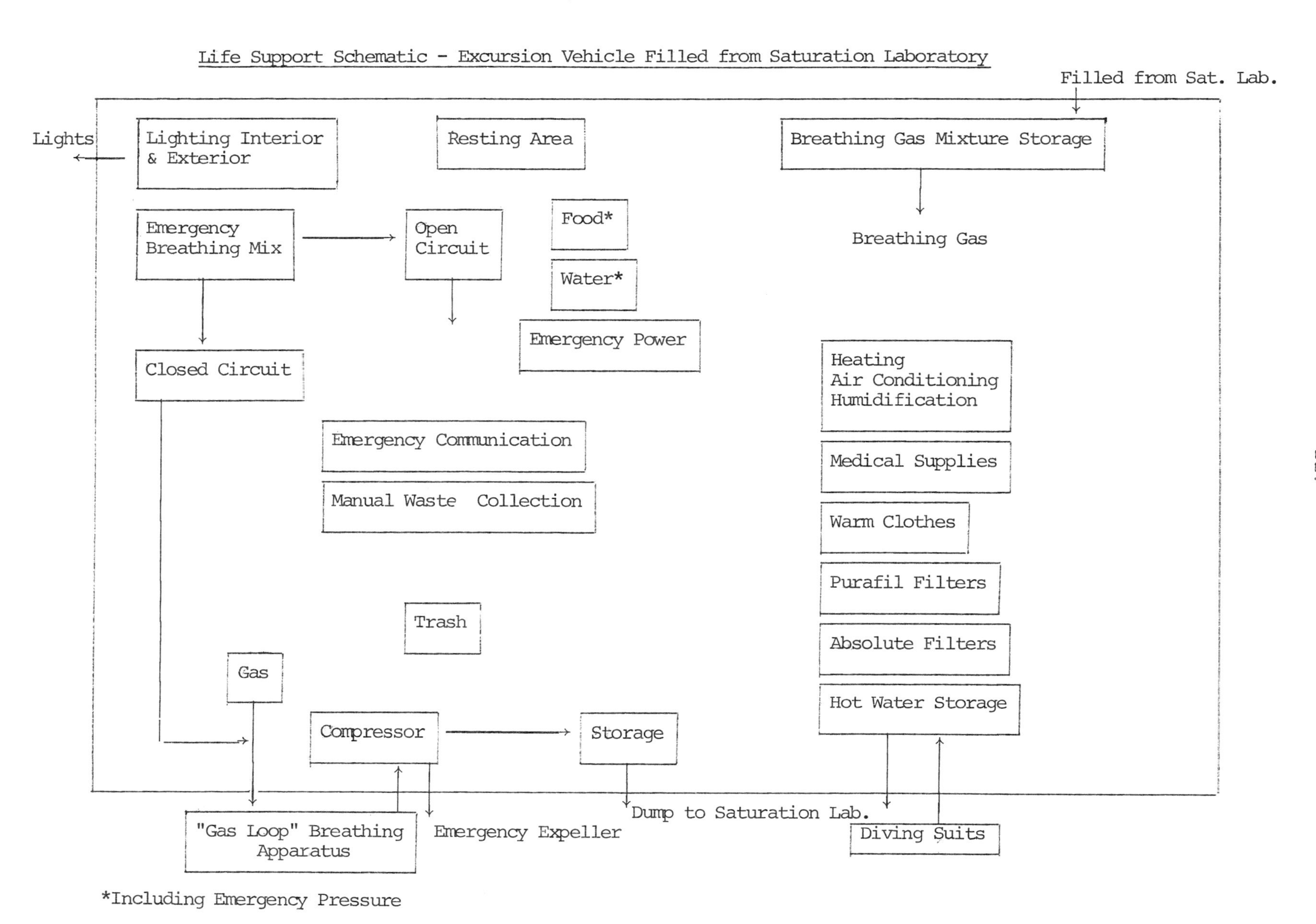

Life Support Schematic - Excursion Vehicle Filled from Saturation Laboratory

Figure 3-10

BIBLIOGRAPHY

1. Concept Development of Manned Underwater Station, Vols. I & II, by General Dynamics Corp., Electric Boat Division, Groton, Connecticut, for Naval Civil Engineering Laboratory, Port Nueneme, California (Cr 67.022), July 1960.

2. Concept for a Manned Underwater Station (u), by Southwest Research Institute Project No. 03-1885, for Naval Civil Laboratory, Port Nueneme, California (Cr 67.022), February 1967.

3. Conceptual Study of a Manned Underwater Station (u), by General Dynamics Corp., Electric Boat Division, Groton, Connecticut, for Naval Civil Engineering Laboratory, Port Nueneme, California (Cr 67.019), April 1967.

4. Underwater Physiology, Proceedings of the Third Symposium on Underwater Physiology, C. J. Lambertson, Williams & Wilkins, Baltimore, 1969.

5. Man's Extension into the Sea, Transactions of the Joint Symposium, 11-12 January 1966, Washington, D.C.

6. The Submarine-Navpers 16160-A. Standards and Curriculum Branch Training Division, Bureau of Naval Personnel.

7. Life Support Requirements. Ocean Industry, December 1967, pp. 24-40.

8. Keeping Man Alive Underwater. Oceanology International, January-February 1967, p. 20.

9. Life Support Systems for Undersea Use, Oceanology International, January-February 1968, p. 31.

10. Automatic Gas-Mixing System Reduces Hazard for Divers. Ocean Industry, January 1968, p. 48.

11. Automatic Control System for Divers' Decompression Chamber. Ocean Industry, April 1968, p. 51.

12. New Undersea System Extends Navy's Salvage Capabilities. Oceanology International, April 1969, p. 39.

CHAPTER IV

OBJECT DELIVERY AND RETRIEVAL

EPIGRAMS

Transfer at sea, in a sea, is an unforgettable experience.

Package preparation is essential for intact delivery.

Knowledge of package and platform motion is an aid to delivery and retrieval; control of package and platform motion assures delivery and retrieval.

Precise knowledge of package location and delivery address has a profound effect on system size and cost.

The critical moment in package delivery occurs when the package changes its coordinate system.

Ad hoc *recovery and delivery mechanisms invite disaster.*

A wide variety of missions encompass the generic category of object delivery and retrieval. The most common, of course is transportation which itself is divided into bulk liquid transport, bulk dry transport, break bulk cargo, containerized cargo and roll-on-roll-off cargo. Equally obvious is salvage where the object is to retrieve valuable objects otherwise lost at sea. Less obvious are missile delivery or weapon launching systems. These are, in fact, simply transportation systems where a package is delivered from one location to another in the face of attempts to impede or resist delivery. Aircraft retrieval aboard carriers is another major function in this generic category, as is catapult or rocket-assisted launch. In each of these retrieval or delivery processes, the same basic sequence of events takes place, as follows:

a) Ascertainment of location and condition of package;

b) Package preparation;

c) Package launch or pickup;

d) Package transportation;

e) Package delivery or retrieval;

f) Post-delivery processing.

The major problems which are encountered in each of these phases constitute the discussion of this subsystem.

Location

The most difficult location and assessment of condition problem is probably associated with salvage of a package lost at sea. In such instance, only the most general idea of location or circumstances of loss and the probable condition of the object is known. The author's experience with the location and recovery of the United States' hydrogen bomb lost at Palomares, Spain, and the location and examination of the United States submarine *Scorpion* underscored the pragmatic problems associated with such a search. The approach adopted in each instance was primarily Bayesian, that is, all of the evidence concerning the events that triggered the loss is considered. Recognizing that much of the evidence itself was uncertain and had a probability of incorrectness, then scenarios were constructed which were consistent with all or part of the evidence. From each scenario a most probable location and a probability distribution about that location is adduced. A board of advisers then makes a subjective estimate of the probability that each of the scenarios is correct. These probabilities should, of course, add up to unity. The probability distributions are then weighted and added to form a total probability map. The search procedure is then optimized so that the product of the probable effective search rate and probable location is a maximum. This technique has proved effective in the limited number of cases in which it has been applied. Something like it occurs in almost every search pattern, whether it be for the location of a lost golf ball or a lost submarine. The movement of a carrier in search of a lost and unretrieved aircraft and the search for a lost container on the pier follow similar but less formal processes. In any instance, the subsystem manager should insist that location procedures and equipment have been provided.

Ascertainment of condition is the next requirement in order to determine the distribution of weight and moment of the object, the distribution of lifting load, the points of attachment, the existence of structural damage, the possibility of further damage due to movement change in ballast, the electrical, chemical and explosive hazards which may exist with the package, dynamic loads which may exist or be initiated, distribution of gas and liquids and manner of containment, etc. The most difficult of these cases is again that of salvage and particularly ship salvage. In this instance, retrieval cannot be accomplished until bottom breakout forces are exceeded. If buoyancy has been used for this purpose, then the immediate excess of force after breakout has taken place will accelerate the body toward the surface. As it rises, expanding air or gas may well increase the buoyancy and further aggravate the imbalance of forces. Such loss of control during the ascent and broaching phases has resulted in disaster for a number of salvage operations. In ship salvage, therefore, techniques for ascertaining the amount of trapped air, its location, the distribution of ruptures in the ship and location of vents, etc., must be part of the system design. Techniques for *in-situ* measurement of bottom characteristics and estimate of breakout forces should also be developed. In aircraft retrieval, radar and visual inspection of the aircraft is required in addition to self-ascertainment by the aircraft of its landing condition. Assurance of "wheels down," knowledge of fuel load, flap conditions, tail hook position and, on occasion, battle or accident damage is vital to setting the configuration of retrieval equipment, damage control equipment and ship heading.

Weight, moment, and condition of cargo is equally important in transportation and, in general, more easily ascertained. It is, however, becoming a major problem with container ships and roll-on-roll-off cargo, since the control of the packaging is now in the hands of a large number of separate sources, some of whom are inexperienced or irresponsible in the containerization of their cargo.

Package Preparation

Package preparation includes the following: Structural containment and assurance of structural integrity, identification of position, establishment of frame of reference, activation of power systems, preparation for launch or breakout pulse. This process is best exemplified in aircraft or missile launching. Package preparation is extensive, consisting of a continuous update from the launching ship of its latitude and longitude together with instructions to the guidance system to insure appropriate address. A means for ascertaining the local vertical (or the horizon) must also be inserted. Unless a solid preinstalled fuel is employed, fueling of missile or aircraft is required, and activation of battery systems. Equipment holddowns and caging is accomplished in preparation for launch or catapult pulse. The use of sabots, shoes, blocks and pads should be considered for launch pulse instigation and for control of the vehicle during the launch phase.

In salvage, package preparation may include the installation of buoyancy foam, tunneling for establishment of chains or for relieving bottom breakout, the attachment of lift devices, pontoons, buoyancy bags, wire ropes, closure of openings and ruptures, sealing of tanks and opening of others, etc. The establishment of position may be accomplished by pingers or transponders located at appropriate points on the object. Establishment of local vertical is generally not a problem because of the gravitational forces which will dominate the operation. For reconstruction of damage scenarios or for other reasons (as in archeological digs), it is useful to locate each object on the sea floor with respect to some fixed coordinate grid. *In-situ* preservation is often desirable since many objects which are preserved when immersed in water will rapidly deteriorate when exposed to the atmosphere.

Package Launch or Pickup

The most crucial phase occurs during launch and retrieval. In general, the launch or retrieval platform will be to some extent coupled to the free surface and will be undergoing extensive motions. Immediately after launch or before retrieval, the object will be subjected to an entirely different set of motions and will operate in a different coordinate frame. Ideally, the object leaves or joins the platform when the relative velocities and relative accelerations are zero. This is rarely the case. For example, the response of a small boat in a given sea to the local waves is very much larger than that of the ship to which it is attempting to return. A most dangerous condition results from the relative motion between both vessels, and there is a possibility of inadvertent contact when the relative velocity is high or that personnel get caught between the boat and ship when transfer is being attempted or that the ship roll will lay a sponson or ladder over on to the boat, etc. In any event, control of the relative motion between platform and package is required if launch and retrieval forces are to be bounded.

Ship gunfire control is a classic example of this problem. A gun located on a fixed earth may utilize a relatively simple gunsight computer when tracking a target such as an aircraft. If rate gyros are employed to determine the rate at which the gunsight is rotated and if range to the aircraft is optically ascertained, then the lead cycle required to compensate for aircraft velocity is easily ascertained. If the same sight is affixed to a ship, then the pitch, roll, heave, yaw and sway of the ship will introduce artificial rates in the sight and will impart motions and accelerations to the projectile which destroy targeting accuracy. This is accommodated by use of gyroscopic stable platforms which align the gun platform to the local vertical and eliminate rotational motions. If required, integrating accelerometers can be employed to determine translational motions. The total effect is to refer the gun platform to an inertial frame of reference and, insofar as possible, to minimize motions in that frame.

Similar approaches are taken in missilery. A guidance platform of modest quality is generally slaved to a ship system which by virtue of size has lower drift rates and requires less time between fixes. The missile thus obtains its initial position from ship coordinates and its direction from its own inertial platform. Progressively less sophisticated techniques are needed for aircraft launch or small boat launch or discharge of cargo at a pier, etc.

With knowledge of motions, minimum launch pulse can be achieved through selection of launch or retrieval times. Nevertheless, substantial accelerations are often required if minimum velocities of the launched vehicle are to be achieved. Considerable effort may have to be expended to insure that the launch or retrieval pulse interval is extended and that the accelerations are uniform throughout this interval. If gas is employed for launching, then constant acceleration can be achieved through use of programmed launch valves (or if a combustible propellant through control of burning configuration). Less uniform but controlled accelerations can be achieved by spring devices for which the acceleration is roughly proportional to the displacement.

Transportation

During the transportation phase the package can be ballistic, as in the case of a gun projectile or ballistic missile, or can be terminally guided by a homing device as a homing torpedo or a ground-controlled approach landing system. It can be continuously under independent control as a manned aircraft, manned small submersible or manned boat; it can be under hybrid control through generation of part of its own force and by cables or connections to the platform. To the extent that control is exerted, the forces, relative motions and positions on landing or retrieval can be minimized, but it is the retrieval process which is perhaps most difficult. In retrieving objects from the water, two basic techniques have been employed, either a snatch at a time when the accelerations and motions of the object to be retrieved are most effective (i.e., at the crest of a wave) or a successive reduction of degrees of freedom with respect to the launching platform. An example of this latter is the proposed surface recovery scheme for the Navy rescue submersible. In this scheme, while fairly deeply submerged, the submersible engages the slack suspension wires of a trapeze. The vehicle is now constrained only by the maximum extension of a tether. The vehicle ballasts negatively until a strain is taken on the tethers, thus constraining the vehicle to follow (more or less) the heave of the ship. The tethers are then drawn up until the submersible is two-blocked against the extended legs of an elevator having nesting shock absorbers. The vehicle is now constrained except for relative rotations about the pitching axis. This final freedom is then restricted and the motions of the submersible are now fully constrained by those of the ship. The submersible is then elevated in a well between the catamaran hulls of that particular design.

In the vehicle snatch scheme employed by Link, a hydraulically-operated articulated crane is extended over the fantail and, while in its flexible articulated configuration, is attached to the vehicle to be recovered. Relative motion is now constrained only by the maximum extensions of the crane. By visual observation the operator waits for the most favorable moment for retrieval. Hydraulic pressure then reduces the freedom of the crane and the vehicle is snatched from the water.

Post-Delivery

Post-delivery problems are usually heavily mission-oriented. They include: in the case of weapons, the assurance of arming and fusing at, before, or after impact; in the case of salvage, the immersion in an appropriate liquid, foam or chemical conditioner to prevent deterioration; in the case of containers, the installation on an appropriate land conveyance, etc. Because of this diversity, no general principles can be adduced vis-a-vis this phase.

A final word should be said about launch and retrieval. The diversity involved is such that systems designers are tempted to rely on *ad hoc* techniques to be improvised on the site when the complete nature of the transfer is first revealed. Such an attitude invites disaster. Any transfer at sea which exceeds control limits has the potential for generating highly destructive forces--at best the package to be transferred is lost; at worst the safety of the platform itself is jeopardized. The cost of preparing for contingencies which never materialize will always be difficult to explain, but must be borne by the manager whose criterion is successful mission performance.

OBJECT DELIVERY AND RETRIEVAL

Nations are developing the inclination and technology to explore, exploit and develop the seabed. Systems, to be operable in the sea, must survive "the perils of the sea," not just the destructive force of surface waves but the hazards due to motion and, therefore, forces near either of the ocean interfaces, the effects of pressure at any depth, and the hostility of living creatures in the marine environment. The sea is vast and objects in it are difficult to locate. Ocean engineering will be costly unless objects become locatable and readily relocatable. The author's suggested salvage system attempts to develop approaches and mechanisms which circumvent the engineering problems of marine salvage, one of the oldest forms of ocean engineering.

The author, T. Gray Curtis, Jr., submitted this case study while on leave from the Office of the Supervisor of Salvage, U.S.N., and studying at M.I.T. for the degree of Ocean Engineer. Mr. Curtis received the Bachelor of Science and Master of Science degrees in Civil Engineering from M.I.T. His proposed salvage system has evolved from his experience in deep-ocean salvage gained while working for the Supervisor of Salvage.

MARINE SALVAGE

by

T. Gray Curtis

As the desire to utilize the seas increases, ocean engineering must develop in a systematic manner. Larger and heavier objects will need to be placed in deeper waters. The capability to deliver, recover and support ocean engineers and their equipment is most appropriately pursued by extending the capability of marine salvage since it is the area of greatest present ocean engineering expertise.

Background

Salvage has progressed considerably since the waters of the wrecking grounds were plied by "wreckers" in their wooden ships and with their steel nerves. These courageous salvors and the coastal inhabitants who acquired bounty from the sea have been maligned by history for their acts of heroism in saving life and cargoes.

The U.S. Navy maintains a fleet of auxiliary ships and equipment to provide naval marine salvage in support of naval operations and subsidiary marine salvage in the private sector of the economy. During World War II, the salvage services dramatically proved their worth by clearing blocked harbor and coastal facilities and recovering valuable cargoes and ships for further use.

The technology which is available for application to marine

salvage problems is expanding at an increasing rate. This is apparent, for example, in the sophisticated deep-diving capabilities being developed. Gone are the days when a salvor had to kick a wire rope to determine the tension in it.

The loss of submarines *Thresher*, *Scorpion* and the H-Bomb off Palomares, Spain, within the last decade has indicated the need for a systematic salvage capability in deep water. Salvage subsystems and components have, in the past, been jury-rigged *ad hoc* solutions to problems unknown *a priori*. As a result they have led to expensive salvage operations of marginal success. The Navy is presently developing a large object-salvage system to partially ameliorate this condition.

Proposed System Summary

It was the author's intent to indicate the generic problems of marine salvage and propose systematic approaches to the technical problems which would not be comparable to opening a walnut with a sledgehammer. He felt that the best solution to a technical problem is a different attack if that avoids the problem and is reasonable.

Subjective Human Purpose

Salvage of a large object from the oceans in depths to 20,000 feet in sea states up to and including six. A large object is defined with regard to depth. In depths to 850 feet, it would include sunken objects weighing up to 100,000 tons. Where depths are greater than 850 feet, but less than 20,000 feet, such an object could weigh up to 100 tons. Salvage will consist of location, recovery and deliverv of an object to an appropriate facility.

Based upon the subjective human purpose, a mission analysis and operation profile were made to determine the variety and extent of the problems. They served to direct the approach made to the operations requirements. The major problems were found to be:

1) Underwater object location and marking

2) Preparation of object for lift

3) Control of relative accelerations due to motions

To solve these and other problems, the system was divided into functional subsets which had dominant technologies and required a minimal amount of interfacing with other subsystems.

A salvage system should be designed around and not through the major sea problems of crushing pressure with depth and control of motions at the ocean interfaces. The major advances of Mr. Curtis' salvage system are stated below.

1) Major logistic support will be supplied from a column-stabilized platform whose motion is uncoupled from that of the sea surface. Dynamic transients in lift forces due to platform motion are small because the platform itself does not move with the seaway. Large underwater hulls allow launch of underwater surveillance and work subsystems from beneath the free surface at sufficient depth to be in quiet waters.

2) Deep lift will be supplied in modular form by hollow

spheres of low ρ/E weight to strength ratios. Lift modules would be ballasted down to the underwater worksite in a clump where they would be placed on the object as needed to provide lift. Thus each salvage job would not require a tailor-made salvage plan.

3) Control pontoons of variable ballast would be attached to the object to provide the marginal lift required to control the object's motion. Properly balanced, the object need not be raised all the way to the surface. Under adverse conditions, the object could be brought near the surface, but held beneath rough surface waters until they are calm.

Lloyd's underwriters are beginning to record the increase in tonnage and value lost at sea due to increased ocean engineering activity. There will be a continuing need for an improved ocean engineering/salvage capability.

What follows is the Object Delivery and Retrieval Subsystem in Mr. Curtis' salvage system. Note that he considered this particular sybsystem of such significance that it was dealt with as if it were a system unto itself.

OBJECT RETRIEVAL AND DELIVERY SUBSYSTEM

MISSION ANALYSIS

I. Inspection

Inspection is undertaken to confirm object identity and to determine the amount of preparation required prior to lift.

A. Object Identification

1. Type

An object is defined as a valuable piece of hardware. Examples: ship, space capsule or booster rocket, bomb, torpedo, airplane, oceanographic equipment, and buoys.

2. Constraints

a. Frangibility. The object must be sound enough not to break up in the lifting process; it must be capable of withstanding moments and lift forces.

b. Dimensions. The ability to salvage an object is highly dependent upon the depth of the object. As the depth increases, fewer salvage lift methods can be applied to the object and, consequently, the object's "dimensions" must be restricted.

Two major regions are considered for salvage which

we shall call shallow water, $h < 850'$ and deep water $850' \leq h < 20{,}000'$.

1) <u>Size</u>

<u>Shallow water</u> - object size shall not be restricted.

<u>Deep water</u> - object size is limited to volumes greater than $2m^3$ distributed roughly as 2x1x1.

2) <u>Weight</u>

<u>Shallow water</u> - object weight shall be no greater than 10,000 t.

<u>Deep water</u> - object weight shall not be greater than 100 t.

3. <u>Obstacles</u>

Object structural integrity <u>in situ</u>.

B. <u>Object Inspection</u>

1. <u>Quantity and Type</u>

The quantity of inspection will be commensurate with that necessary to ensure definition of the problems to be encountered when using a given lift method.

a. <u>Objective</u>. Because a lost object on the ocean bottom will probably bear little resemblance to the same object before it was lost, it must be examined for

1) Structural integrity
2) Geometric form similitude

b. <u>Environmental</u>. For sophisticated work systems to be employed it will be necessary to quantify the behavior of the environment in which work is to be done.

1) <u>Oceanic</u> - measurement of sea properties

a) Temperatures
b) Currents

2) Foundation - measurement of submarine soil conditions

a) Bearing capacity

b) Porosity

c) Cohesiveness

2. Constraints

a. Systemic

1) Power and frequency required for object visibility

2) Mobility of reconnaissance component

b. Environmental

1) Depth, i.e., hydrostatic pressure - subsystem must be able to operate under an ambient pressure of 9000 psia

2) Submarine currents V = 3 kts

3) Submarine topography - operate in salt water environment

3. Obstacles

a. Time-power is presently in short supply

b. Operation of underwater work systems

II. Object Preparation

A. Quantity and Type

Depends upon the object to be lifted and its sensitivity to moments, forces, and chemical attack. Preparation will be restricted to that necessary for the success of the salvage operation.

1. Objective

The object must be prepared to accommodate the loads and moments of lift.

a. Structural

1) Deformation control - only limited deformation, depending upon the object, will be allowed

deformation results from yielding as a result of excessive moments due to load.

2) Load-bearing point production - the object must be strengthened where concentrated lifting loads are to be applied.

3) Cutting - because of weight limitations imposed on deep-depth lifts, unnecessary weight may have to be cut away.

b. Metallurgical

Attack by marine growth, chemicals in the sea and physical force may require preparation of the object's surface.

1) Cleaning - removal of adverse material to permit desired quality bonding.

2) Coating - protection of surface from elements.

2. Environmental

The conditions around the object may need to be improved for a successful operation.

a. Soil Stabilization. Improve the resistance of the submarine soils in the area to disturbance.

b. Breakout Force Reduction. Lower the level of force required to pull an object off the bottom.

1) Dredging - reduce embedment and therefore increase flow of water around the object.

2) Cohesion reduction - reduce the holding power of submarine soils by increasing flow of water through the soil.

B. Constraints

1. Objective

a. Reaction (Moments and Forces) Tolerance. Normally not known due to condition of a lost object in situ. It may be possible to make rough estimates from reconnaissance.

b. Quality. The level of acceptable moments, forces and

deformations will vary with the sensitivity, here called quality, of the object.

2. Environmental
   a. Submarine currents V = 3 kts
   b. Marine life

3. Systemic
   a. Weight of work systems
   b. Power acquired for work systems

C. Obstacles

1. Environmental
   a. Submarine Topography
   b. Submarine Soil Conditions
      1) Cohesiveness
      2) Consolidation
      3) Shear strength
   c. Hydrostatic Pressure

2. Objective
   a. Geometry - ease of access
   b. Structural integrity

III. Lift

Salvage lift methods generally must be applied in combinations in order to meet the wide range of possible salvage requirements.

A. Types of Lift

1. Added Buoyancy
   a. Internally Added
      1) Water-pumped deballasting

a) Constraints

(1) Pressure - head which has to be pumped against

(2) Geometry - a link with the atmosphere by which air can displace water. Object usually has to be in a certain attitude.

(3) Sea state - cofferdams sensitive to physical forces of the sea - must keep air passage clear.

b) Obstacles

(1) Stability control of rising object upsetting weight of cofferdam.

(2) Preparation of object for dewatering by patching holes.

2) Air-blown deballasting - air is forced down a hose or other link to the object's interior.

a) Constraints

(1) Hydrostatic pressure determines pump capabilities required.

(2) Flow rate from compressor.

(3) Geometry - object attitude determines effectiveness of air blowing.

b) Obstacles

(1) Air delivery

(2) Stability control - object will have a waterplane and free-surface effects.

3) Generated-gas deballasting - gas is produced in the object.

a) Constraints

(1) Rate of reaction
(2) Density } f(Pressure)

(3) Geometry - object attitude determines effectiveness of gas-generated deballasting - object's airtightness.

b) Obstacles

(1) Chemical storage and handling chemicals highly reactive and toxic.

(2) Stability control due to free surface.

4) Foam deballasting - injection of catalyst and resin which forces water out of object displacing it with a lighter material which solidifies.

a) Constraints

(1) Density
(2) Rate of reaction } f(pressure)
(3) Chemical delivery, power needed to pump chemicals

b) Obstacles

(1) Chemical storage and handling
(2) Cost

Foam deballasting has the advantage over the other methods of internal added buoyancy of

No free surface. The foam sets solid after displacing the water.

Constant buoyancy. Once the void space is filled, reduced pressure will not change the foam's volume appreciably.

Expanding gas, on the other hand, can markedly alter buoyancy.

b. External Pontoons

Buoyancy can be added to the object externally in the form of a pontoon.

1) Variable gas supply - gas can be introduced at depth to create the buoyancy only when it is desired.

a) Chemical production

(1) Constraints

(a) Rate of chemical reaction
(b) Density } f(pressure)
(c) Attachment loads - concentrations of stress
(d) Submarine currents - external pontoons subject to environmental forces

(2) Obstacles

(a) Chemical storage and handling

(b) Gas generator delivery and control

b) Physical transport - blown gas

(1) Constraints

(a) Density - f(pressure)

(b) Attachment loads

(c) Currents

(d) Power for pumping

(2) Obstacles

(a) Pressure losses in delivery

(b) Interface penetration

c) Change of phase delivery - cryogenic gas supply

(1) Constraints

(a) Insulation of supply; bleedoff/time

(b) Rate of reaction

(c) Gas density } f(p) => depth limit

(2) Obstacles

(a) Weight of cryostat

(b) Handling of cryostat

2) Deballasting pontoons - constant fluid mass pontoons are ballasted down - attached - deballasted in order to attach lift.

a) Constraints

(1) Pontoon wall strength a function of pressure

(2) Ballasting material

(a) Density

(b) Quantum

(3) Attachment stress concentrations

(4) Submarine currents

b) Obstacles

(1) Transport of pontoons - weight of ballast and pontoons must be carried by platform.

(2) Deballasting procedure and effect on environment and object.

2. Applied Force

Lift can also be accomplished by attachment to the object of a dynamic lift.

a. Surface Supplied

1) Constraints

a) Strength of line

b) Natural frequency of the line

c) Attachment stress concentrations

2) Obstacles

a) Line storage

b) Dynamic loads - resonance of the line due to excitation from platform and sea greatly reduce load capability.

b. Submersible Supplied

1) Constraints

a) Submersibles reserve buoyancy

b) Submersible moment carrying ability trim control

c) Currents - affect coupled motion depending upon area perpendicular to flow

d) Manipulation load limits and moment limits

2) Obstacles

a) Attachment method

b) Transfer of object near surface

c. Diver Supplied

Swimmers can pick up objects dead lift or dynamic (air lift).

1) Constraints

a) Excursion depth
b) Currents
c) Weight of object

2) Obstacles

a) Environmental working conditions: light, temperature
b) Transfer near surface

B. Object Control

In order to regulate the motions of the object, the forces and moments acting upon it must be controllable. Different regions of the flow will be dominated by different forces and require different methods of control.

1. Region

Bottom - breakout forces required to lift an object off the bottom are the result of pressure forces due to potential flow about the object which are time-dependent.

a. Methods of Control

1) Reduction of applied high-level force directly after breakout eliminates object accelerations.

a) Constraints

(1) Lift control - rate of lift reduction
(2) Attachment strength

b) Obstacles

(1) Observation of breakout-time delay
(2) Lift line (length and construction)

2) Low-level applied force

a) Types

(1) Long-term solution - depends upon time-dependent nature of object behavior.

(2) Environmental influence - water jetting to increase submarine soil porosity and flow. Water jetting was selected after considering

water jetting

explosive shock - too dangerous

vibration - not technically feasible

structural substitution - creates artificial failure plane

b) Constraints

(1) Reaction forces

(2) Momentum transfer

(3) Cohesiveness of submarine soils   Water jetting

(4) Power available

(5) Time availability   Long-term solution

c) Obstacles

(1) Portability of jetting equipment

(2) "Visibility" once jetting is started

2. Mid-Depth

Inertial forces are due to buoyancy accelerations in a dissipative fluid, i.e., one in which friction is present.

a. Control of Rising Motion

1) Methods of control

a) Reduction of rising motion

(1) Drag increase - terminal velocity of an object falling through a fluid is proportional to the area normal to the velocity => drag chutes to increase the area of concern.

Drag chutes - effectiveness limited in velocity range $0 < V < 4$ kts. At this rate another form of control must be used. Effective in velocity range $4 < V < \infty$ because the drag force is proportional to the velocity squared.

(2) Lift decrease

(a) Control pontoon of variable ballast

(b) Surface lift line tension variation

b) <u>Increase rising motion</u>

(1) Lift increase - same methods as for decrease
(2) Weight reduction
(3) Jettison drag system

2) <u>Constraints</u>

a) <u>Drag increase</u>

(1) Area perpendicular to flow
(2) Shroud and chute material strength
(3) Attachment strength

b) <u>Lift adjustments</u>

(1) <u>Ballast</u>

(a) Weight of quantum
(b) Amount available

(2) <u>Gas generation</u>

(a) Rate of delivery and rejection
(b) Amount of gas available

(3) <u>Lift line</u>

(a) Strength
(b) Handling rate

3) <u>Obstacles</u>

a) <u>Drag</u>

(1) Currents
(2) Descending object motion

b) <u>Lift Adjustment</u>

(1) Hydrostatic pressure
(2) Ballast control

b. <u>Equilibrium Control</u>

Requires fine adjustment of lift forces.

1) <u>Methods of control</u>

a) <u>Lift control</u>

(1) Ballast pontoon

(2) Lift line

b) <u>Weight reduction control</u>

2) <u>Constraints</u>

Buoyancy adjustment

a) <u>Ballast</u>

(1) Weight of quantum

(2) Amount of ballast

b) <u>Gas generation</u>

(1) Amount of gas available

(2) Rate of delivery and rejection

(3) Fuel resupply capability

3) <u>Obstacles</u>

a) Depth - hydrostatic pressure

b) Sea water

IV. <u>Object Protection</u>

Once an object has been raised to the surface, it is subject to more destructive force than at depth, due to increased chemical and physical action near the surface.

Protection must start before a change of environment initiates deterioration by chemical and biological action.

A. <u>Chemical Action</u>

1. <u>Types</u>

a. Oxidation

b. Marine growth effects

2. Methods of Protection

   a. Isolation
   b. Protective coating

3. Constraints

   a. Oxygen levels
   b. Temperature

4. Obstacles

   a. Object geometry - size and arrangement
   b. Hydrostatic pressure
   c. Sea water composition

B. Physical Action

1. Types

   a. Wave action
   b. Dynamic rigging loads

2. Methods of Protection

   a. Controlled submergence. Placed below the free surface at an appropriate depth, the wave forces on the object will be minimized.
   b. Removal from the free surface.

3. Constraints

   a. Objective

      1) Weight
      2) Size - water plane area
      3) Salvaged stability

   b. Systemic

      1) Control pontoon gas resupply
      2) Line loads allowable

   c. Environmental

      1) Sea state
      2) Swells

V. Object Delivery

An object recovered from the ocean must be removed to an appropriate port and turned over to the harbor forces for further processing.

A. Transport

Object is taken from the salvage scene to the address of delivery.

Method

1. Carry Object on Board

a. Constraints

1) Weight
2) Volume
3) Frangibility
4) Time
5) Storage capability

b. Obstacles

1) Platform stability
2) Weather and sea state
3) Tie down

2. Tow

a. Constraints

1) Range
2) Time - 2 months
3) Sea conditions
4) Power available
5) Towline tension

b. Obstacles

1) Objective

a) Towing characteristics
b) Salvaged stability

2) Systemic

a) Attachment of external buoyancy
b) Tow linkage

3) Environmental

a) Sea state
b) Weather

4) Logistic - reprovision underway

B. Transfer

The suitably-prepared salvaged hardware is turned over to the harbor forces.

1. Preparation

The salvaged object is rigged for transfer.

a. Constraints

1) External buoyance
2) Sea conditions

b. Obstacles

1) Salvaged stability
2) Compatibility with harbor forces' handling capability
3) Propinquity of facility

2. Address of Delivery

Harbor forces of U.S. naval shipyards or other alternatively designated facility precision of delivery (environs).

a. Constraints

1) Environmental

a) Sea state
b) Harbor configuration

(1) Shipyard facilities
(2) Harbor depth

2) <u>Systemic</u>

a) Port facility capabilities

b) Condition of object

b. <u>Obstacles</u>

1) Sea conditions

2) Harbor navigability

DESIGN

I. Approach

A. Subsystem Breakdown

1. Rigging. The rigging subsystem* is responsible for handling large forces and moments associated with the salvage operation occurring during lift and delivery.

Attachment mechanisms

Application mechanisms

2. Work. The work subsystem* is responsible for the manipulation of tools and hardware required in the course of the salvage operation.

Underwater preparation for lift

Surface

1) Preparation for transport

2) Preparation for delivery

3. Control. The control subsystem* is a redundancy required by the object retrieval and delivery subsystem manager to assure his subsystem success. It will be responsible for the fine adjustment in lift forces required to control object motion. This will require some environmental sensing for which the control manager will be responsible.

B. Depth Regimes

This system analysis and design will deal primarily with the problems of deep ocean salvage. It is felt that the approach to shallow-water salvage will be facilitated by the use of a properly-designed salvage platform. Most of the methods of shallow-water salvage have not changed and therefore do not need to be expounded upon here. Where a

---

*Subsystems are indicated by *

large lift is needed in shallow water, it is felt that the deep-ocean salvage system can be applied either in toto or partially. For example, the control pontoons can be utilized in shallow water as well as in deep, and the attachment of pontoons can be done by divers or submersibles.

II. Subsystems*

A. Rigging

The application of large forces in lift and tow is complicated by the effects of the free surface. Because the dynamic loads in lines between two objects, one of which is in coupled motion with the free surface, are capable of being many times the steady line tension, it has been decided to uncouple all forces from interface effects whenever possible.

The implications of such uncoupled motion are major. Such motion control could be gained by two basic approaches:

1) Submerged operation. Away from the free surface, platform and object motion are not influenced by the interface;

2) Platform design which would minimize platform motion.

The rigging subsystem will rely upon submerged applied lift in the form of pontoons whenever conditions dictate. There will be times when surface-supplied lift is necessary, such as with the removal of a submersible, lift-support system such as the diver's PTC, or small object from the water entirely. When surface lift is applied, it should be uncoupled.

During tow operations the tow-line tension will be subjected to the dynamic loads of the sea. As these cannot be eliminated except perhaps by a sophisticated arrangement of ram tensioners, which would be costly, bulky and not in keeping with the philosophy of engineering around a problem instead of through it, controlled submergence of the object will be utilized when necessary.

1. Systems Implications of Uncoupled Motion

a. Platform

The salvage platform must give limited response to all sea conditions up to and including state 6 seas. The sea condition will be defined by the superposition of 6-foot swells and the Pierson-St. Denis Sea Spectrum.

1) Limited Response

The motions imply accelerations of the handling system.

Heave - the natural frequency in heave must be less than 0.3 rad/sec. Magnitude of heave will not exceed one foot.

Roll - magnitude of roll will not exceed ±3°.

Pitch - magnitude of pitch will not exceed ±3°.

Arrangment will be so made to permit lifting and lowering at the C.G. of the platform so as to eliminate coupled effects of roll and pitch on heave.

b. Navigation and Location

The navigation system depends upon computers to correct for ship motions. The limited motions of the platform are desirable design limits for the navigation and location subsystem.

2. Methods of Applied Lift

The relative capabilities of various methods of applying lift are shown graphically in the following plot. Some methods, although not presently state of the art, are not technically unfeasible and ought to be available within three years. Note that the cross-hatched regions indicate a decreased efficiency.

Methods of lift are greatly restricted by depth, i.e., more correctly, hydrostatic pressure. In shallow water, $h < 850$ feet, the salvage officer has at his disposal 1) air-blown deballasting, 2) foam deballasting, 3) synchronous, and 4) dynamic lift as well as diver capabilities. With these added capabilities, the restrictions on weight retrieval can be greatly eased.

The condition of the object may have considerable bearing on which lift method is employed. A compact ocean engineering instrument may be readily "snatched" from the ocean bottom and placed on the platform by the use of simple lift, i.e., for example, by wire rope. The object has well-designed attachment points and its behavior subsequent to "lift off" is approximately known. When salvaging equipment lost at sea, its weight and configuration are not known ahead of time. Its response can only be estimated. Attachment and lift may also be a big problem. As the scale of the operation increases, lift will have to be supplied by adding buoyancy.

a. Added Buoyancy

External pontoons appear to be the most generally applicable. For large weight lift it will probably be necessary to apply other methods as well, such as foam and air-blown deballasting.

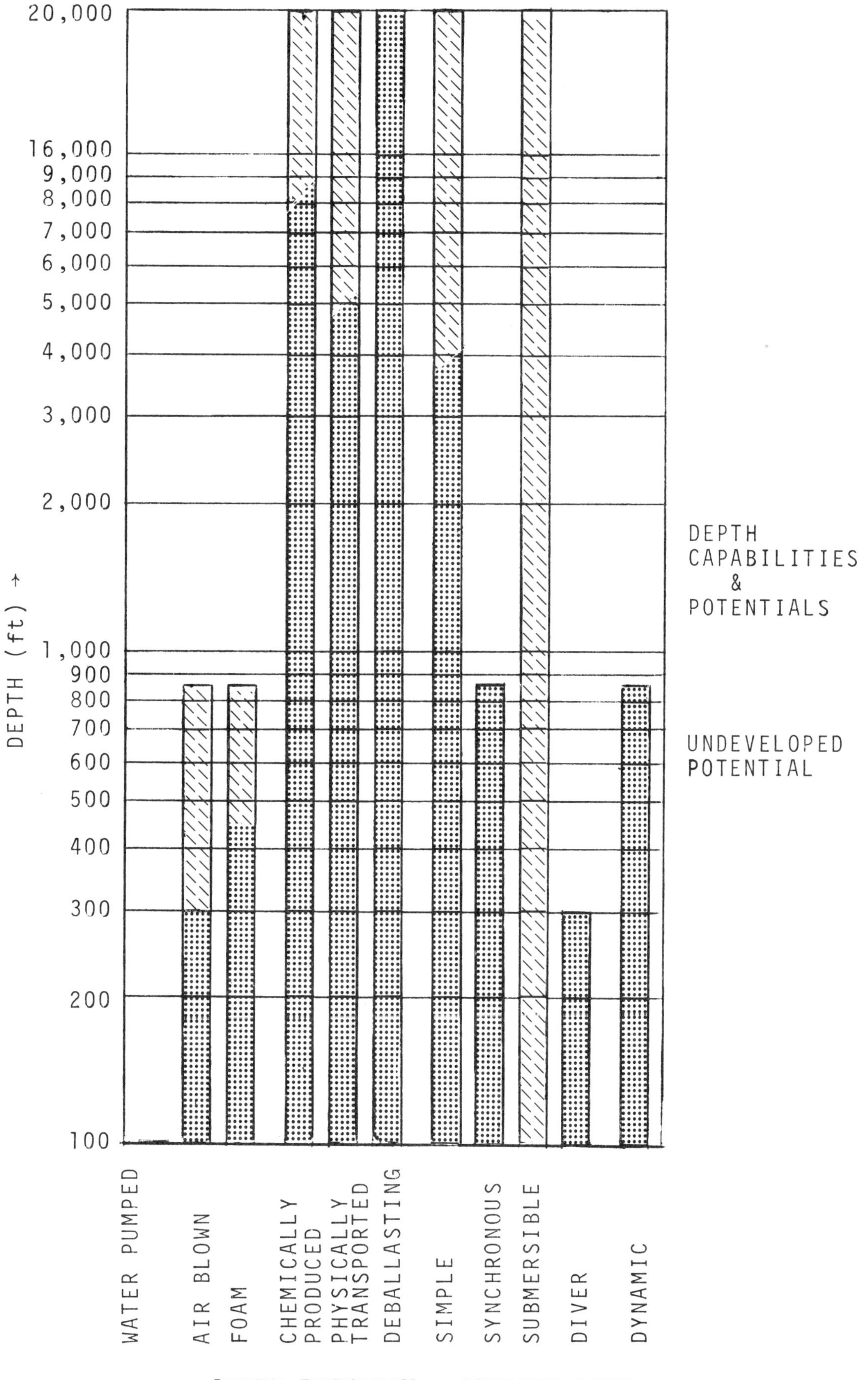
20,000
16,000
9,000
8,000
7,000
6,000
5,000
4,000
3,000
2,000
1,000
900
800
700
600
500
400
300
200
100
DEPTH (ft) →
DEPTH
CAPABILITIES
&
POTENTIALS
UNDEVELOPED
POTENTIAL
WATER PUMPED
AIR BLOWN
FOAM
CHEMICALLY
PRODUCED
PHYSICALLY
TRANSPORTED
DEBALLASTING
SIMPLE
SYNCHRONOUS
SUBMERSIBLE
DIVER
DYNAMIC
ADDED BUOYANCY - APPLIED LIFT

Figure 4-1

At present, salvage pontoons are 1) inflatable and 2) deballasted. Gas is either generated chemically or pumped down to the pontoon from the surface.

b. Surface Lift

The platform will be able to raise and lower objects into the ocean.

1) Simple Lift

Capability - 20-ton lift capability to 20,000 feet with a safety factor of 5.

Equipment

a) Two simple lift lines of 6 x 37 monitor AA wire rope 3-1/2 in. diameter with 1/4-in. taper 25,000 ft weight ≃ 100 tons.

b) Two simple lift lines of 2-7/8 in. diameter nylon line - 20,000 ft.
   Weight ≃ 70 tons
   Volume 4,000 cu ft

c) ONR instrumentation - communication - lift cable. Rolling capstans can be used to hold tension in lines while they are stored on reels not under tension. This will improve wire life.

   In this manner it will be possible to use a variety of lines without requiring respooling of a main-line winch.

2) Synchronous Lift

Capability - 400 tons from 850 ft.

Equipment - 4 synchronous lift engines and lifting lines.

c. Submerged Lift--Lifting Pontoons

1) Inflatable Pontoon

The inflatable pontoon will consist of a flexible inflatable bag which is blown with gas at depth to gain buoyancy.

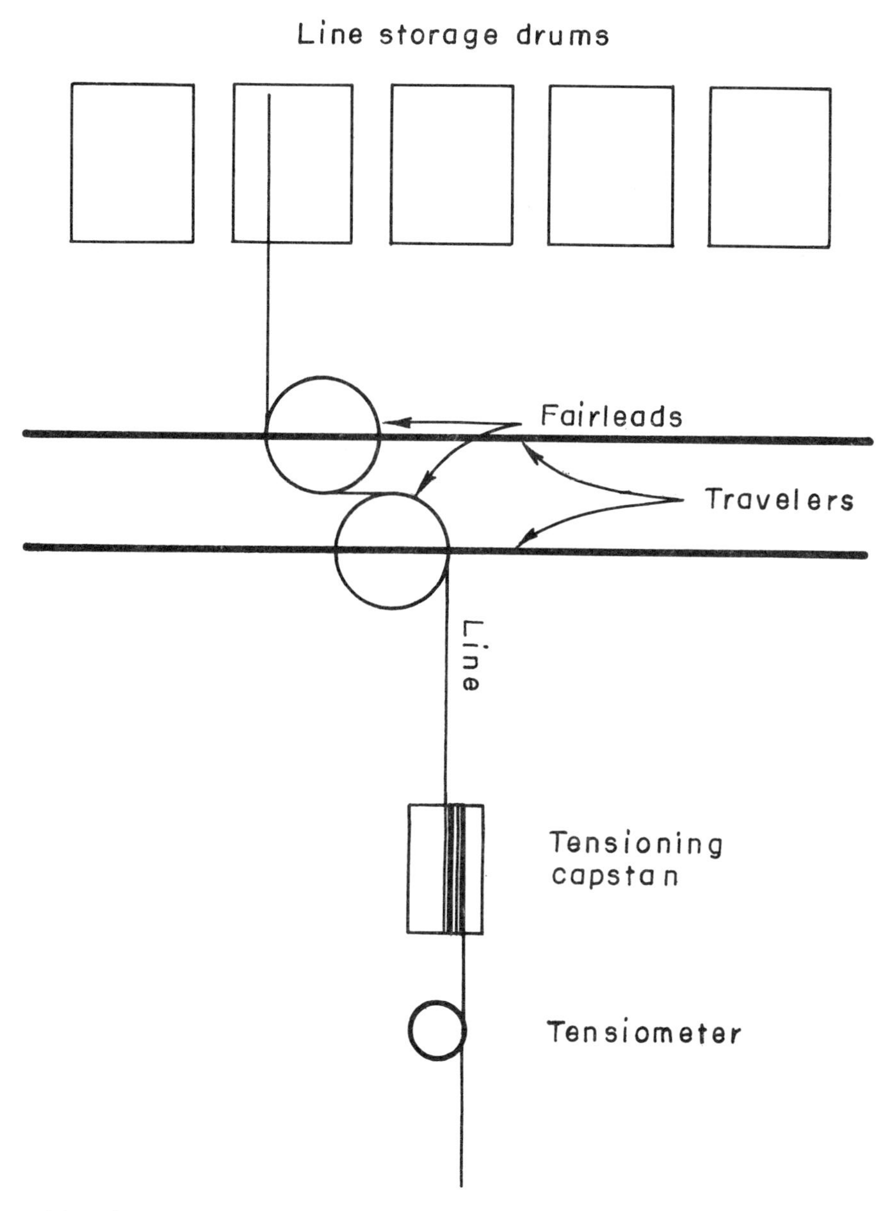
Line storage drums
Fairleads
Travelers
Line
Tensioning capstan
Tensiometer
Lift line layout

Figure 4-2

a) Shallow Water Operation

The inflatable bag will be blown with nitrogen supplied in cryogenic form to depth. Nitrogen in liquid form is fairly inexpensive. Safe to handle compared to the more reactive LiH and hydrazine, N_2H_4, N_2 does not perform as well at depth and will not be used at depths greater than 850 feet (see Figure 4-3).

b) Deep Water Operation

Below 850 feet, a lighter chemical must be used to deballast the inflatable pontoon. The cryostatic gas generator will be replaced by a chemical reaction gas generator. Now under study, hydrazine is closer to being operationally available than any other chemical for use in gas generation. Hydrazine is a good chemical bottle for hydrogen (see Figure 4-4). So also is lithium hydride, as Figure 4-4 would imply. A solid, lithium hydride has not been developed for gas generation. Its potential warrants an investigation.

c) Attachment

The inflatable pontoon is to be attached to the object in various configurations depending upon whether the object is being lifted or towed. Since the bag is flexible, there is no advantage to a restricting, rigid attachment. Attachment will be through a flexible link, a short tether. This will allow some freedom-of-attachment location.

d) Constraints

(1) Small and light so that it can be readily handled by the work system.

(2) Actuated by command.

(3) Safe--will not endanger equipment in the immediate area.

(4) Strength of attachment capable of accommodating the load.

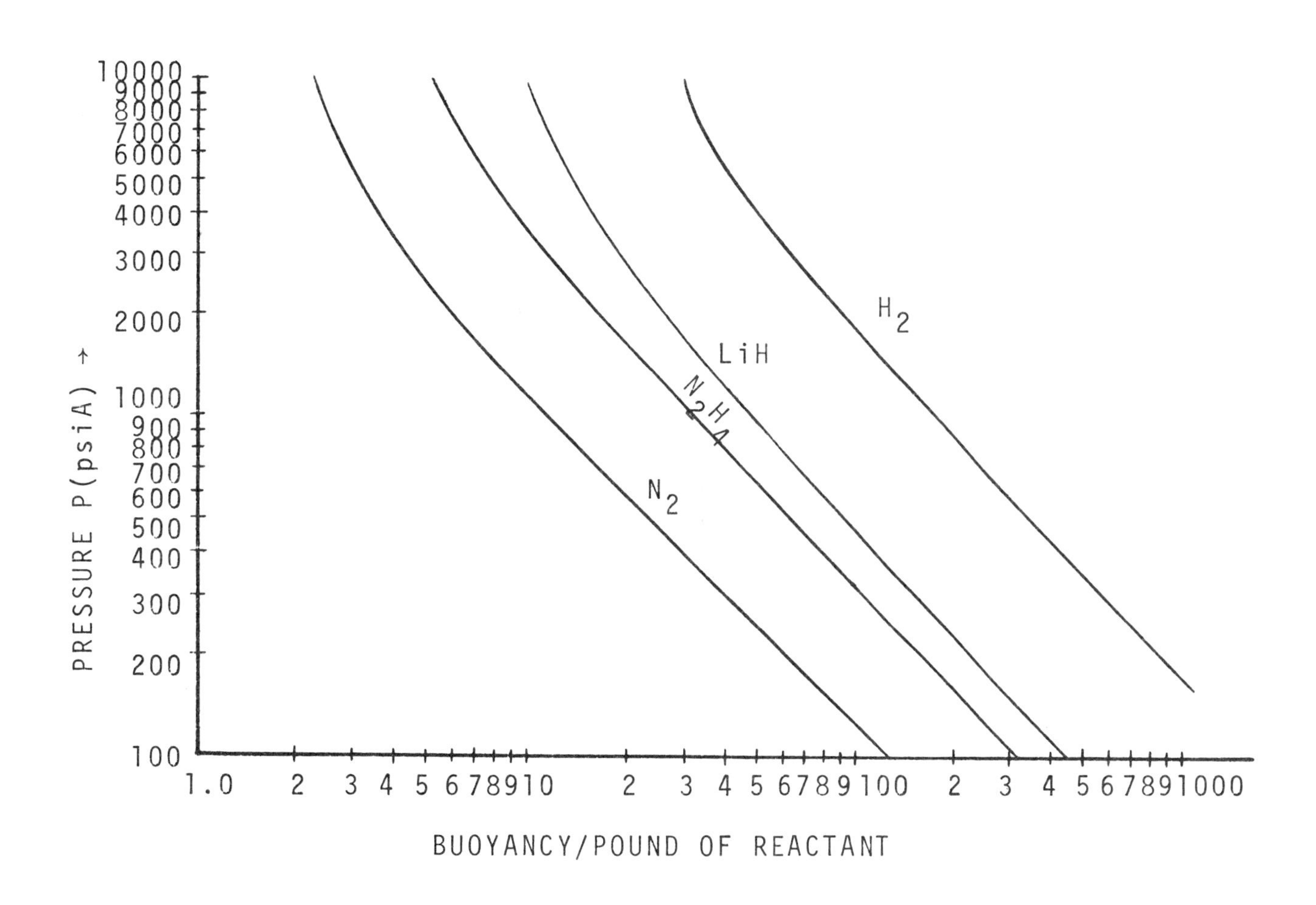

Figure 4-3. PRESSURE EFFECT ON BUOYANCY

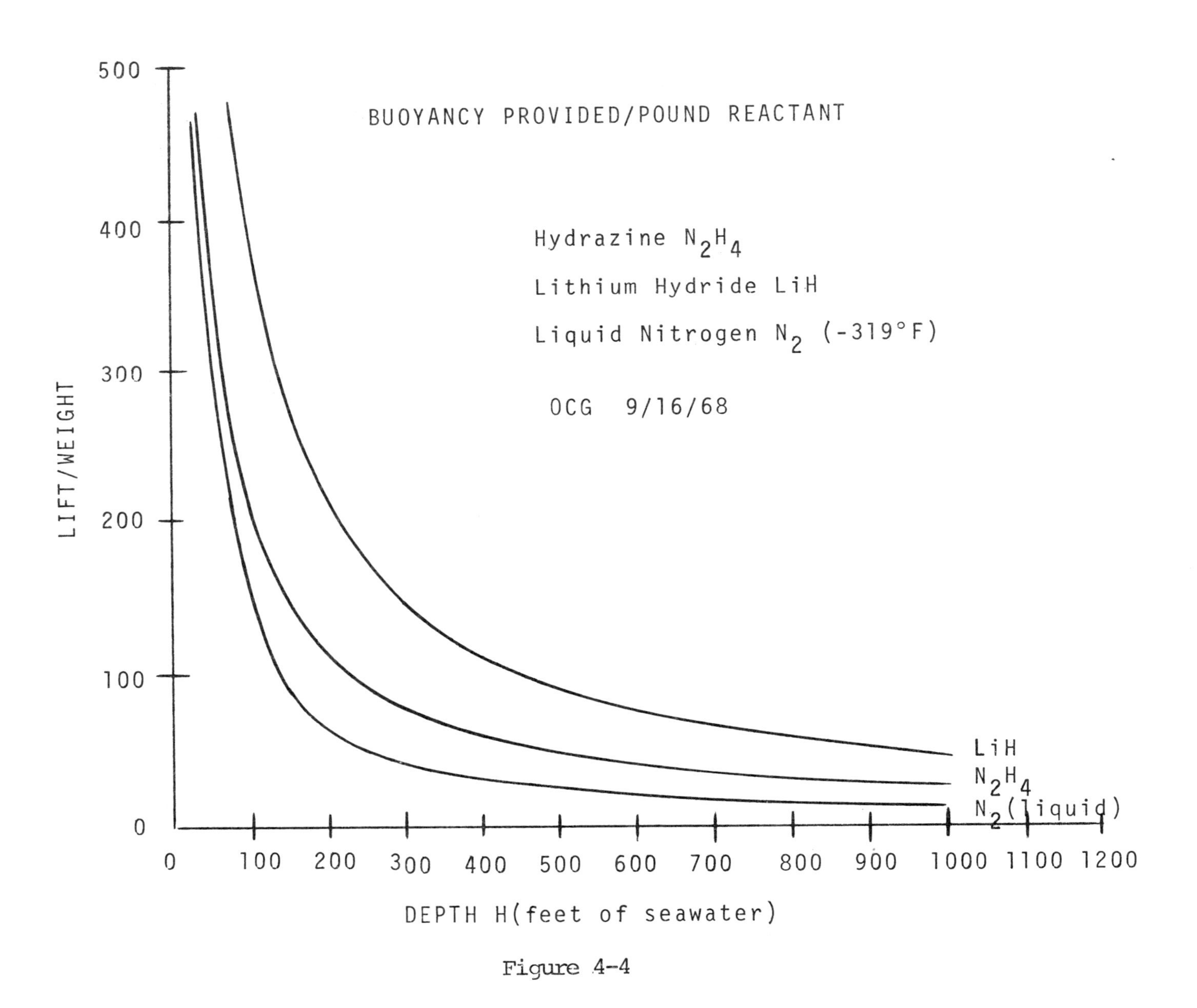

BUOYANCY PROVIDED/POUND REACTANT
Hydrazine N_2H_4
Lithium Hydride LiH
Liquid Nitrogen N_2 (-319°F)
OCG 9/16/68
LIFT/WEIGHT
500
400
300
200
100
0
LiH
N_2H_4
N_2(liquid)
0
100
200
300
400
500
600
700
800
900
1000
1100
1200
DEPTH H(feet of seawater)

Figure 4-4

e) Obstacles

The attachment device will have to work under high hydrostatic pressure => explosive device.

(1) Explosive podeyes
(2) Explosive welding

The shock from explosion must not endanger the other components of the system.

If explosive methods are used, other components must be protected from the magnitude of the pressure and, more importantly, its transient behavior. Methods of protection:

(1) Insulation - surround objects with a material with ρc different from that of the object to be protected.
(2) Since $P \propto \frac{1}{R}$, separation usually can mean protection.

2) Constant Buoyancy Pontoons

A pontoon can be constructed of a constant mass of material which has a density less than that of sea water. The pontoon can be ballasted to make it negatively buoyant. Once sunk, pontoons can be attached to the object to be lifted. By carefully deballasting the pontoons, they can be brought to the surface with the object attached.

a) Modular Lift Concept

Given enough ants, the largest picnic lunch can be carried off. This analogy is the basis for the modular buoyancy lift concept. Various sizes of object can be lifted off the ocean floor by attachment of one or more lift modules,

b) Module Design

Hollow containers with high void ratios provide the most desirable lift to displaced volume ratio, as is indicated in the following table. The container must be strong enough to withstand the external pressure of depth. A spherical shape was chosen for the Ant because its symmetry makes it the strongest shape under uniform loading. Glass has the lowest weight to displacement ratio of container materials for appropriate spherical containers.

Note in comparison: The inflatable pontoon is subjected to ambient pressure, the gas density increases with depth (pressure), and the inflatable pontoon loses efficiency.

The basis of each Ant will be a glass sphere. Each sphere is located in a ballast tank filled with soluble ballast. The Ant is attached to the object to be raised by means of an attachment device on the end of a tether connected to the Ant. The attachment device is similar to that for the inflatable pontoon. The tether permits freedom of attachment.

TABLE 4-1

APPROXIMATE BUOYANT FORCE OF BALLAST MATERIALS

Material	Buoyant Force (lb/cu ft)*
Gasoline	21
Kerosene	14
Ammonia-water (70% NH_3)	17
Lithium	33
Aluminum spheres (4-inch diameter)	20
Petroleum jelly	12
Waxes	15
Woods	21
Glass spheres (4-inch diameter)	16
Glass cylinders (4-inch diameter, S-inch length	23
Microballoon-filled plastics	24

*Measured at the surface and neglecting container weight.

The included graph (Fig. 4-5)relates the size of glass spheres to buoyancy as determined from consideration of the weight of the sphere alone and its displaced water volume weight for a sphere designed for 20,000 ft depth capabilities.

Based upon the present state of the art, development, the glass sphere design will be based on a 7-foot outside diameter glass sphere. This will provide roughly 3.8 tons of lift per ant. To get the required lift, the surface platform will have to carry 15 such ants and their ancillary equipment.

The ballasting material to be used will be a soluble, dense salt. It must be kept isolated from the water in the environment in order to prevent dissociation.

Module - assumed near cubic in form.

Displaced ant volume 174 ft^3 → 3.8 ton lift

ballast ≃ 3.8 tons

ballast volume | sodium chloride salt block | 118 ft^3

c) Constraints

(1) Platform

(a) A module must weigh less than twenty tons in order to permit handling by the platform crane and handling system.

(b) The platform must be able to accommodate 60 tons of buoyancy at depth, in storage.

(c) The platform must be able to provide storage space and access for the ants 10'x10'x10'.

(2) Work Systems

(a) The underwater work system will have to be able to maneuver the ants in ballasted condition in a 3-knot current.

(b) All handling and deployment of ants will be done by unmanned work systems.

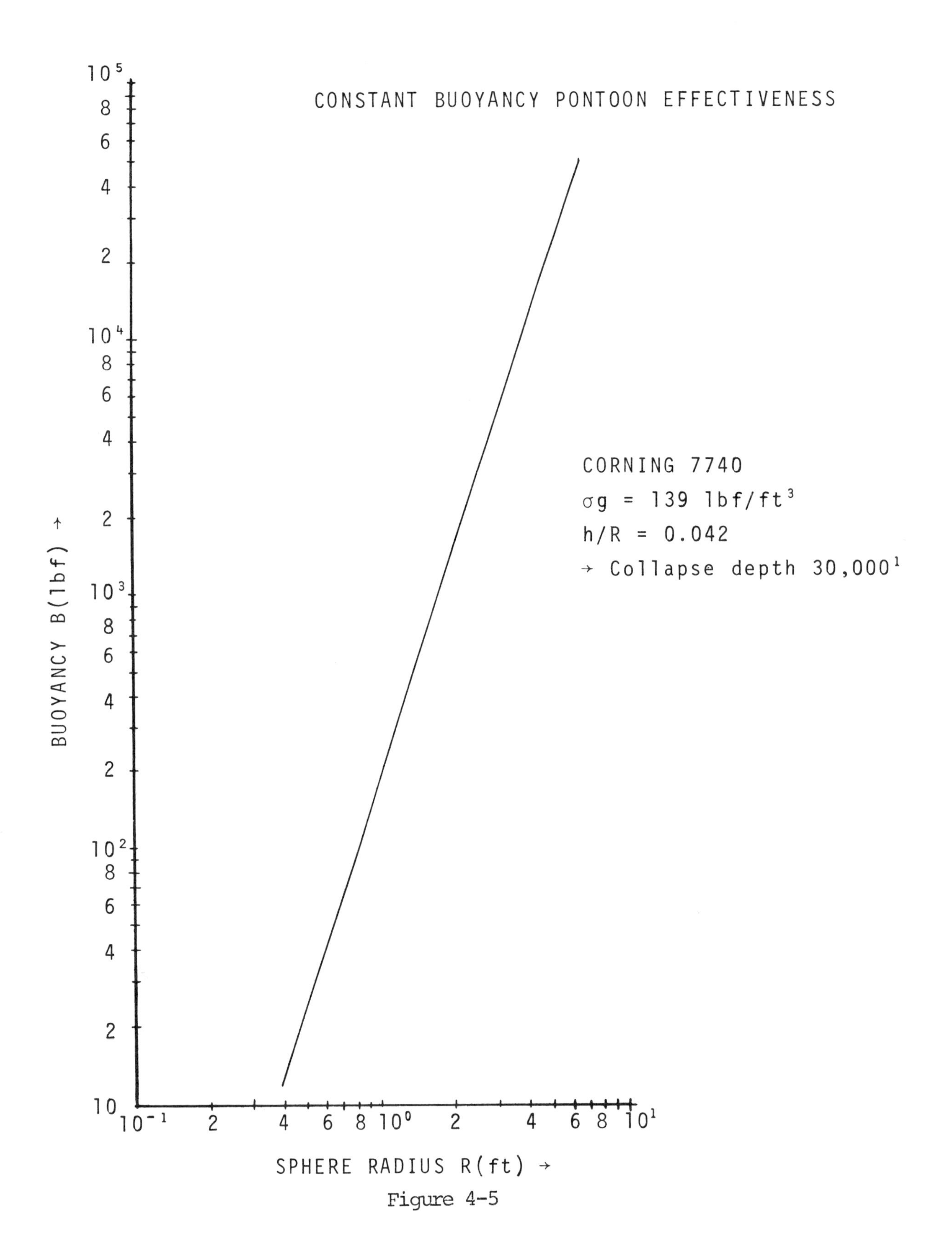
CONSTANT BUOYANCY PONTOON EFFECTIVENESS
CORNING 7740
σg = 139 lbf/ft3
h/R = 0.042
→ Collapse depth 30,000'
BUOYANCY B(lbf) →
SPHERE RADIUS R(ft) →
10
2
4
6
8
10^2
10^3
10^4
10^5
10^-1
10^0
10^1

Figure 4-5

(3) Performance

The module will have to survive a cyclic loading to 10,000 psi for 100 cycles without catastrophic failure with 80% probability and 80% confidence.

d) Obstacles

(1) Ballasting Material

It is desirable to ballast the ant down with a dense, soluble material which will go into solution when exposed to sea water. In this manner, the ant can be deballasted without littering the bottom or burying the object to be recovered.

In order to provide a neutrally-buoyant pontoon, the total pontoon weight must equal that of the displaced water. Thus, the size and weight of the sphere determine the amount of ballast needed. The density of the ballasting material will determine the total ant size.

$$\text{Net buoyancy} = \sum_{i=}^{N} \forall_{B_i} (\gamma_{m_i} - \gamma_w)$$

$\forall_{B_i}$ = displaced volume of water due to i^{th} component

γ_{m_i} = specific weight of ith component

γ_w = specific weight of water

The salt to be used will most likely be sodium chloride ($\rho_m/\rho_w = 2.165$). Its form will depend upon the rate at which the salt is required to go into solution. In block form the ballast volume required would be 118 ft^3, while in granular form it would be approximately 231 ft^3. Tests should be conducted to determine rates of dissociation as a function of temperature and granulation. The advantage of a gradual change in pontoon buoyancy is that there will never be any large jerk-induced forces to pull the attachment device from the object to be raised.

(2) Massive Glass Production

The production of large glass spheres for use in the sea has been plagued by production difficulties and technological inertia subsequent to initial failures. Fabrication of massive glass requires appropriate equipment and not jury-rigged capabilities. Therefore, fabrication will not be cheap. Done well, sphere sphericity can probably be held within acceptable limits of ±0.002 in. This may require grinding of spheres by random rolling over an abrasive field with an abrasive material inside.

A key component of the deep-lift capability, the subsystem success cannot be allowed to hang on massive glass. Although probably available within four years, an alternative will be developed concomitantly, consisting of titanium spheres of the same size as the glass spheres.

Titanium sphere development

Titanium does not have as desirable a weight to displaced volume ratio as does the more brittle glass,

$$W_s/\forall_s\gamma_w = 0.38 \text{ for titanium } \sigma_y = 150{,}000 \text{ psi} .$$

Thus, a titanium sphere the size of a glass sphere will deliver less net buoyancy and, as a result, require less ballasting material to ensure neutral buoyancy.

A seven-foot diameter titanium sphere will provide a net buoyancy of only 3.18 tons. Fabrication difficulties do not plague the construction of somewhat larger titanium spheres. A ten-foot diameter titanium sphere weighs 6.1 tons and provides a net buoyancy of 10.1 tons and will be developed because of its increased lift capability.

Constraint

Compatibility - it will be necessary to ensure that both spheres fit properly into the clump.

(3) Module Delivery

A method is required for placing the ants on the bottom so that they could be readily deployed without wasted effort.

(a) Types of Delivery

1. Unit delivery - each ant is lowered from the surface

   a. Control-pontoon lowering
   b. Lift-line lowering

2. Batch delivery

   a. Control-pontoon lowered
   b. Lift-line lowered

Unit delivery is not desirable because it is wasteful of stored energy which will be required for control during delivery and it is wasteful of time. Batch delivery will be used unless the buoyancy required is too small to warrant that, in which case unit delivery will be undertaken. Modules will be delivered to the bottom in a "clump" which will act as the center for the underwater work site. As such, the clump will provide power, storage and protection for the underwater work systems.

(b) Systems Implications of "Clump"

1. Responsibilities

   a. Platform - The clump must be carried aboard the platform.

      Constraint - Weight and compatibility with platform arrangement.

   b. Control - The control of the clump in the process of locating it on the bottom.

      Constraints

      1) Weight space, power, for equipment required for control.
      2) Clump status and command signals through the location - navigation and communications subsystems.

c. Object Retrieval and Delivery

1) Rigging

a) Lowering the clump through the free surface

b) Emergency recovery

2) Work systems - Compatibility of clump arrangement and design with work tools, power supply constraints.

a) Arrangement

(1) Tether pickup - The work system will require clear access to the attachment tether of each ant which will protrude one foot into an unobstructed flow region.

(2) Space and weight for work system components aboard the clump.

(c) Clump Design

The clump will consist of a chassis into which can be mounted the individual ant modules and a central control, communications, power and work section. When the ant modules are collected in the chassis, the clump can be lowered into the free surface where it floats, barge-like, until flooded down and sunk to the work site.

Clump Delivery

In keeping with the philosophy of uncoupled motions, the clump will be "flown" and not lowered to the work site. Being, roughly, neutrally buoyant, an integral set of control ballast tanks in the stem and stern of the clump can so control the clump weight distribution and attitude that the ants can be glided into position. Note that, prior to glide-down, the environmental currents shall be measured.

Emergency Recovery

In the case of an ant implosion and the loss of buoyancy, the release of that ant's ballast will minimize

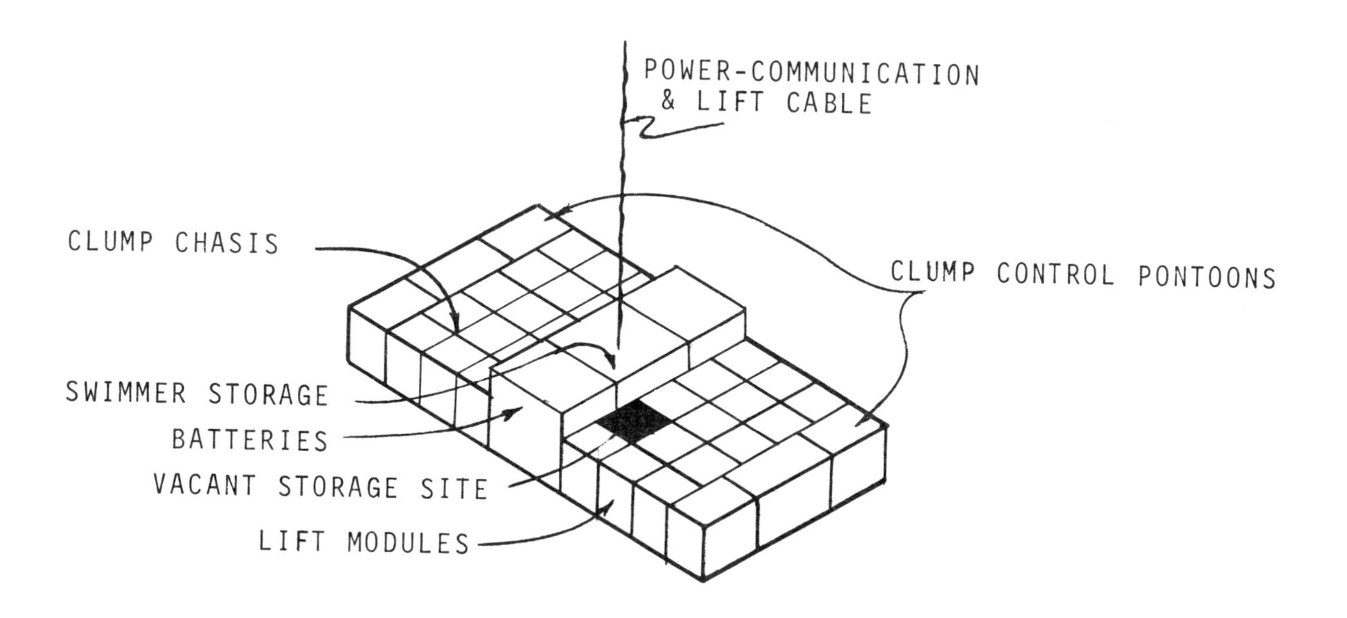
SUBMERSIBLE "CLUMP"
POWER-COMMUNICATION
& LIFT CABLE
CLUMP CHASIS
CLUMP CONTROL PONTOONS
SWIMMER STORAGE
BATTERIES
VACANT STORAGE SITE
LIFT MODULES

Figure 4-6

the net loss of buoyancy to be compensated for in the ballast and trim tanks.

When reserve buoyancy of the ballast tanks is insufficient to effect recovery, the "nerve" center of the clump will be detached from the chassis. Lift of this most valuable section of the clump will be effected by grappling to and lifting with communications and power cable, one end of which floats near the surface, in order to shorten the acoustic link between the clump and surface and by which electrical power can periodically be transmitted.

Constraints on Clump Design

1. Platform - The clump will be suspended beneath the lowest deck of the platform and lowered into the sea from that position.

   a. Attachments between the chassis and deck must carry the load of the clump and admit to ready lowering.

   b. The platform must permit access to the clump so that maintenance and replacement of ants and communications, power and work gear can be effected.

2. Object Retrieval and Delivery

   a. Rigging - Lowering equipment must be properly arranged within the platform relative to the clump to facilitate lowering.

   b. Work - Clump must provide space and weight for power source and work system* components.

3) Control Pontoon

The control pontoon is a variable ballast, external, added buoyancy pontoon which will permit correction of the magnitude of force applied to the object and, thereby, control of object motions. If partially full when in force equilibrium with an attached object, further blowing will supply the force required for breakout. Once off the bottom, the control pontoons can be flooded to return the object and pontoons to force equilibrium. The pontoons are initially set at half-flood equilibrium in order to allow for variations in object weight, which will probably be inevitable. The reserve volume in the pontoon acts as a buffer.

a) Modes of Operation

(1) Automatic Rise

During rising motion, the time rate of change of pressure will be sensed and the buoyancy of the pontoon automatically adjusted so that $dP/dt \leq 2.225$ psi/min, i.e., 5 ft/min.

(2) Automatic Hold

Adjustment of dP/dt, say, $dP/dt \rightarrow 0$. The pontoon will establish a null table at a given depth. The pontoons will be flown and ballasted, as needed, and automatically to keep an object at a given depth.

(3) Manual Override

Control personnel will be able to interrupt automatic operation to provide forces of buoyancy and systems response which the automatic operation could not foresee.

b) Control Pontoon Interfaces

(1) Object Retrieval and Delivery

(a) Rigging - Pontoon structural design.

(b) Control - A control subsystem* has been set up to handle the crucial problem of control of this special pontoon.

i) Mechanisms to control pontoon buoyancy and sense pressure variation automatically as well as manually must be coordinated.

ii) Make pressure signals of control available to the system's command and control subsystem.

(c) Work

i) Provide power for control subsystem.*

ii) Provide for attachment of pontoon to object.

c) Pontoon Structural Design

Lift per pontoon ≃ 20 tons

Volume displaced for lift ≃ 805 cu ft

Cylindrical shape - ring-stiffened

Cylinder dimensions - 10 ft diameter, 12 ft length

Weight ≃ 3 tons

Attachment and instrument weight constraint - 1 ton

(1) Constraints on Pontoon Design

(a) External attachment mounts and penetration in cylinder.

(b) One-ton instrument and control package.

B. Work Subsystem*

The requirement of the work subsystem* is to provide all the underwater work mechanisms and underwater support needed to carry out the responsibilities of the object retrieval and delivery subsystem. Although the task is large, it is all based on a common technological expertise in high-performance mechanisms.

1. Analysis

a. Functions

1) Observation

Requirement - provide a reconnaissance platform for

a) Search and location

b) Inspection

2) Manipulation

Requirement - provide

a) Mechanisms for the performance of tasks

b) Platform base for manipulation of mechanisms

3) Support

Requirement - provide for power and its distribution in support of underwater work

b. Platforms

The basic philosophy behind platform choices was: A mechanical, unmanned system can be designed to do the same tasks as manned systems and they can usually do them better and at less cost.

This is also true in the deep ocean. More and more, the manned submersibles in use are relying upon technologically-derived, as opposed to human, links for his knowledge about his pressure hull. A properly-designed communication - and command and control - link will permit real-time control of undersea systems and the added comforts of man-safe equipment, i.e., unmanned.

To perform the various tasks previously outlined, three different types of unmanned platforms are proposed.

1. Unmanned search and location sled
2. Inspection - unmanned swimmer
3. Unmanned manipulator platform

These were settled upon in order to avoid conflicting performance requirements.

1) Unmanned Search and Location Sled

a) Function - Facilitate the location of sunken objects by carrying search sensors deep within the ocean for extended periods of time and over large areas. Location of sensors near the bottom reduces the resolution required for location verification or, to say it another way, increases the probability of identification.

b) Performance Constraints

1. Endurance - unlimited
2. Speed - 5 knots
3. Stability - the object will not roll, pitch or sway more than 1/2° when towed at 5 knots in still water
4. Payload capability

   a. Weight of sensors - 500 lbs
   b. Arrangement

      1) Magnetometer isolation 20 ft
      2) Sonar attitudes - side-looking sonar - side-mounted
      3) Optical field unobstructed

Based upon the desirability of uninterrupted, unlimited endurance, the unmanned search and location platform must need be a towed sled. The sled must be designed to be towed in a stable orientation at five knots with (another constraint) a tow-line tension at the bitter, i.e., submerged end of less than 1000 lbs.

2) Manipulation Platform

a) Function - Facilitate the performance of tasks underwater required in preparation of object lift. Tasks will consist of objective and environmental observation, alteration or manipulation.

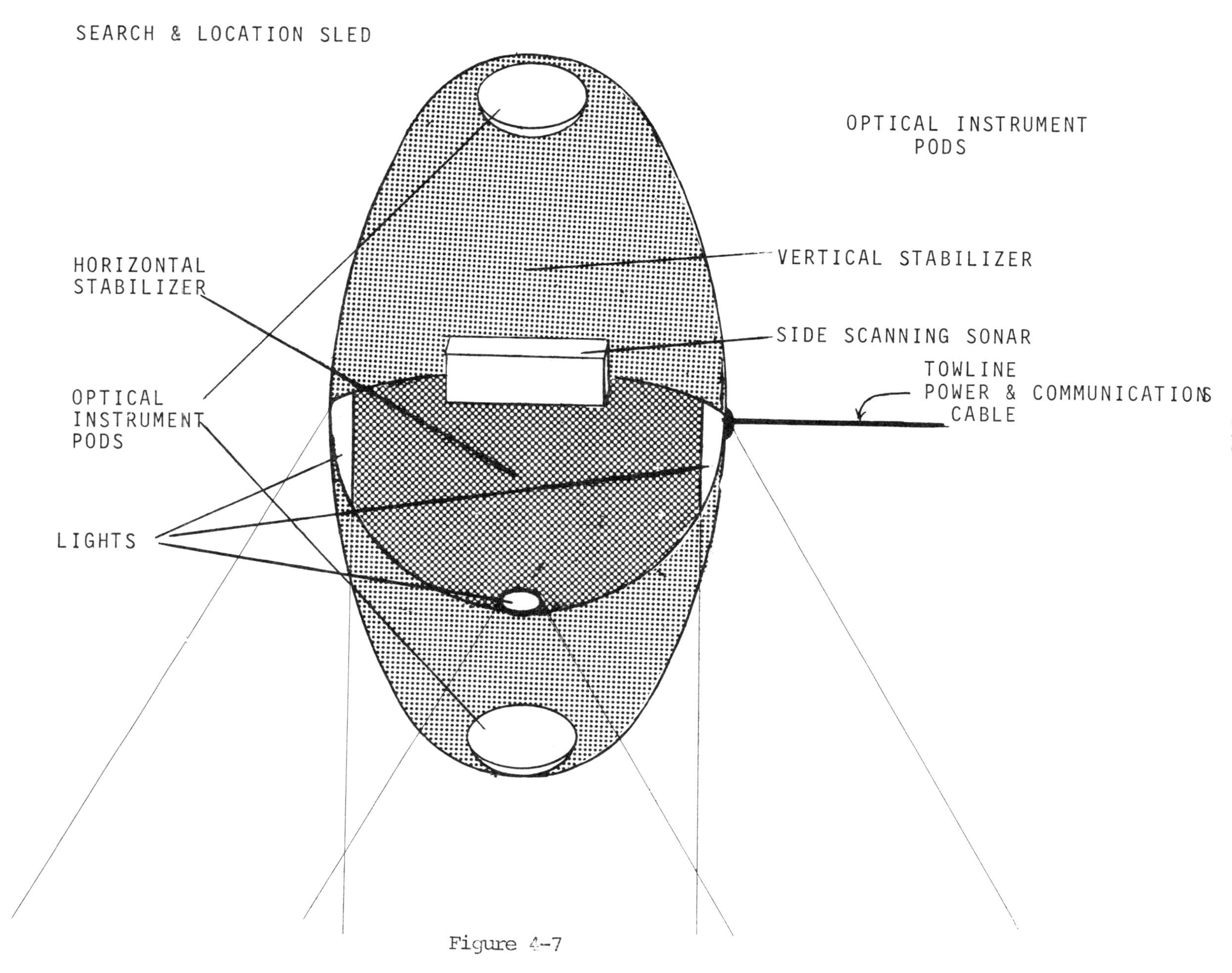
SEARCH & LOCATION SLED
OPTICAL INSTRUMENT PODS
VERTICAL STABILIZER
HORIZONTAL STABILIZER
SIDE SCANNING SONAR
TOWLINE
POWER & COMMUNICATIONS CABLE
OPTICAL INSTRUMENT PODS
LIGHTS

Figure 4-7

b) Performance Constraints

(1) Maneuverability

Stationkeeping - ±2 inches in a five-knot current

(2) Power - 1500-lb thrust from 0° and 180°

(3) Application of Force-Moments

(a) Lift
(b) Push-pull
(c) Torque

(4) Application of Control - Surface-Controlled

(a) Function

i) Operation of tools
ii) Remote control of other components

Based on these requirements, the manipulator platform must be equipped with

i) Manipulators
ii) Propulsion system
iii) Sensing and controllable mechanisms

(5) Signals

(a) Control Signal

i) Form digital electric
ii) Strength - 0.6 volts

(b) Platform acceleration sensed and transmitted to surface

Frequency
Strength 0.6 V
Form digital electric

Manipulators

The platform will be equipped with two manipulators of 500 lbs thrust each when statically tested. They will provide unobstructed access to hemisphere of radius 15 feet, excluding a concentric hemisphere of 3 feet radius. At maximum extension, the manipulator will be able to resist a moment of 7500 ft lbs.

Propulsion System

Due to the range of speeds in which the platform will operate, a tandem propulsion system was chosen. This system permits operation in the velocities of 0-4 knots without loss of control.

Thrust delivered - 1500 lbs
Weight - 1000 lbs

3) Underwater Swimmer Platform

a) Function - Provide reconnaissance in those regions that are inaccessible to or are considered hazardous for the manipulator platform.

b) Performance Constraints

(1) Control

(a) Ability to hover with positioning error of less than 2 inches in a five-knot current.
(b) Ability to make 1 knot against a 5-knot current.

(2) Sensing

In order to properly command and control an underwater operation, a display of orientation of the various components and their status is required. For this orientation information, there has to be a relatively high-resolution sensing instrumentation package.

Types of Sensors Used

1. Acoustic - high-resolution sonars (hulographs) are eventually to be used for object identification and orientation at close range.

2. Optical - T.V. is to be used in an hulographic form to indicate depth of field and object orientation to surface command and control personnel.

Constraints

1. Arrangement - clear access - protected from abuse
2. Power requirements lighting strobe 10,000 watt/sec
3. Weight and volume - 500 lbs
4. Control - initiation and termination signals
5. Signal transmission
   Communications problem - base signal strength from platform

2. Platform Designs

a. Search and Location Platform

Whenever possible, when desirable, instruments shall be protected inside glass instrument spheres. Arrangement of spheres is the problem.

Constraint - Spheres 10 inch OD
Weight → buoyancy per sphere = 11 lb
Submerged - neutrally buoyant

Intersecting lensatic shapes make faired sled structure

Constraint - Penetration of spheres will be made in equatorial collar at two points 180° apart on the circumference. Penetrations 1/16 in. in diameter

Power Supply to Sled

Power to instruments on the sled and to communications will be supplied from the surface. The two cables will be 1/2 in. instrument cable which will be capable of carrying a 1000-volt current to power the instruments. The losses involved will be high but the unlimited endurance capability is worth the price.

b. Manipulator Platform

1) Propulsion - Tandem-propelled

2) Hull Structure - Titanium cylinder

8 ft long
4 ft diameter

3) Control - Manipulator platform will be surface-controlled.

c. Unmanned Swimmer

1) Propulsion - tandem-propelled

Powered by an electronic link with the clump, controlled from the surface.

2) Hull Structure

All sensors except acoustic and pressure transducers will be sealed inside glass instrument spheres of 10"I.D. Hemispheres will be joined at the equator with a titanium band through which penetrations of no greater than 1/16 in. will be permitted.

The configuration of the swimmer will be that of a body-centered cubic-packed array of spheres. The controls will be nestled in a 16-in. diameter Corning 7740 glass sphere. To this sphere the leads from the instruments will be fed.

a) Constraints

(1) Lead line diameters

(2) Lengths

(3) Tether (s) must be neutrally buoyant so that it will not sink the swimmer

b) Line Control - Neither lead line nor tether must be permitted to tangle in propellers. This may require a takeup and payout mechanism.

c) Control

Underwater swimmer (unmanned) will be surface-controlled interface with control subsystem.

(1) Control Signal

Form - digital electric
Strength - 0.6 volts

(2) Sensors give accelerations of swimmer signal

Form - digital electric
Strength - 0.6 volts

3. Tools and Support

In order to do work underwater, mechanisms must be designed to do it. Even in shallow waters in which divers can be employed, they need assisting mechanisms. The slightest physical exertion for a diver underwater becomes a greater problem with depth.

Assistance needed underwater is normally in one of the following areas:

a. Observation
b. Connecting and disconnecting (attachment and cutting)
c. Lifting
d. Pushing

a. Observation

1) The unmanned underwater swimmer is an observation assist.

2) Sea-floor soil condition can be observed by stirring up the silts but, more desirably, conditions can be quantized by use of an in-situ Vane-shear device and penetrameter. Based on penetration bests and shear tests, it will hopefully be possible to determine soil conditions and predict break-out forces.

b. Connecting and Disconnecting

1) Attachment

It will often be necessary to attach the tether of a pontoon to an object. Toward this end, work must be directed to produce an explosive bolt or welding process.

As envisioned, the manipulator platform would transport a lightweight attachment package over to the object, use a stopgap attachment method until it could move of sufficient distance to fire remotely at a safe distance the explosive charge.

2) Design

3) Cutting

This capability is limited at this time to somewhat shallow depths. Although oil companies have cut drill pipe at 18,000 ft, this has been a very special application.

Types Applicable

Chemical
Explosive
Hydraulic

Chemical cutting has the drawback of some very nasty by-products which are quite corrosive. Explosive cutting is not controllable enough to prevent the shock from perhaps damaging the object in an undesired manner.

Hydraulic cutting is achieved by accelerating a particle of water to very high velocities so that it shoots holes in the material to be cut. This method has none of the harmful effects of chemical cutting, nor the drastic loss of efficiency of conventional gas compression methods.

At this time the hydraulic cutting methods are not available. Development of hydraulic and chemical cutting should be continued concomitantly in order to get a workable tool subject to the following constraints:

1. Weight < 500 lbs
2. Size < 80 ft^3
3. Control - mechanical control at pressure
4. Pressure 9800 psi

4) Lifting

a) Diver's Lifting Assist - inflatable lifting bag.

(1) Constraints

(a) Underwater inflation control by diver

(b) Load capability - 1 ton

(c) Stowage - life

(2) Gas Generator

Blowing of inflatable pontoons, control pontoons and clump ballast tanks requires the displacement of water at depth by gas. The work subsystem* will be responsible for the mechanism by which the gas is delivered.

(a) Chemical Production

A gas generator to produce H_2 from disassociation of hydrazine is to be procured subject to the following:

(b) Constraints

1. Performance

   a. Production rate 1 cu ft/min at 20,000 ft
   b. Efficiency - 80% dissociation at 2,000 ft

2. Control - electrical - wire size and length

3. Geometric compatibility - attachment volume

4. Weight - 500 lbs

(c) Obstacle - Fuel Resupply

Care must be taken to ensure that the gas generator is not exposed to sea water directly.

(d) Cryogenic Supply

Because of the relatively low cost of air when compared to hydrazine, it will be desirable to replace the hydrazine gas generator with a cryogenic gas delivery system when the object is in depths less than 1000 ft.

5) ± Pushing

The removal of debris and the transfer of an object from one point to another, such as would be required in the attachment of an ant tether is to be fulfilled by the manipulator platform.

4. Underwater Power Source

The most desirable power source for use in the ocean would be the source with the highest energy density and reliability at the desired life (see Figure 4-8).

Prime desirable source - nuclear reactor

Desirable - fuel cells

Least desirable but available - batteries.

The work subsystem* is proceeding in investigation of a cost and feasibility at using a derivative of the snap 8 reactor to supply power at 20,000 feet.

Prime Systems Problem

Efficient energy converter - on the way

Goal - 50 kw continuous power

Advantages

Continuous surveillance

Multiple submersible operations, i.e., several underwater unmanned platforms could be operated continuously

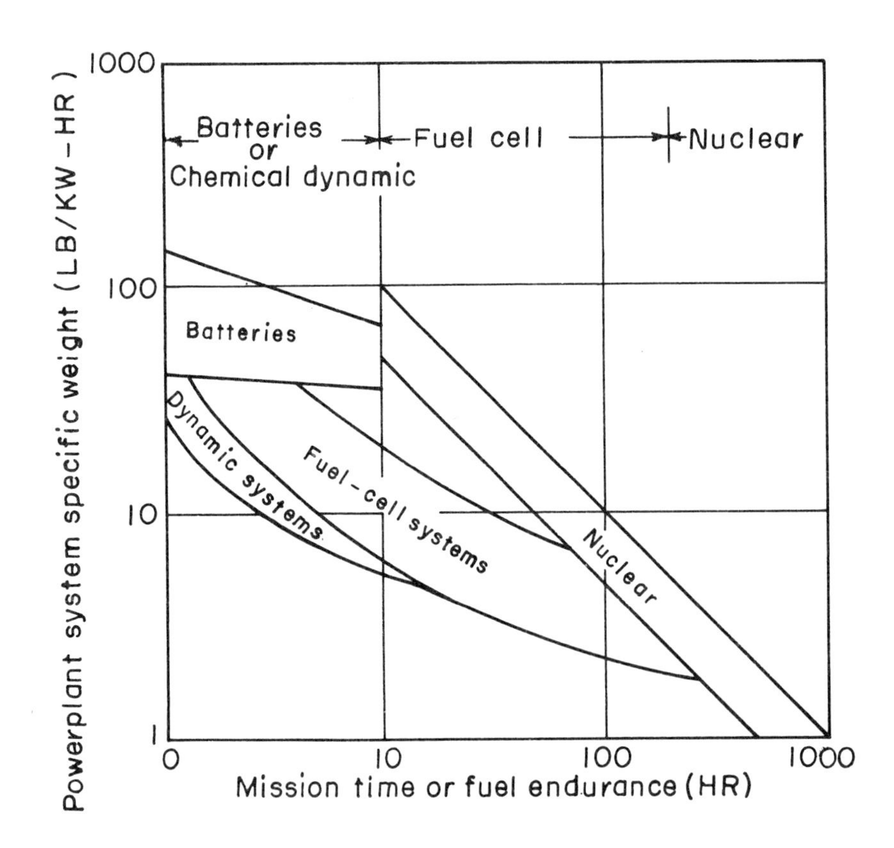

Figure 4-8

The alternative to the more desirable energy sources is batteries. The table below gives the rough cost of using various battery forms in terms of weight, volume, cost.

Type	Dry Weight (lb/kwh)	Volume (ft^3/kwh)	Wet Weight (lb/kwh)	Cost ($kwh)
Lead-acid	50-115	0.25-0.6	17-80	30-1000
Silver-Cadmium	30-37	0.21-0.23	17-22	800-8000
Silver-Zinc	12-50	0.11-0.35	5-35	500-6000

a. Performance Requirements

1) Power output 15 kw

2) Endurance 72 hours

b. Constraints

1) Lift Subsystem* - clump design

a) Weight - 5400 lbs

b) Volume - 110 cu ft

c) Cost - $2,700,000

2) Work Subsystem*

a) Battery Recharge

Surveillance must be done on an intermittent observation basis unless the system is willing to recharge batteries frequently. Placed in the clump and sunk to the worksite, the power leads to the instrument platforms, tools and ancillary equipment can be kept relatively short. However, once on the bottom, the batteries will have to be recharged on the bottom. During charging, there will have to be a physical link with the surface.

b) Connections

Battery to sink-lines

Maximum length - 200 ft
Weight - cable weight neutrally buoyant in water

Battery recharge - from surface

Power - communication-lift line of clump
Capacity - 1000 volts

C. Delivery Subsystem*

Has been dissolved because it is felt that the rigging subsystem* can handle the object on the surface. One of the lifting lines can be used as a tow line, for example. The line payed out around the rolling capstan is handled much the same way as a towing engine handles a tow line.

D. Control Subsystem*

The requirement of the control subsystem is to provide the sensing and controlling signals for ballasting and blowing the control pontoons.

1. Types of Control

a. Rise

Automatic Control

1) Senses hydrostatic pressure and its rate of change with time.
2) Compares data.
3) Signals buoyancy corrections to control pontoons.
4) Operates control devices.

Manual - same as for automatic except that topside control interrupts command cycle.

b. Drag - Deployment of drag chutes.

2. Pressure Sensing

Pressure transducers provide electrical signal to an object-mounted control center. Transducers are located on each control pontoon as part of its control package.

a. <u>Constraints</u>

1) Signal strength

2) Power requirements 100 watts

3) Sensitivity of transducers

±1 psi level
1/2 psi/min rate of charge

b. <u>Data Comparison</u>

1) <u>Trim Control</u>

Signals from the control pontoons are balanced on Wheatstone bridges. Signals of unbalanced strength $> V_o$ will actuate control mechanism on pontoon.

2) <u>Rise Control</u>

Maximum rate - inductance coils in the control center will generate signals of sufficient strength to actuate correction if the signal from the control pontoon is changing too rapidly. The polarity of the current will indicate rising or falling motion. A minimum signal V_1 will do the same to prevent the object from stopping its motion and hanging neutrally buoyant where the object pleases.

3) <u>Equilibrium Rise Control</u>

Placed in equilibrium mode by surface signal, the induced current tolerances are reduced so that the control pontoons will seek $\frac{dp}{dt} = 0$.

4) Constraints

a) Communications

(1) Signal strengths

(2) Form and frequency

b) Work Subsystem

(1) Gas generator control signal

(2) Form and strength

electrical 1 volt

c. Signal for Buoyancy Correction

1) Automatic

Illation of control modification occurs when V_1, the induced voltage magnitude, exceeds one volt. The sense of the current initiates standpipe or gas generator.

2) Manual

Surface signal override tells the nerve center to feed voltages $|V_1| < V$ into the control loop to initiate proper subsystem response.

d. Design Constraints

1) Lift Subsystem

a) Pontoon penetration for standpipe

b) Equipment down (see Lift section)

2) Work Subsystem*

e. Drag

Drag chutes will be attached to the control pontoons and be deployed upon signal from the nerve center.

1) Constraints

a) Signal - acoustic
Frequency

b) Communications - transmission of command and control signal

c) Command and control signal

BIBLIOGRAPHY

1. Craven, John P., "The Design of Deep Submersibles," Soc. of Nav. Arch and Mar. Engs., No. 9, June 1968.

2. Craven, John P., "Recent Contributions under the Bureau of Ships Fundamental Hydrodynamics Research Program," Journal of Ship Research, October 1958.

3. Edwards, Russel N., Floating Power Plant to Support Submerged Offshore Operations, Offshore Technology Conference, Paper No. OTC 1131, 1969.

4. Frink, Donald, Arthur Coyle, and Donald Hackman, Battelle Memorial Institute. Private communications.

5. Hansen, R. O. and D. Uhler, Office of the Supervisor of Salvage. Private communications.

6. Arthur D. Little, Inc., Stress Analysis of Ship-Suspended Heavily Loaded Cables for Deep Underwater Emplacements (ADL, August 1963).

7. Waters, O. D., The Ocean Engineering Program of the U.S. Navy, U.S. Government Printing Office, 1968.

CHAPTER V

COMMUNICATIONS SUBSYSTEM

EPIGRAMS

Communication reception is one form of environmental sensing.

Communication transmission is one form of environmental control.

The most appropriate environment to control is the one in which you are immersed.

The major problem of a fish out of water is getting back across the interface.

Communication, even if it is only physical, exists at every interface.

General Purpose

1. Communications is the transfer of intelligence from one point in space and time to another point in such a manner that the intelligence aids in the operation of the system and the achievement of system goals. In most sea systems communications is not an end in itself and, in fact, often constitutes the entire interface between subsystems which are not physically connected. As a result the communications requirements are uniquely determined by the nature of the other subsystems if command and control are included as a subsystem. Indeed, if communications are confused with command and control, then this subsystem becomes a system's master rather than a system's servant.

2. Since subsystems may be operating in differing environments and since communication in general utilizes the environment as the medium for communication, the problem of communication subsystem design is not only that of design of a transmitter and receiver, but includes transducing requirements in order to transmit the signal from one environment to

another. Subsystem managers will desire to communicate in a format which is most compatible with the physical generation of the information which must be communicated. For example, if an environmental sensing instrument obtains its information in analog form, there may be a strong desire on the part of the subsystem manager to transmit this information as modulation on a continuous wave signal. On the other hand, if a subsystem manager having responsibility for an inertial guidance system receives his information in the form of pulses from a pulse pendulum, then he would most likely desire to send his information in a digital format. Subsystem managers will also desire to utilize the most convenient form of the environment for transmission of the signal. For example, a subsystem manager having responsibility for a satellite subsystem will desire to transmit his information in the form of high-frequency electromagnetics, whereas a subsystem manager having responsibility for a deep-ocean subsystem will prefer to transmit his information acoustically. It is the responsibility of the communication subsystem designer, therefore, to select the most appropriate communication means which can provide a direct link, if possible, between the two subsystems or, in the alternative, to select transmission schemes which most suit the needs and requirements of each subsystem together with an appropriate transducing equipment, generally located at some environmental interface. With this philosophy, the format and content of information becomes the interface between two subsystems not otherwise physically connected and each of these subsystems has, in turn, an interface with the communication subsystem for the injection of the previously-determined type of message into the communication system.

The communication subsystem manager has the responsibility for transmitting the information in appropriate format or assuring that it arrives in the appropriate format. He should have the responsibility for the design of the transmitter, transducer and receiver which will be employed between the subsystems.

Major Design Considerations

Intelligence is received and transmitted by controlled perturbations of the environment. The energy form used for perturbation of the environment which is chosen will depend upon the nature of the environment and the system needs, such as information rate, security, antijam requirements and communication delay time. Figure 5-1 is a matrix indicating the energy forms which have been shown to be most appropriate above and below the free surface and for communication between various types of platforms.

A number of generalities can be drawn which explain this diagram, as follows:

a) High Data Rate requires high frequency. The maximum amount of information which can be transmitted in a channel is given by

$$\frac{dI}{dt} = W \log \left(1 + \frac{P}{N}\right)$$

Here $\frac{dI}{dt}$ is the information in bits per unit time

W is the bandwidth

P is the signal power

N is the noise power

Only at high frequency can a wide bandwidth be obtained without filling up the spectrum. For example, AM broadcast at frequency in the range of from 540 to 1600 kc can only transmit the sound spectrum of 0 - 10kc

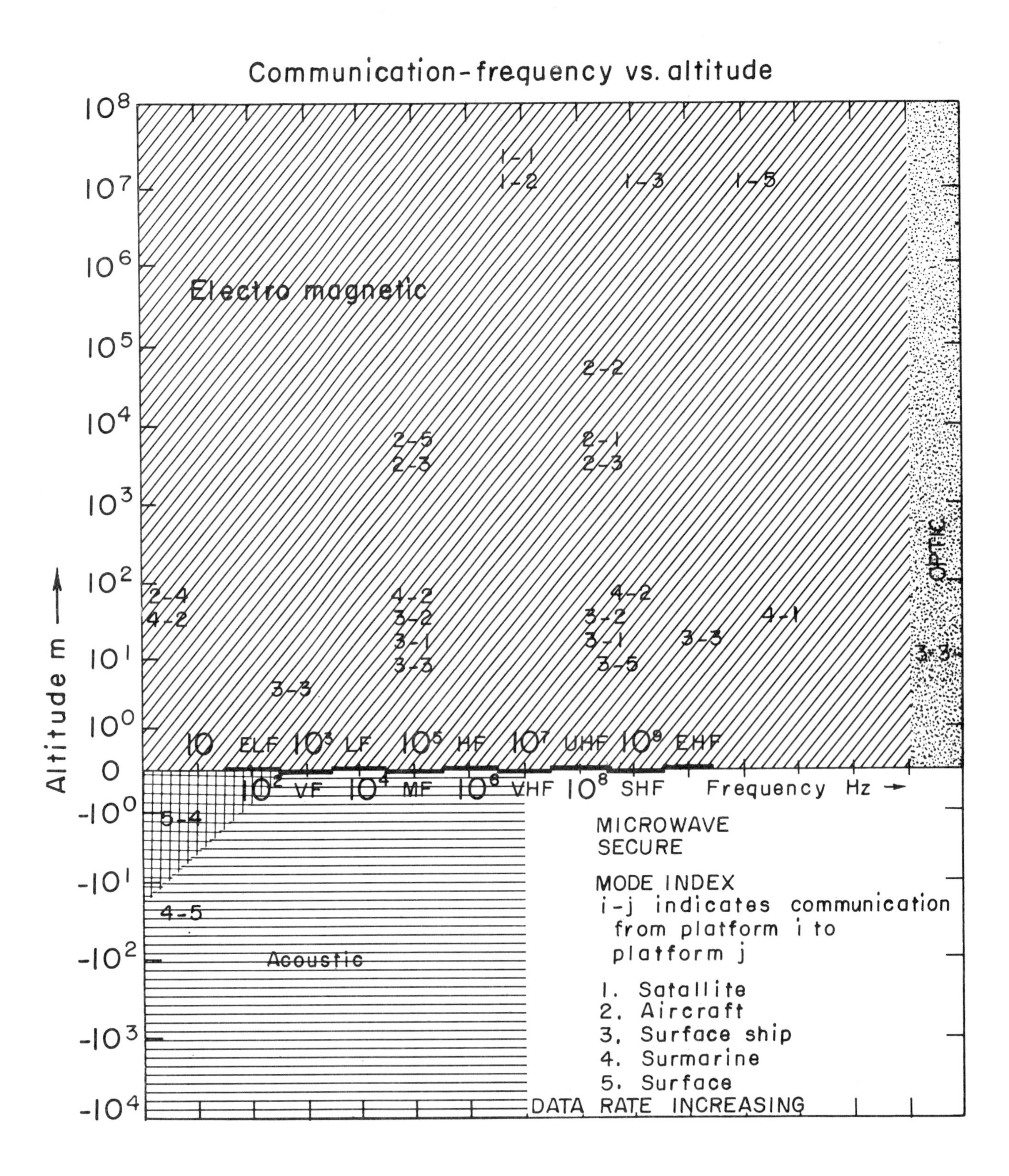

Communication-frequency vs. altitude
Altitude m
Electro magnetic
OPTIC
Acoustic
Frequency Hz
ELF
VF
LF
MF
HF
VHF
UHF
SHF
EHF
1-1
1-2
1-3
1-5
2-2
2-5
2-3
2-1
2-4
4-2
3-2
3-1
3-3
3-5
4-1
5-4
4-5
MICROWAVE
SECURE
MODE INDEX
i-j indicates communication from platform i to platform j
1. Satallite
2. Aircraft
3. Surface ship
4. Surmarine
5. Surface
DATA RATE INCREASING

Figure 5-1

without substantial interference between neighboring stations. On the other hand, FM broadcast in the megacycle range of 90 - 105 megacycles can transmit audio in the 0 - 20kc and "commercial music" on a carrier of 88kc with a bandwidth of about 20kc. The total bandwidth of about 100kc is only about 1% of the total FM spectrum.

b) High-frequency communication requires high power for significant range since the attenuation varies as an inverse function of the wavelength.

c) High-frequency communication is distorted,diffracted or scattered by physical objects or phenomena, having a scale which is large compared with 1/4 wavelength (roughly).

d) High-frequency or optical communication benefits powerwise from collimation and such collimation permits high security and freedom from jamming.

e) High frequency requires small antennae for transmission but may require large dishes or collectors for reception.

f) Low-frequency communication requires low power for long range since attenuation varies inversely with wavelength.

g) Low-frequency communication is highly reliable since it is almost unaffected by physical objects or phenomena, having a scale with 1/4 wavelength or less.

h) Low-frequency communication requires large antennae for transmission, but can be received on small or modest-size antennae.

i) Low-frequency communication would require large arrays for directional transmission and large arrays for directional reception. Antijamming capabilities are therefore difficult and transmission security is low.

The generalities coupled with environmental characteristics have constrained oceanic communication to the parameter range shown in Figure 5-2.

The environmental medium in which communication occurs drastically changes the ability to use electromagnetics, acoustics, light or other physical phenomena. When communicating with a subsystem which is completely immersed in the ocean, the important physical characteristics to be considered are salinity, temperature and density gradients, the character of the ocean bottom in terms of its reflectivity and physiography, and its absorptive and dispersive characteristics with respect to magnetic fields, electromagnetics, and light. The energy forms for communication in water are listed in Table 5-1. Of all these energy forms, only acoustics has the properties which permit transmission over long ranges in the water environment, and this may be seen by noting that the power for transmission determines the range communication possible at a given frequency and for a given energy form. As an empirical law the range r is pragmatically limited to a value given by $r_{lim} = 10\ dB/\alpha$ where α is the attenuation coefficient. This rule is based on the fact that large power increases beyond state-of-the-art reasonableness will occur for ranges in excess of this value. For short-range surface communication between ships, the traditional means of communication has been optical. In the past, this has been accomplished through the use of flags and semaphor by day and by lights at night. It would be only natural to hope that optical techniques would work equally well underwater. Unfortunately, not only is light attenuated in water, it is readily scattered by the particles in the

TABLE 5-1

METHODS OF INFORMATION TRANSFER

A. High Energy Particles

B. Cosmic Rays

C. Electromagnetic Radiation

1. Ultraviolet
2. Optic
3. Infrared
4. Extremely High Frequency } Microwave
5. Super High Frequency } Microwave
6. Ultra High Frequency } Microwave
7. Very High Frequency
8. High Frequency
9. Medium Frequency
10. Low Frequency
11. Very Low Frequency
12. Voice Frequency
13. Extremely Low Frequency

D. Sonic Radiation

1. Acoustics
2. Seismic

E. Hard Wire

F. Physical Delivery

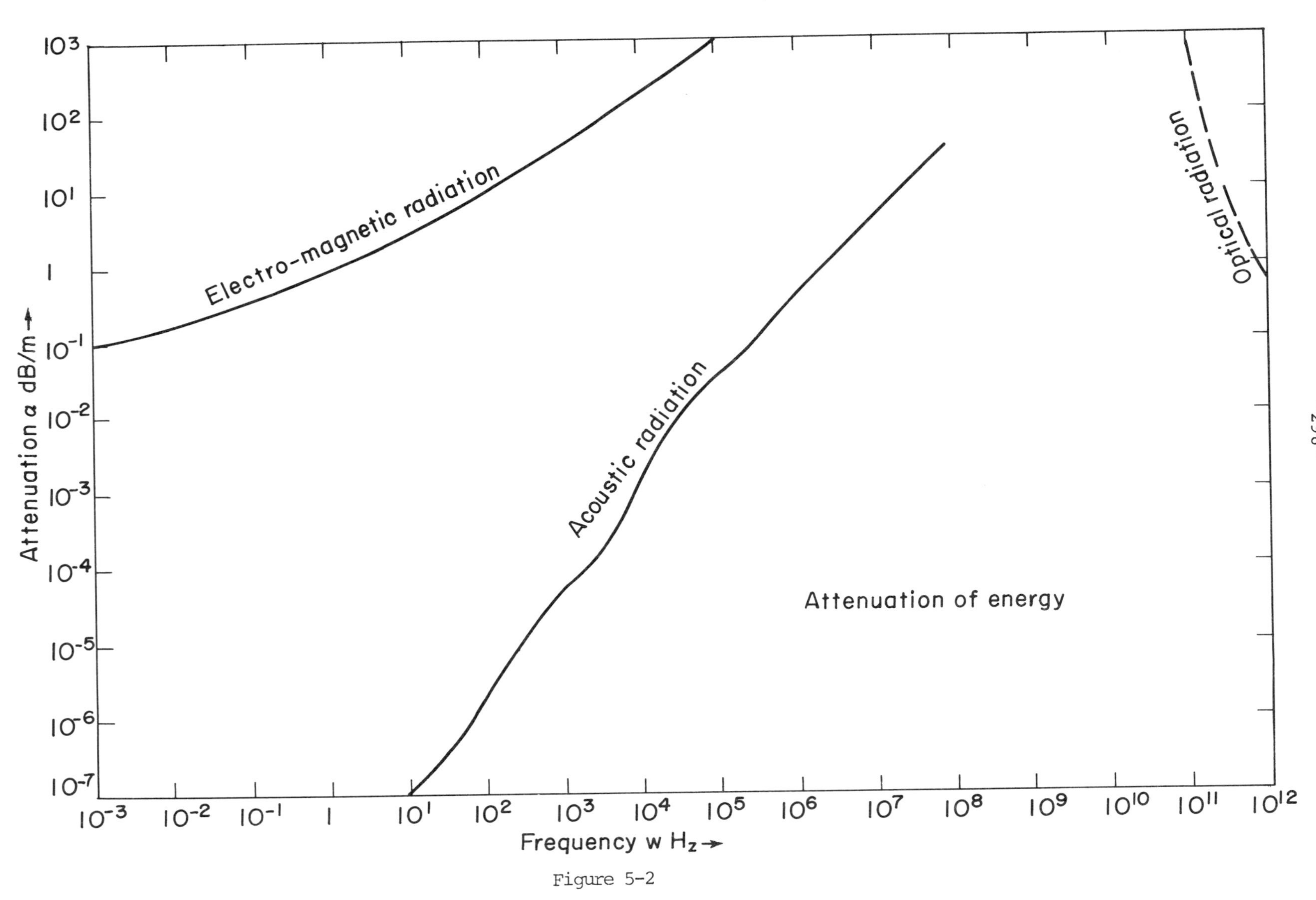

Electro-magnetic radiation
Acoustic radiation
Optical radiation
Attenuation of energy
Attenuation α dB/m →
Frequency w Hz →
10³
10²
10¹
1
10⁻¹
10⁻²
10⁻³
10⁻⁴
10⁻⁵
10⁻⁶
10⁻⁷
10⁻³
10⁻²
10⁻¹
1
10¹
10²
10³
10⁴
10⁵
10⁶
10⁷
10⁸
10⁹
10¹⁰
10¹¹
10¹²

Figure 5-2

water producing a phenomenon known as backscatter. The forward attenuation is drastic when compared to attenuation of acoustic energy, that is,

$$I_{(optical)} \cong I\,e^{-x}$$

$$I_{(acoustic)} \cong I\,e^{-0.1x}$$

where I is the intensity at a distance x from a source of intensity I_0 at x = 0. The sea does act as a bandpass filter for blue-green frequencies because conductivity of energy increases as the polar water molecule rotates at frequencies greater than 10 KHz. Many system designers are misled into believing that this results in a "window in the sea." In fact, the attenuation coefficient is reduced only about one-half of that for light in the red portion of the spectrum. Particles of matter or bubbles of gas in the water and turbidity can increase the attenuation rate fourfold. Backscattering quickly decreases the signal-to-noise ratio so that in typical sea water the backscattering will frequently limit visibility to distances of less than 40 feet.

If the designer recognizes that optical communication underwater is limited to extremely short distances such as those used by a diver or by small vehicles in close proximity, then it can indeed by a useful means of communication. A number of techniques exist for improving the range over which a light beam can be observed. These include such things as the use of electronic shutters which gate out the backscattering and admit only the light which arrives at the same time as the reflected pulse. Other techniques can maximize signal identity; for example, objects should be of maximum color contrast with their surroundings. It has been shown that the maximum range of visibility occurs when objects are colored international orange or yellow in natural underwater sunlight and are colored black in artificial or murky light. Colors which are the same as the background, either blue-green or white, are generally indistinguishable, either because they blend in with the water or because they blend in with the backscattered light. While such techniques for increasing the power of the light and the sensitivity of the receivers can be considered, or such things as shifting to observation at blue-green frequencies might increase the range, these improvements in visibility range are generally expensive and of marginal effect.

Electromagnetic Communication

The salinity of the water makes it a highly-conductive medium insofar as electromagnetics are concerned. The net result is that electromatic waves which are propagated through the atmosphere will tend to regard the ocean as a flat conducting plate. The penetration of the medium is therefore confined to the skin depth which is a very small distance except for extremely low-frequency waves. The use of electrodes in the water itself will produce a highly-localized field. At distances which are large compared to the spacing of the electrodes, this field will be in the nature of a dipole. This attenuates much more rapidly than a pure source and will be highly inefficient except in the "near field" locations. This suggests that electromagnetic or electrical communication within the sea water itself is limited as is optics to very "near field" conditions and should be employed only for bodies or subsystems that are in close proximity to each other and where the inefficiency is traded off against some system simplicity.

Acoustics

The clear superiority of acoustics for long-range propagation makes this the desirable type for underwater communication even at the expense of the low propagation rate, the average velocity in ocean water being approximately 1400 meters per second. When considering long-range acoustic means of communication, the system designer should be aware of the fundamental properties of sound propagation in the ocean. Reflections and refractions of acoustic energy in the sea trap sound in what might be conceived of as oceanic wave-guide. When so trapped and redirected the sound spreads cylindrically and not spherically. Consequently, signals as a first order will travel further than they would before they are reduced to background noise level but, on the other hand, the signals will be heard only in the channel and can be directed away from significant portions of the ocean.

Shallow Water

In shallow water reflections from the surface and bottom redirect the energy. Consider a sound source, as shown in Figure 5-3, located at point S. In general, a receiver located some distance away will receive signals reflected from the free surface and bottom, as well as the signal transmitted directly to it. Wave-guide effects become significant at ranges greater than 10 times the depth of the water. At these ranges signal strength can only be maintained by near-total internal reflection and negligible transmission. The condition of the bottom, therefore, exerts controlling influence on signal strength. Strong signals are the result of hard, dense bottoms. On the other hand, sound channels do not exist in shallow water when the bottom is "slow" because energy

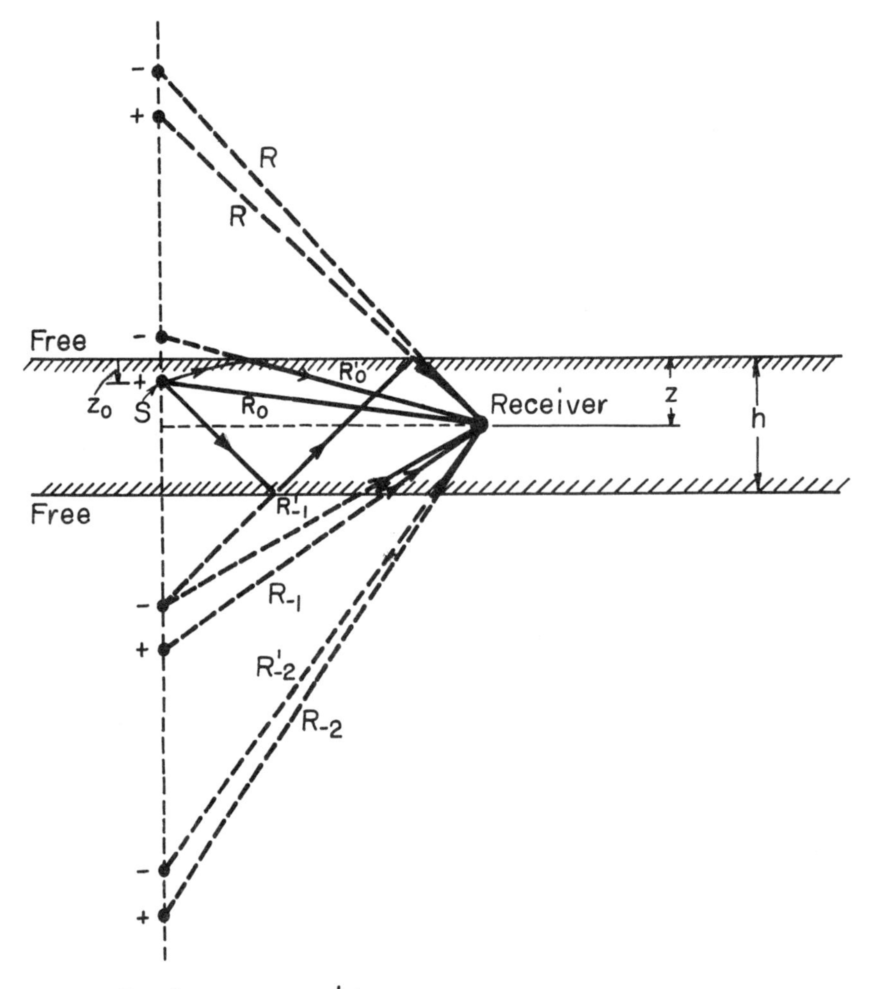

Figure 5-3

is refracted into the bottom. Indeed, as a general rule, one can expect that on continental shelves or other shallow water areas signal propagation for extensive distances will be poor and multipath, i.e., the repetition of signal will be large.

Deep Water

While for most shallow-water sound channels the thickness can be considered such that the velocity of sound is approximately constant, in deep water the spatial variation with temperature, salinity and depth becomes significant. Sound is continuously refracted into regions of lower propagation velocity where its extinction is only a function of its attenuation. The velocity of sound is normally expressed as

$$C = K \frac{P}{d\delta/\delta}$$

where K = bulk modulus

P = pressure

δ = density

An empirical relationship has been determined by A. B. Wood which reflects the effects of variation of the temperature, depth and salinity upon density and pressure, and therefore the speed of sound in the sea. It is

$$C = 1410 + 4.21\ T - 0.37\ T^2 + 1.145\ S + 0.018\ D$$

where

C = velocity in m/sec

T = temperature in °C

S = salinity in parts per thousand

D = depth in meters

Depth effect--The velocity of sound will increase with depth because the pressure increases in the compressibility of the fluid. The rate is roughly 0.018 m/sec per meter of depth.

Temperature effects--The surface waters of the ocean act as a thermal buffer between the gaseous atmosphere above and the liquid one below. Variations in temperature, due to seasonal and diurnal fluctuations in solar radiation result in thermal stratification in the upper ocean (see Figure 5-4). Wind waves work to destroy the stratification by inducing

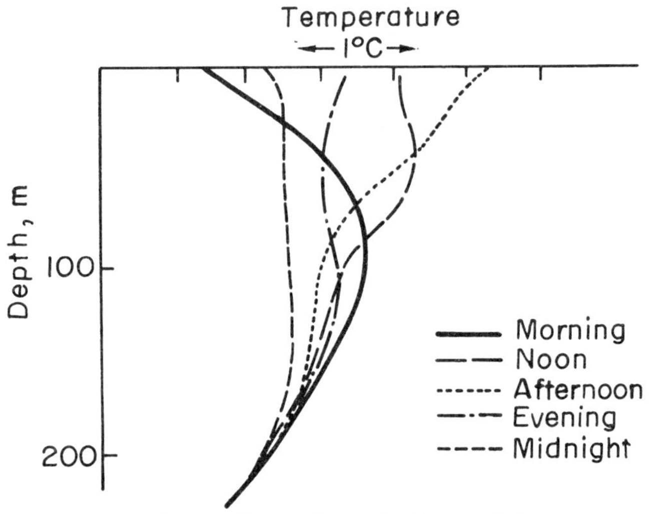

Typical diurnal variation of temperature in the surface layer of the ocean.

Figure 5-4

mixing. Isothermal layers as deep as 200m may be created. Below such layers, the temperature gradually decreases to an isothermal state. The thermal lapse rate varies considerably, but for warm surface water the effect is roughly

$$\frac{1}{C}\frac{\partial C}{\partial T} \sim 1.8 \times 10^{-3} \ (°C)^{-1}$$

Salinity effects--The velocity of sound increases with salinity because the density of water does. At the surface, where evaporation occurs, the component of velocity due to salinity variation is at its greatest. However, even then it is of small import. Changes in S due to evaporation amount to one part per thousand. In estuary mixing of fresh and salt water, the greatest influence on S and therefore C will be felt.

Typically, temperature and depth effects dominate the sound velocity profile in the sea. This can be seen graphically in Figure 5-5.

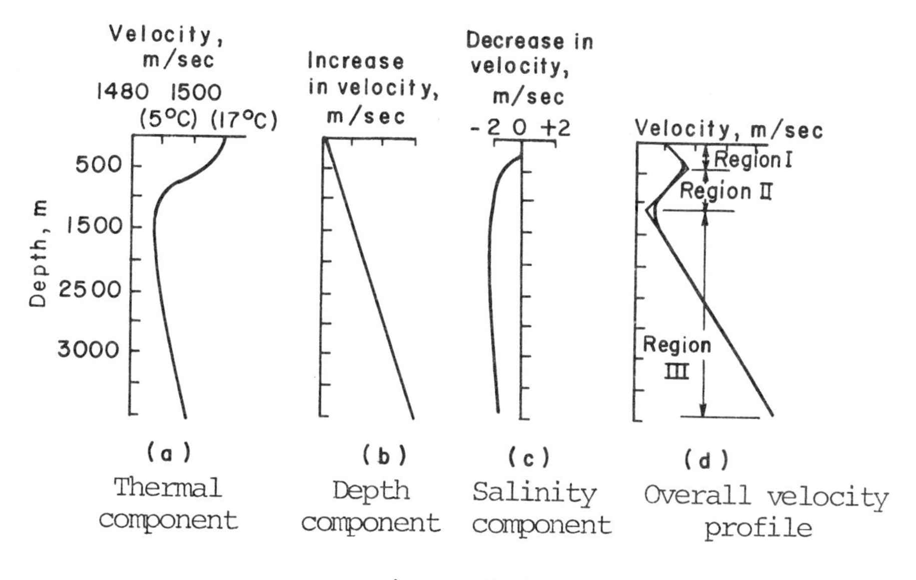

Figure 5-5

Near Bottom

Sensing and transmitting signals near the ocean bottom may become difficult for a variety of reasons. Sensing requires recognition of a contrast between the desired signal and signal noise reflected from the bottom. The previous analysis of signal reflection, and transmission remains valid for this bottom interface.

It becomes more difficult to identify the interface, particularly after it has been disturbed. Optical visibility can be decreased drastically as the attenuation coefficient increases as much as fourfold. Once the bottom sediments, normally a diatomaceous or globergerena ooze, are placed in suspension, it may take weeks or even months for the fines to settle out, particularly if there are no currents in the region.

Acoustic visibility depends greatly on the composition of the bottom. The Atlantic and Pacific Ocean bottoms exhibit greatly different characteristics because of their relative densities. Calcium carbonate is a constituent of the Atlantic diatomaceous ooze. This results in sufficient density difference so that $m = \rho_2/\rho_1$

where ρ_1 = density of sea water

ρ_2 = density of sea bed

is greater than unity and the bottom acts as a reflecting surface by Snell's Law. In the Pacific, calcium carbonate has redissolved because of the increased pressure before it can reach the bottom. The sediment deposited is mostly a red clay which is so unconsolidated that ρ_2 is too small for m to be greater than one. Thus, much of the Pacific Ocean bottom scatters and dissipates acoustic energy in diffusion with negligible reflection. The consequence of this is that much of the Pacific Ocean bottom, unlike that of the Atlantic, is invisible to acoustic imaging.

When there is a stable thermal stratification in the sea, there are two regions of low-speed sound and, consequently, two sound channels, one at the surface (region I) and one at the main thermocline (region III). If the surface water should become isothermal, through mixing, the surface sound channel would disappear (see Figure 5-5). In the high latitudes, the minimum sound velocity may be at the surface, in which case the deep sound channel would not be present. In general, one should expect both sound channels to be present.

The major implication of sound channels to system design is that a sound source at any range can only be heard when a receiver transducer is located at the appropriate depth in the proper sound channel. Note that the fundamental difference between shallow and deep sound channels is that, in the former, the acoustic energy is reflected from the free surface and refracted upward from the depths while, in the latter case, the channel results solely in the convergence of energy by refraction. Sound originating in the lower sound channel will remain primarily in that channel, while sound originating in the surface channel will leak into the lower channel so that the signal is more rapidly reduced to the ambient noise level (see Figure 5-6). Due to refraction, there are shadow zones, regions in which acoustic energy will not penetrate, which must be avoided by transducers if signals are to be received.

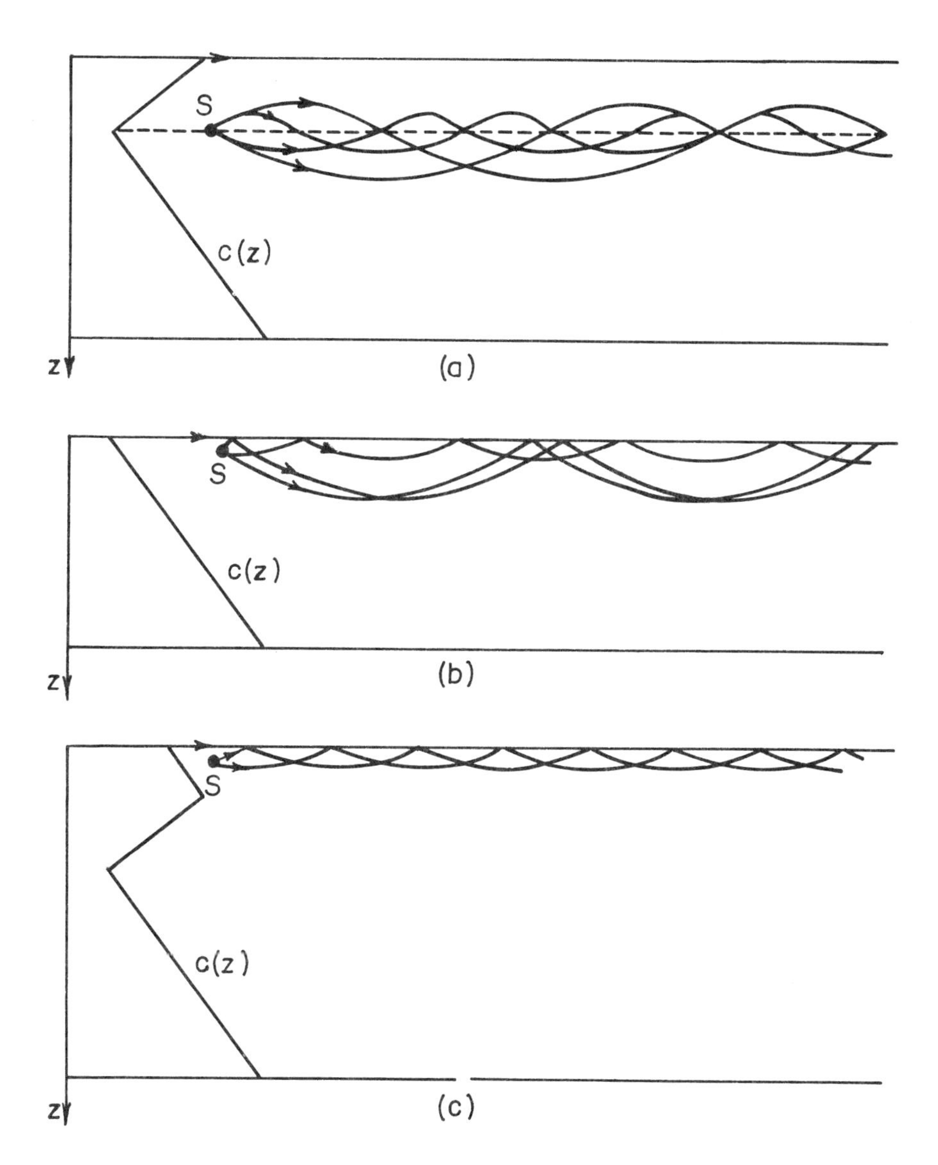

Main types of ocean stratifications displayed by c(z) (compare with Fig. 5-5.) Also shown are schematic ray diagrams illustrating the waveguide effect for a source S in the deep (a) and in the surface (c) channels. Case (b) corresponds to the Arctic type of sound-velocity stratification.

Figure 5-6

As with other subsystems, special problems are associated with operation above the free surface, at the free surface, and near the sea bed. In the atmosphere a marked diurnal variation in signal reception occurs as a result of the change of the ionosphere such that high-frequency communications are quite unreliable during daylight hours and during periods of high sunspot activity. The systems which are relying on high-frequency ship-to-ship or ship-to-shore transmissions over considerably large ranges should adjust their mission profile to ensure that the majority of relevant communication takes place during the nighttime hours. The use of low frequency or very low-frequency transmissions will alleviate this problem, but the antenna size makes such transmission from ship to shore or from aircraft to shore or from aircraft to ship exceedingly difficult to accomplish. At the free surface line-of-sight type of communications, light, visual observation, infrared and microwave communication are, of course, horizon-limited. In addition, optical systems are limited by the conditions of the atmosphere. Line-of-sight acquisition systems are motion-limited as a result of sea state and they require rather elaborate gimbaling and directing techniques to assure receipt of communication. These problems are particularly aggravated in the case of buoy systems which are close to the free surface and exceedingly limited in their horizon.

Through the Free Surface

For all practical purposes, transmission through the free surface is unfeasible via radiant energy, but is possible with hard wire or transducer links. Buoys may be located on the surface which receive the signal electromagnetically, or by other appropriate above-water communication techniques and then transduce the signal and readmit it acoustically into the water. The advantage of the hard-wire antenna technique is that no transducing is required and the hard wire can be attached to a submersible which is moving. The reception capability is, of course, a function of the orientation,size and disposition of the antenna which is at the free surface and the signal strength obtained there. The disadvantage of the hard-wire system is that the depth to which the submersible can operate may be constrained by the wire. The maneuverability of the submersible may be constrained by the wire and the survivability of the wire may be constrained by the maneuvers of the submersible required by the mission. The use of transducing buoys, on the other hand, usually requires that the submersible be in the proximity of the buoy. At the same time, the submersible is greatly freed to sound to great depths and to maneuver without fear of jeopardizing the communication system. The design of the buoys is, of course, a major ocean component problem, including, as it must, appropriate power sources, appropriate receiving antenna, appropriate hydrophones, abilities to maintain position on station either by anchor or by dynamic systems, the need for low maintenance, high reliability and long life and the jeopardy to which the buoy is exposed as a result of even well-intentioned mariners who might interfere with or retrieve such objects on the high seas.

CASE STUDY

LONG-RANGE ANTISUBMARINE WARFARE SYSTEM--COMMUNICATIONS

The communications subsystem example was provided by Lieutenant Commander Gerald Sedor, U.S.N. His design of an extensive ASW system relies heavily upon an appropriate and effective sensing and communications subsystem to detect and transmit intelligence of clandestine submarine movements to control centers in real time.

Author's Background

During the academic year of 1969 Lieutenant Commander Sedor was attending M.I.T. under the Navy program for officer postgraduate education when he designed this system. A Naval Academy graduate of 1957, he attended the Navy's school for submarine training and has seen service on conventional and, after being trained in nuclear power, nuclear submarines. As the communications and sonar officer aboard the USS SABALO (SS 302) in the western Pacific, Lieutenant Commander Sedor worked with the problems of communications on a daily basis. His practical knowledge of underwater communications is excellent.

Location and monitoring of submarine activity is primarily a sensing and communications problem. An event, the movement of a submarine, must be perceived above background environmental noise and transmitted to appropriate subsystems for consequent action.

Subjective Human Purpose

Improve the national defense posture of the United States against ballistic missile submarine (SSBN) of actual or potential adversary nations. The system accomplishes this purpose by providing the capability to search for, locate, and identify such submarines within an outer barrier approximately 2500 miles from the coastline of the continental United States and to track and trail such submarines within an inner barrier approximately 1000 miles from the United States coastline. Ocean areas are limited to the Atlantic and Pacific.

Mission analysis has indicated that sensing must be done at fairly long range with sufficient resolution for positive object identification. The present location of a target submarine inside the outer barrier must be communicated to a coordinating center in real time and updated frequently so that submarines can be located, tracked and trailed by a variety of appropriate platforms (see Figures 5-7 and 5-8).

Systems Implications

1. The size of the oceans dictates that magnetic anomaly detectors and acoustic sensing stations be strategically located about the oceans. Power requirements for continuous monitoring of the seas indicate that active sonars are unfeasible. For short-range interrogation for short times, as for tracking and trailing, active sonars can be used effectively.

2. Long-range sensing can most effectively be done in the sound channels of the seas. Thus, acoustic transducers must be located at the depth of these channels. Acoustically-quiet shadow regions must be avoided because inordinate amounts of power are required to hear signals in the

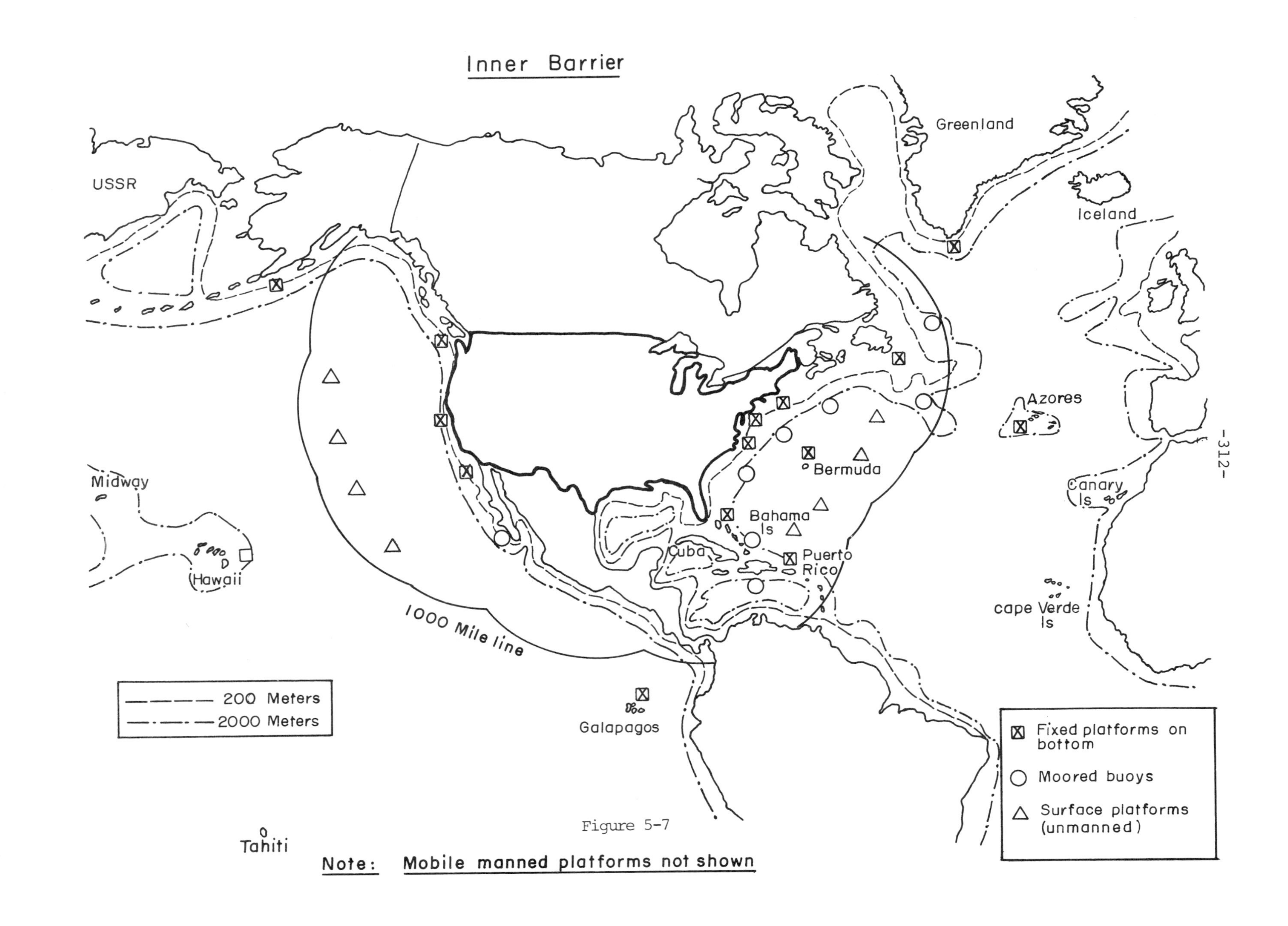
Inner Barrier
USSR
Greenland
Iceland
Azores
Bermuda
Bahama Is
Puerto Rico
Cuba
Canary Is
cape Verde Is
Midway
Hawaii
Galapagos
Tahiti
1000 Mile line
200 Meters
2000 Meters
Fixed platforms on bottom
Moored buoys
Surface platforms (unmanned)
Note: Mobile manned platforms not shown

Figure 5-7

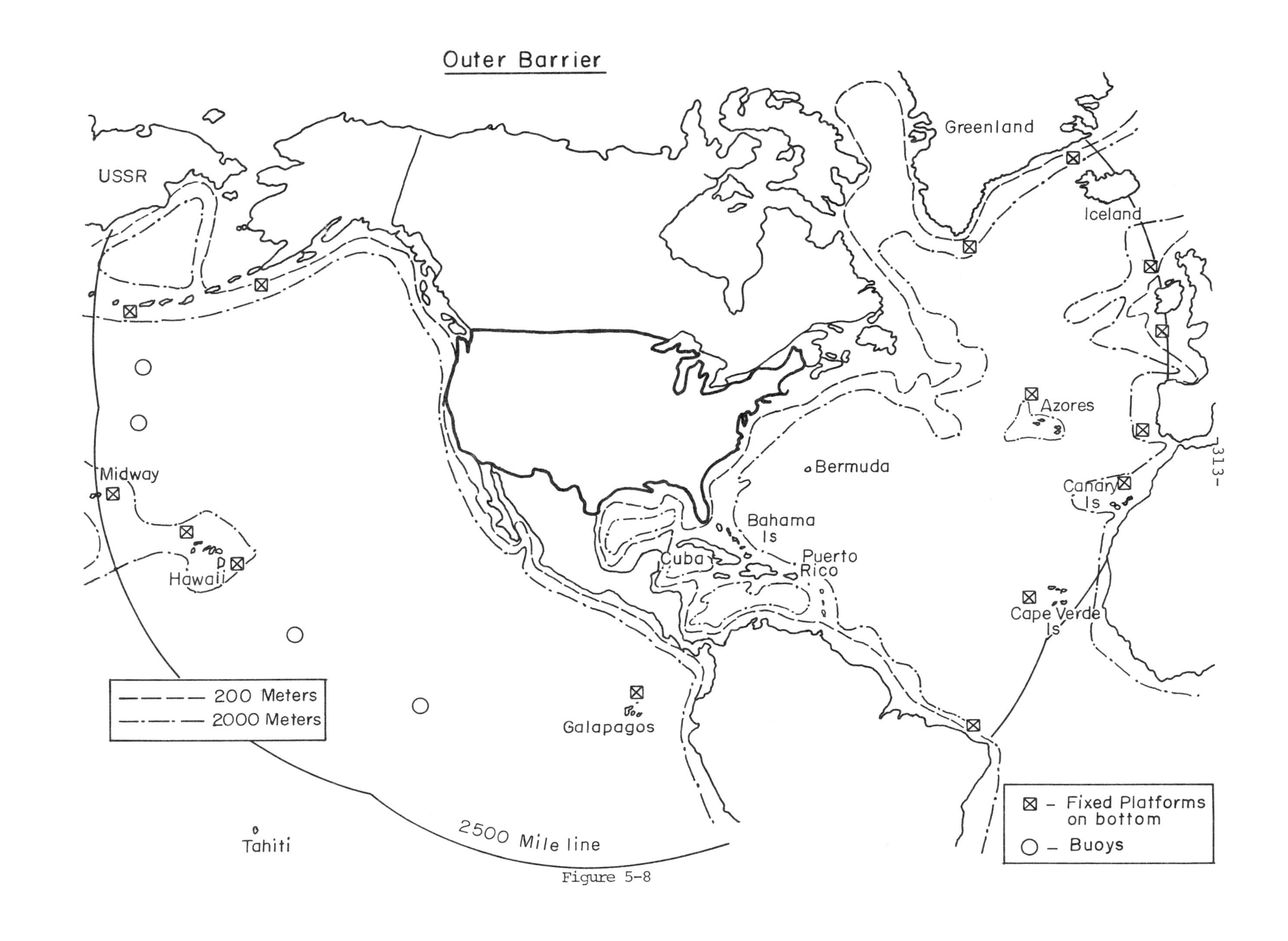

Outer Barrier
USSR
Greenland
Iceland
Azores
Bermuda
Canary Is
Bahama Is
Puerto Rico
Cuba
Cape Verde Is
Midway
Hawaii
Galapagos
Tahiti
2500 Mile line
200 Meters
2000 Meters
– Fixed Platforms on bottom
– Buoys

Figure 5-8

zone which originated outside of it.

3. Real-time communication requires that acoustic links be as short as possible and that electromagnetic signals be used wherever possible because of the relative speed of signal propagation. Major signal transduction is needed.

4. Signal transmission through the air-sea interface should be by fixed wire.

What follows is basically Lieutenant Commander Sedor's design and engineering analysis data with annotations dispersed throughout. Lieutenant Commander Sedor first reviews the General Characteristics, General Descriptions of Communication Systems, and Typical Methods for Communicating between Various Terminals. From this review he makes his subsystem choices.

The nations of the world are more extensively employing the sea for deployment of their strategic missile systems. Proliferation to nations other than those currently involved in the strategic confrontation could result in instabilities. It is anticipated that in the future such systems will be operating pursuant to arms control agreements. The ability to monitor the oceans to ensure peaceful use implies the existence of antisubmarine surveillance systems to maximize peacetime knowledge of deployments and for tracking and trail of violators of any agreement. The system published by Lieutenant Commander Sedor is entirely of his own creation and is not drawn from classified material, nor is it intended to represent any system now in being or in comtemplation by the United States.

SUBSYSTEM ENGINEERING CONSIDERATIONS

OCEAN SYSTEMS COMMUNICATIONS

I. General Characteristics of Communications Systems

  A. Factors Determining System Requirements and Selection

  B. Location of System Terminal Points

  C. Methods of Transmission

  D. Major Components

  E. General Design of System

II. Description of Systems

  A. Electromagnetic Systems

  B. Acoustic Systems

  C. Optical Systems

  D. Fixed Wire Systems

III Typical Communication Methods between Various Terminals

IV. Bibliography

OCEAN SYSTEMS COMMUNICATIONS

I. General Characteristics of Communications Systems

A. Factors Determining System Requirements and Selection

1. Locations and/or operation areas for terminals
2. Distances between terminals
3. Types and quantity of information to be exchanged
4. Desired rate of information exchange
5. Quality of signal required
6. Reliability of system required
7. Security of system required
8. Existing means available for transmission of information

B. Location of System Terminal Points

1. Underwater
2. Sea level atmosphere
3. Low atmosphere
4. High atmosphere
5. Space

C. Methods of Transmission

1. Electromagnetic radiation
2. Acoustic
3. Optical
4. Fixed wire
5. Physical Delivery

D. Major Components (see Figure 5-9)

1. Information source--person or device generating the original information to be transmitted by the system to the user (e.g., human voice, scene in TV studio, informa- stored in compute memory, photo from satellite).
2. Source transducer--any device capable of being actuated by waves or a source of energy, and of supplying waves or

energy related to the input. Output energy produced may be of same type or different from input (e.g., microphone in telephones, video camera, photoelectric cell).

3. Channel encoder--device which can accept the signals from the source transducer and transform them into a form suitable for transmission through the selected communications channel (e.g., relay repeater, amplitude modulator).
4. Channel decoder--device which accepts signal delivered by communications channel and transforms it into a form suitable to the user transducer (e.g., biased polar relay, voice channel separation network, frequency detector).
5. User transducer--device which transforms the signal delivered by the channel decoder and delivers the type of energy output (optical, electrical, acoustical, etc.) called for by the information user (e.g., loudspeaker, TV receiver, magnetic tape recorder).

MAJOR COMPONENTS OF A COMMUNICATIONS SYSTEM

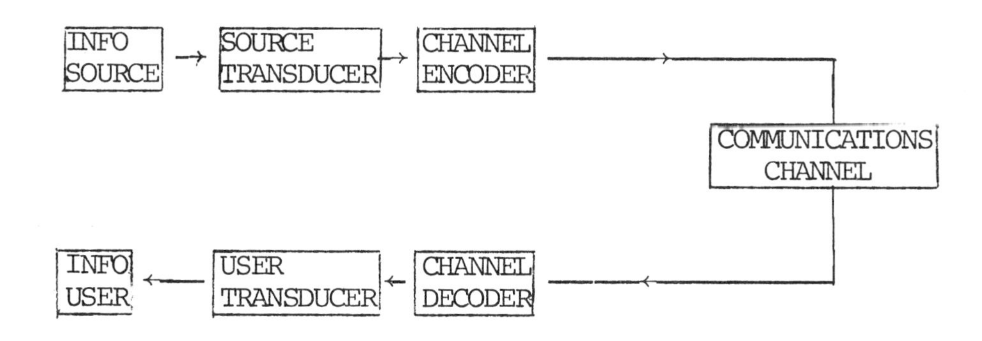

FIGURE 5-9

E. General Design of System

1. Parameters of System Design

a. Intelligence to be moved--communications objectives

b. Traffic analysis--volume and loading requirements

c. Criterion of acceptability--quality and grade of service

d. Operating requirements--service schedule

e. Initial survey--geography of system

f. Facilities survey--available services, transmission medium

g. Survey of communication techniques and equipment

h. Service quality and reliability vs. cost

i. Serviceability and personnel requirements vs. cost

j. Volume growth, new services, future prospects vs. cost

k. Obsolescence

2. System Design Flow Chart (see Figure 5-10)

II. Description of Systems

Electromagnetic communications methods should be utilized for transmission through the air atmosphere. This means that signals can be transmitted over relatively long range in real time.

A. Electromagnetic Systems

1. Standard Radio Bands

BAND NO.		DESIGNATION	FREQUENCY RANGE
2	ELF	Extremely low freq.	30 - 300 Hz
3	VF	Voice freq.	300 - 3000 Hz
4	VLF	Very low freq.	3 - 30 kHz
5	LF	Low freq.	30 - 300 kHz
6	MF	Medium freq.	300 - 3000 kHz
7	HF	High freq.	3 - 30 mHz

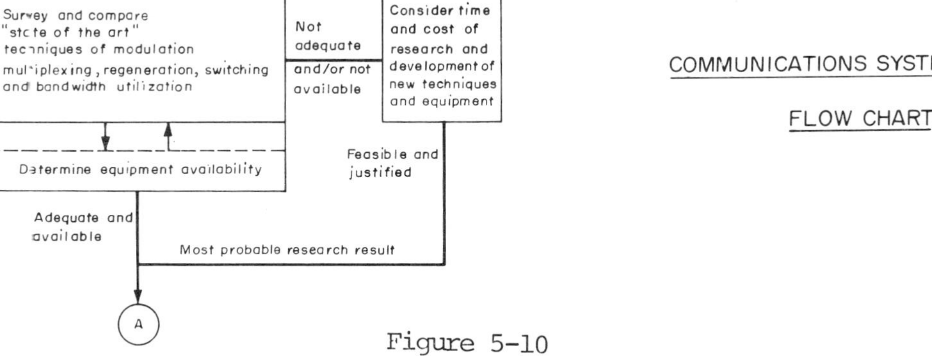

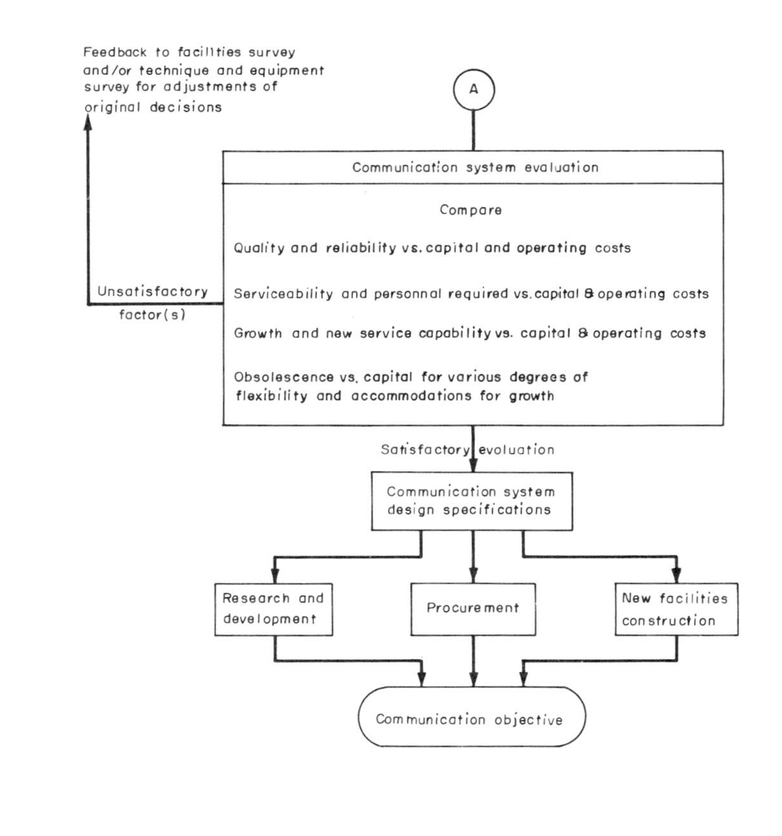

COMMUNICATIONS SYSTEMS DESIGN

FLOW CHART

Figure 5-10

1. Standard Radio Bands (continued)

BAND NO.		DESIGNATION	FREQUENCY RANGE
8	VHF	Very high freq.	30 - 300 mHz
* 9	UHF	Ultra high freq.	300 - 3000 mHz
*10	SHF	Super high freq.	3 - 300 GHz
*11	EHF	Extremely high freq.	30 - 300 GHz
*12	-		300 - 3000 GHz

(*--Microwave Bands)

2. Terminal Distance Considerations

a. Short Distances--line of sight up to 50 miles

BANDS: VHF/UHF used, based on

(1) Relatively large bandwidths

(2) Available spectra

(3) Low power requirements

PROBLEMS:

(1) Multipath reflections from waves or other surfaces causing distortion

(2) Ducting or layering effects from inversion layers of atmosphere

b. Medium Distances--50-500 miles

BANDS: MF/HF ground wave or surface wave, based on

(1) Less fading than sky waves

(2) Moderate power requirement

(3) Good antenna system

(4) Discrimination against atmospheric noise

(5) Minimum number of channels required

PROBLEMS:

(1) Bandwidths and available frequencies relatively limited

SURFACE WAVE PROPAGATION:

(1) Transmission along curved surface of earth

(2) Daylight broadcast and medium range communication in band of 1.5-30 mHz

(3) Sensitive to conductivity of surface over which transmitted (moist solid and salt water offer best conditions)

c. Long Range--over 500 miles

BANDS: HF skywave or ionospherically reflected propagation; LF; VLF

PROBLEMS:

(1) Several channels, widely separated in frequency, required for continuous 24-hour use

(2) Limited bandwidth and available frequency

(3) Ionospheric disturbances and fading

(4) Slow data rates (digital data rates limited to 500 bits/sec, due to multipath effects)

SKYWAVE PROPAGATION: Generally follows several predictable cycles--daily variation, annual variation, sunspot cycle (11 years)

3. Operating Frequency Selection

a. Best frequency determined by calculations from available data. Limits set by

MUF - Maximum Usable Frequency
FOT - Frequency of Optimum Traffic
LUF - Lowest Usable Frequency

b. Predictions used to determine FOT

(1) Long range--used for planning of facilities and allocation of frequencies

(2) Short range--determine daily FOT's closest to those allocated; monthly ESSA publication gives computer-based predictions

c. Allocation of frequencies

(1) Nongovernment users--FCC rules based on international agreement

(2) Government users--Interdepartment Radio Advisory Committee

d. Frequency allocations applicable to ocean systems

(1) Aeronautical mobile (ground-air; air-air)

General: 200 kHz - 136 mHz

Distress: 500 kHz (telegraph); 156 mHz (telephone); 243 mHz (survival craft equipment)

Telemetry: 216 - 220 mHz

(2) Broadcast

Standard AM: 535 - 1605 kHz

FM: 88 - 108 mHz

TV: 54 - 890 mHz

Int'l AM: 5.9 - 26 mHz

(3) Government: 510 kHz - 38.6 mHz

(4) Maritime mobile

General: 110 kHz - 3.4 mHz

Distress: (see Distress above)

Ships and coast stations:

Telegraph and facsimile: 2 - 22 mHz
Telephone: 2 - 162 mHz

Ship calling:

Telegraphy: 4 - 22 mHz
Double SB telephone: 8 - 22 mHz

(5) Meteorological aids

Radiosondes: 400 - 406 mHz; 1660 - 1700 mHz

Radars (ground): 5.6 - 9.5 GHz

(6) Satellites

Communications: 3.7 - 8.4 GHz

Meteorological: 137 mHz - 7.75 GHz

Radio navigation: 149 mHz - 14.4 GHz

4. Antenna Selection

a. Physical dimensions usually constrained by space shortages on ship, or other platform.

b. Shipboard installation usually determined by convenience of location rather than performance, usually resulting in an interaction problem between antenna and ship structure.

c. Directivity pattern--depends on

   (1) Frequency
   (2) Loading condition of nearby antennas
   (3) Movement of nearby objects
   (4) Height of antenna above water

d. Directivity losses on ocean platforms--large structures above deck tend to act as parasitic radiating elements and cause undesired directivity effects which distort radiation pattern of antenna. Typical losses--at least 10 db for 10% of the time, 15 db for 1% of the time.

e. Types of antennas used:

   (1) Inductively loaded short verticals (VLF)
   (2) Nonresonant structure-to-structure wires (VLF)
   (3) Ground plane verticals (HF, VHF, UHF)
   (4) Discone (HF)
   (5) Half-wave horizontal doublet (HF)
   (6) Multi-element horizontal or vertical array (VHF, UHF)
   (7) Parabolic reflectors (UHF - high end of band)

5. <u>Transmission of Information--Modulation</u>

<u>Modulation</u>--the control process used to modify the generated radio waves so that the wave carries with it the desired intelligence. May consist of

a. Simple interruption of the wave (CW telegraphy)

b. Interruption of a tone-modulated wave (modulation CW)

c. Continuous modulation in accordance with a telephonic current (voice)

d. Continuous modulation in accordance with scanning of a fixed image (facsimile radio transmission)

e. Continuous modulation in accordance with the scanning of a moving image (TV)

Method of Modulation--by variation of amplitude, frequency, or phase of the AC producing the radio wave.

6. Radio Noise

a. Limiting factor of almost every communication system

b. Noise power at a given point depends on

(1) Geographic location

(2) Time of day

(3) Season

(4) Operating frequency

(5) Bandwidth of receiving system

c. Signal/noise ratios

(1) International Radio Consultative Committee (1956) recommended ratios of 11 db to 32 db, depending on service.

(2) Must add fading allowance for reliability--recommendation vary up to 33 db for SSB.

d. Interference--caused by proximity of electrical machinery to radio equipment. Need careful planning of systems and analysis of environment to minimize.

(1) Noise reduction techniques should be applied to electrically-operated devices on platform.

(2) Environment should be measured to determine total noise content and individual noise sources.

(3) Spectrum signature of possible noise contributors--can aid in future detection of interference problems.

(4) Military specifications used by USN to establish acceptable environmental noise levels.

7. Underwater Radio Wave Transmission

a. Limited by conductivity of salt water (approximately 4000 times that of fresh water).

b. Attenuation--varies with frequency, with a propagation "window" in the visible light frequency range (see Figure 5-11 below).

c. Reception of VLF possible at shallower depths.

ATTENUATION OF RADIO WAVES IN WATER
AS A FUNCTION OF FREQUENCY

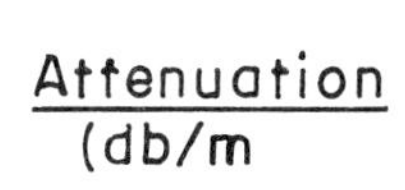

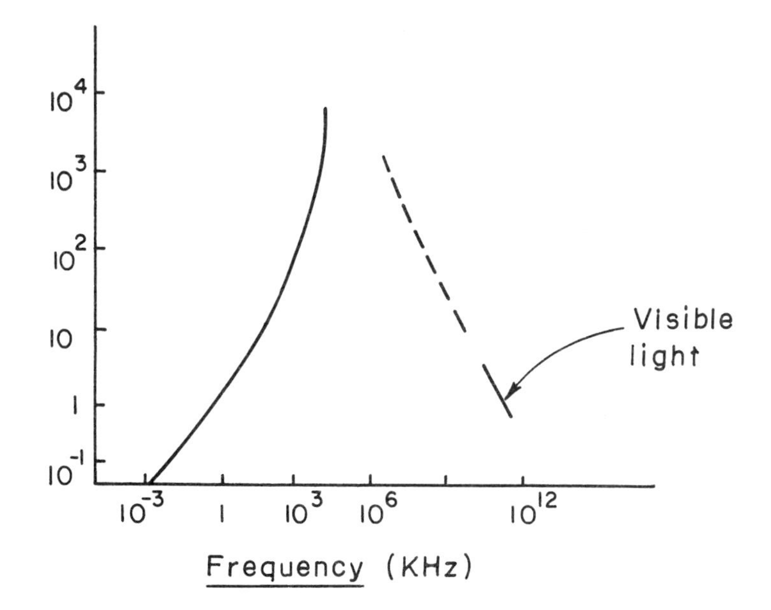

Figure 5-11

Based on the design principle

$$r = \frac{10}{\alpha}$$

it is obvious that electromagnetic energy is attenuated too rapidly to be useful as a form of communications underwater.

B. Acoustic Systems

1. Basic Definitions and Units

I = Acoustic intensity of sound wave, representing the average rate of flow of energy through a unit area normal to the direction of wave propagation (watt/m^2)

$I = P_e^2/\rho_0 c$

P_e = Effective (RMS) pressure (dynes/cm^2)

ρ_0 = Density of medium of transmission

c = Velocity of sound in medium

IL = Intensity Level = $10 \log (I/I_0)$ db

SPL = Sound Pressure Level = $20 \log (P_e/P_0)$

1 microbar = 1 dyne/cm^2 = 0.1 Newton/m^2 = 10^{-6} atm.

Sound Levels in db

MEDIUM	TYPE	I_0 (w/m^2)	P_0 (microbar)	Rel. Intensity of 0 db	Rel. Level of 0 db
Air	Intensity	10^{-12}	--- 1	1	0
Air	Pressure	---	.0002	1	0
Water	Pressure	---	.0002	.00027	-35.5
Water	Pressure	---	1.0	6700	38.5

2. Velocity of Sound in Salt Water

Empirical relationship:

$$c = 1449 + 4.6t - .055t^2 + .0003t^3 + (1.39 - .012t)(s-35) + .017d$$

where
c = speed of sound (m/sec)
t = temperature in °C
d = depth of water (m)
s = salinity of water (0/00)

Temperature effects are the most significant and subject to the widest fluctuations.

TYPICAL SOUND-SPEED PROFILES

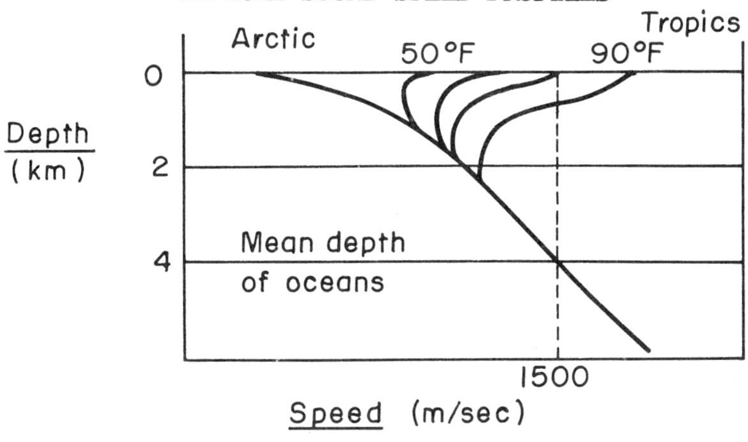

Figure 5-12

3. Sound Transmission Losses in Salt Water

Ultimately all acoustic energy in any medium is degraded into some form of heat energy. The sources of dissipation may be classified as follows:

a. In transmitting medium:

(1) Viscous losses
(2) Heat conduction losses
(3) Molecular exchange of energy

b. At boundaries of medium

Relationships:

S = Sound pressure level of source, in db, relative to one microbar generated at the reference position

S = 20 log 10 P = 71 + 10 log W db

where W = sound power output (w)

$P^2 = \rho_0 cW/4\pi$

H = One-way transmission loss, referenced to 1 m. distance

H = 20 log r + ar + db

where r = distance from source (m)

a = absorption const. (db/m) = a(freq,temp)

A = transmission anomaly, due to refraction, diffraction, scattering

α = Attenuation coefficient = (a)(const) $\cong$ .01 f^2 db/km (f<30kHz)

VARIATION OF ATTENUATION COEFFICIENT WITH FREQUENCY

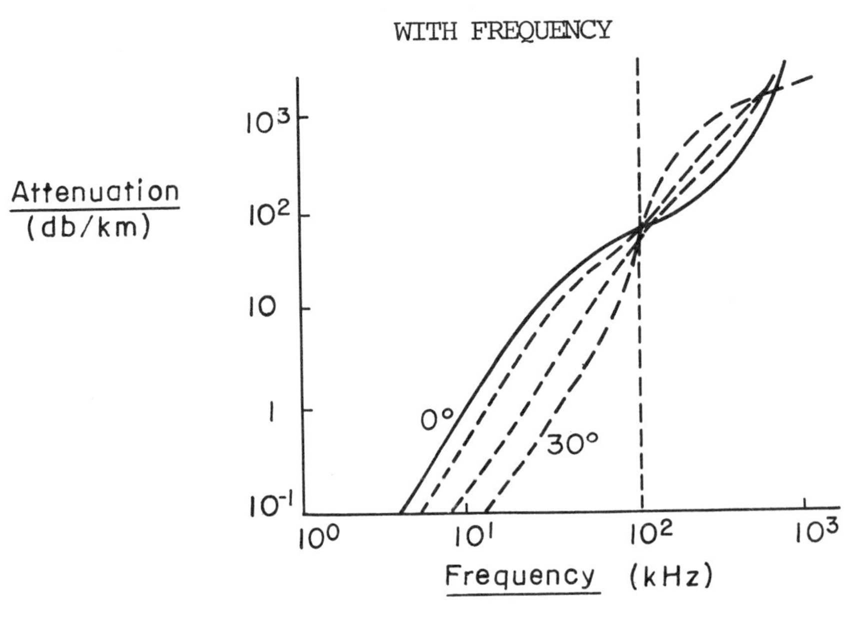

Figure 5-13

Longer range implies lower frequencies. The low attenuation rate of sound in sea water permits ranges on the order of tens of kilometers for the appropriate frequency.

$$r = \frac{10}{\alpha} \qquad \alpha = 0.01 f^2$$

$$1000 = \frac{10}{0.01f^2}$$

range (km)	frequency (Hz)
1	1000 Hz
100	30 Hz

$$f^2 \sim \frac{1000}{1000} = 1$$

$$f \sim 1000 \text{ Hz}$$

4. Refraction of Sound Waves

Transducers must be located in a sound channel if long-range reception is to be made at reasonable power levels. Submarines may operate in either the deep sound channel, sofar, or the shallower RSR, refracted surface reflected, channel. This implies the need for the variable depth sonars.

Due to variations in sound velocity, sound waves may be refracted during transmission, creating "Shadow Zones" in which the sound pressure is theoretically zero.

a. Refraction always toward direction of lower sound speed.

b. Usually decreases detection range of listening devices near surface.

c. When sound velocity minimum occurs in Sound Velocity Profile (SVP), sound rays become "trapped" in a channel which can extend the range of transmission.

d. SOFAR (SOund Fixing And Ranging) channel between depths of equal sound speed acts as a wave-guide to increase range of transmission.

e. Convergence zone--refocusing of limiting rays at large distances from source. Spacing depends on depth of SOFAR axis. Varies from about 25 km in subarctics to about 75 km in tropics.

Shadow Zones--make it impractical to attempt to transmit out to ranges beyond the limiting ray because penetration into the shadow zone can only occur by volume scattering phenomena similar to those of tropospheric scatter propagation in the upper atmosphere, or by bottom reflection, or by diffraction. Several orders of magnitude increase in power is required to obtain adequate signal-to-noise ratio at a receiver in the shadow zone. The propagation path is severely distorted by multipath effects which severely degrade the intelligibility of a communications link.

EXAMPLES OF REFRACTION PATTERNS

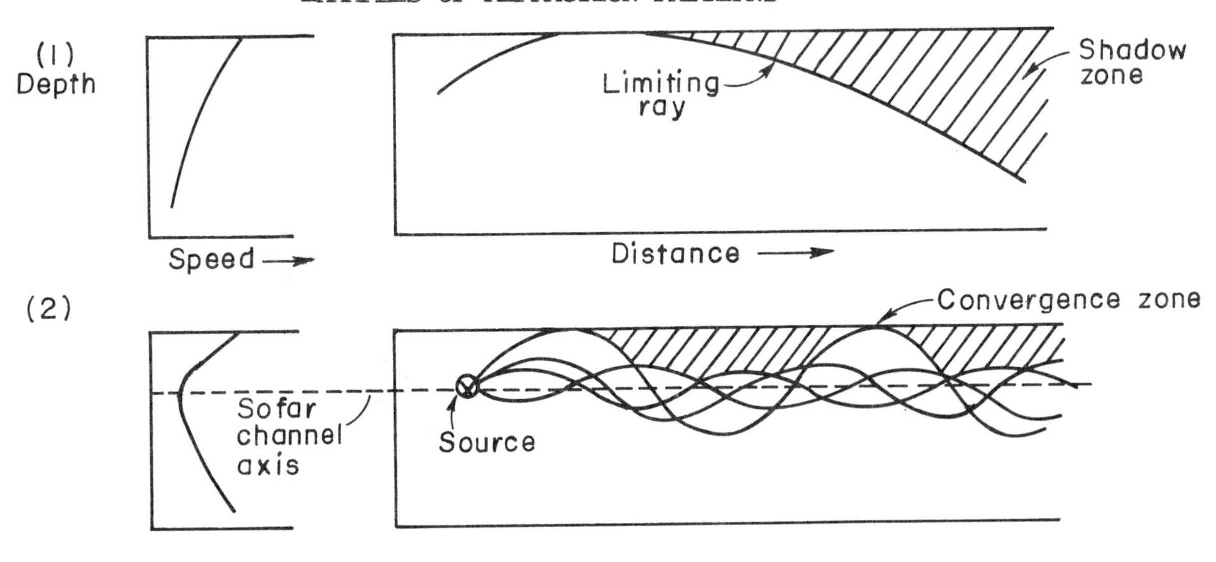

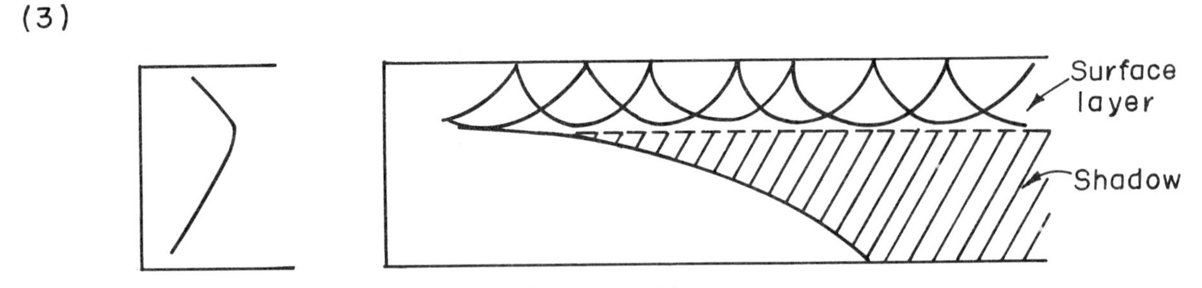

Figure 5-14

<u>SOFAR Channels</u>

a. Conditions

(1) Thermal gradient where the temperature decreases steadily toward a nearly constant 4°C, where sound velocity is at a minimum.

(2) At deeper depths, velocity increases due to the increase in pressure. At shallower depths, velocity increases due to increase in temperature.

(3) All rays originating in the region of minimum velocity which make a small angle with the horizontal will tend to curve back toward this level without reaching either the surface or the bottom, forming a deep sound channel.

b. Transmission loss: $10 \log r_1 + 10 \log r$. Loss is less than for spherical divergence.

c. Absorption: Small over large distances. For low frequency signals, reception up to 3000 km.

5. <u>Passive Listening Devices</u>

a. Ability to detect signal depends on

(1) Nature of sound signal radiated by source

(2) Magnitude of one-way transmission loss

(3) Nature of masking noise

(4) Directivity index of receiving hydrophone system

(5) Recognition differential and bandwidth of detector

A submarine is a hard nonresonant object whose target strength is roughly

$$T_s = 20 \log \frac{\ell}{4r_o}$$

In order to backscatter enough energy to make observation possible the lower limit on frequency must be

$$f_{\ell\ell} = \frac{c}{2\pi a\sqrt{3}}$$

if $a = 5m$

$c = 1500$ m/sec

$$f_{\ell\ell} \sim \frac{1500}{2\pi 5\sqrt{3}} \cong 28 \text{ Hz}$$

Below that frequency, power requirements increase rapidly. Thus, if frequencies less than $f_{\ell\ell}$ are required, the sonar system must be a passive one.

b. Signal Interference--most common cause of interference is sound produced by fish and other sea life. Interfere in three ways.

(1) Produce sufficient noise in signal frequency to change radically the signal-noise ratio.

(2) Produce sounds sufficiently like signal to create false signal.

(3) Interposing themselves as signal attenuators or reverberators.

Problem is usually greatest near continental shelves, where concentration of marine life is greatest.

6. Active Underwater Acoustic Devices

a. Signal Interference--often more serious than with passive devices. Problems caused by

(1) Sound production by marine life.

(2) Acoustic interference by fouling of transducers.

(3) Oasis effect--organisms (algae) attracting sound-producing or sound-attenuating fish.

(4) Acoustical characteristics of marine life.

Laboratory experiments have demonstrated that sonic energies in high frequencies may produce resonances in very small organisms, which may result in cavitation. Temporary formation of "clouds" of acoustic attenuators or reverberators.

b. Signal Transmission--Any of the carrier modulation and multiplexing techniques common to conventional

EM communications may be used in underwater communications, including amplitude, frequency, or phase modulation; time-division or frequency-division multiplexing. Information may be encoded in analog or digital format, transmitted in real time or stored for return.

c. Voice Transmission

Usual Technique--8-11 kHz suppressed carrier AM, SSB; ranges up to several km.

Comparison of AM and FM:

FM: works particularly well in deep water where effects of multipath interference not as severe as shallow water; intelligibility of FM voice transmission equal to or better than AM.

AM: generally preferred over FM for voice, due to power efficiency, ability to perform in high multipath environment encountered in long-range horizontal transmissions and in shallow water.

d. Acoustic Telemetry--Two systems currently under development by Bendix Corp.

(1) System to transmit speed, depth, course of deep research submersible to surface support vessel.

(2) System using several acoustic/RF telemetering buoys arrayed in circle about planned impact point of air dropped weapon. Information on weapon's arming and firing sequence telemetered acoustically to buoys and relayed via RF link to central station.

Modulation--FM selected for above systems due to

(1) Wideband noise improvement.

(2) Ease in obtaining high degree of linearity.

(3) Amplitude modulation introduced in medium can be removed by circuits in receiver.

(4) FM subcarriers capable of transmitting DC data and provide advantages in cross-talk noise improvement.

C. Optical Systems

1. Above-Surface Systems

Techniques Used--Flashing light, UV, IR, Lasers

Limitations:

a. Short ranges within LOS
b. Relatively low security and reliability
c. Relatively low signal quality
d. High attenuation of signal by clouds, fog, etc.

RELATIVE ATTENUATION (db/km) IN DIFFERENT MEDIA

SYSTEM	MICROWAVE	MILLIMETER	SUBMILLIMETER	OPTICAL
FREQ (Hz)	4.2×10^{9}	4.2×10^{10}	4.2×10^{11}	4.2×10^{14}
Clear Dry Air	.01	.1	10	.03
Light Haze	--	--	--	3.3
Dense Haze	--	--	--	16.0
Fog	.02	2.0	200	560
Light Rain	.01	0.4	--	--
Heavy Rain	.02	5.1	--	--

Quantity of Information to be Conveyed:
Large bandwidth of optical systems (e.g., lasers) provides extremely high capacity

INFORMATION CARRYING CAPACITY FOR VARIOUS FM CARRIERS
(Assuming a bandwidth of 4200 Hz per Channel)

SYSTEM	RADIO	TV	MICROWAVE	OPTICAL
Freq (Hz)	4.2×10^{4}	4.2×10^{7}	4.2×10^{9}	4.2×10^{14}
Usable Bandwidth	10%	10%	1%	0.1%
Bandwidth (Hz)	4.2×10^{3}	4.2×10^{6}	4.2×10^{7}	4.2×10^{11}
Assumed Freq per Channel (Hz)	4.2×10^{3}	———————	———————	———————→
No. Channels	1	1000	100,000	100,000,000

2. Underwater Systems

   a. Attenuation of Light by Spreading, Scatter, Absorption

      (1) Spreading--inverse square law

      (2) Scattering--deflection from normal path by particles in water (minimum in green region of spectrum).

      (3) Absorption--by water molecules, colored particulate matter, colored dissolved matter. Varies with wave length of light. Gives water its characteristic color.

      Minimum Attenuation--In blue-green region, wavelength of 4700 Angstroms. No lasers presently available at this frequency. Current USN projects investigating new laser materials for this frequency.

   b. SONOPTOGRAPHY--"Sonic-Optical-Graphy," the process of constructing a visual image of an object without use of lenses and no equipment other than a sound source, recording medium, and coherent optical source. Two-stage process:

      (1) Recording made of interference pattern created when sound wave front created by object is biased by coherent sound wave. Record is sonoptogram.

      (2) Realistic 3-D visual image created in second stage of process. Sonoptogram illuminated with coherent light source, optimally from a laser.

      Note that here is an attempt to convert one form of energy to man's favorite energy from optical frequency electromagnetic radiation.

      Current Status--Still undergoing development by McDonnell-Douglass Corp. In early experiments, objects of sizes less than two sound wavelengths (equivalent to 1 micron on optical scale) resolved in reconstructed images.

Use in Ocean System--Need large mosaic array of sound sensors that can be scanned electronically to display sound pattern falling on them. Each element in array similar to each dot in a halftone newspaper engraving.

High resolution requires large acoustic arrays. The length of an array in a given direction is (from general design principles),

$$L = \frac{c}{f\Theta}$$

where

$$\Theta \sim \frac{dR}{R}$$

Size of Array--For 10 ft resolution at range of 3000 ft, total aperture 366 wavelengths for 100 kHz system results in 15 ft diameter array. Narrow field of view--1000 sensing elements; 180° field--420,000 sensing elements.

Environmental Effects--Less sensitive to turbidity and turbulence than conventional imaging systems. Theoretically infinite depth of field.

D. Fixed Wire Systems

1. Advantages

a. High reliability

b. Excellent security

c. Signal not affected by environment. Fixed wire provides a pipeline through which energy can cross the air-sea interface.

d. Simplicity of components required

e. Low power requirements

f. Fixed-wire communication increases the speed of propagation of the signal to that of light and therefore reduces communication lag time.

2. Disadvanges

a. Relatively limited capacity

b. Limited flexibility of operation with mobile platforms

c. Cost of installation of fixed wire

d. Maintenance requirements of system.

III. Typical Communications Methods between Various Terminals

FROM \ TO	UNDER WATER	SEA LEVEL ATMOS.	LOW ATMOS.	HIGH ATMOS.	SPACE
UNDER WATER	Acoustic Optical Fixed Wire	Acoustic Fixed Wire	——	——	——
SEA LEVEL ATMOS.	Acoustic Optical Radio (VLF) Fixed Wire	Acoustic Optical Radio Fixed Wire Phys. Deliv.	Radio Optical	Radio	Radio
LOW ATMOS.	Radio (VLF)	Radio Optical	Radio Optical	Radio	Radio
HIGH ATMOS.	——	Radio	Radio	Radio Optical	Radio
SPACE	——	Radio	Radio	Radio	Radio

As a consequence of his review and analysis of the problems of communications, Commander Sedor has made the following subsystem selections.

SYSTEM COMPONENTS

COMMUNICATION AND ASW SENSORS OF VARIOUS PLATFORMS

PLATFORM TYPE	ASW SENSORS	TRANSMISSION OF INFORMATION
Fixed ocean bottom	1. Acoustic	1. Fixed wire
Deep-moored buoys	1. Acoustic 2. MAD	1. Acoustic relay 2. Radio relay (UHF) via surface antenna
Unmanned surface platform (sea pallet)	1. Acoustic	1. Radio relay (UHF) 2. Acoustic relay
Surface ships	1. Optical 2. Acoustic	1. Radio (UHF, HF, MF, receive VLF) 2. Acoustic 3. Optical
Submarines	1. Acoustic	1. Acoustic 2. Radio (UHF, HF, MF, (receive VLF)
Manned Airborne	1. Optical 2. Monitor acoustic and MAD sensors	1. Radio (UHF, HF, MF)
Satellite	None (serves only as relay station)	1. Radio (UHF)

FIXED BOTTOM PLATFORMS

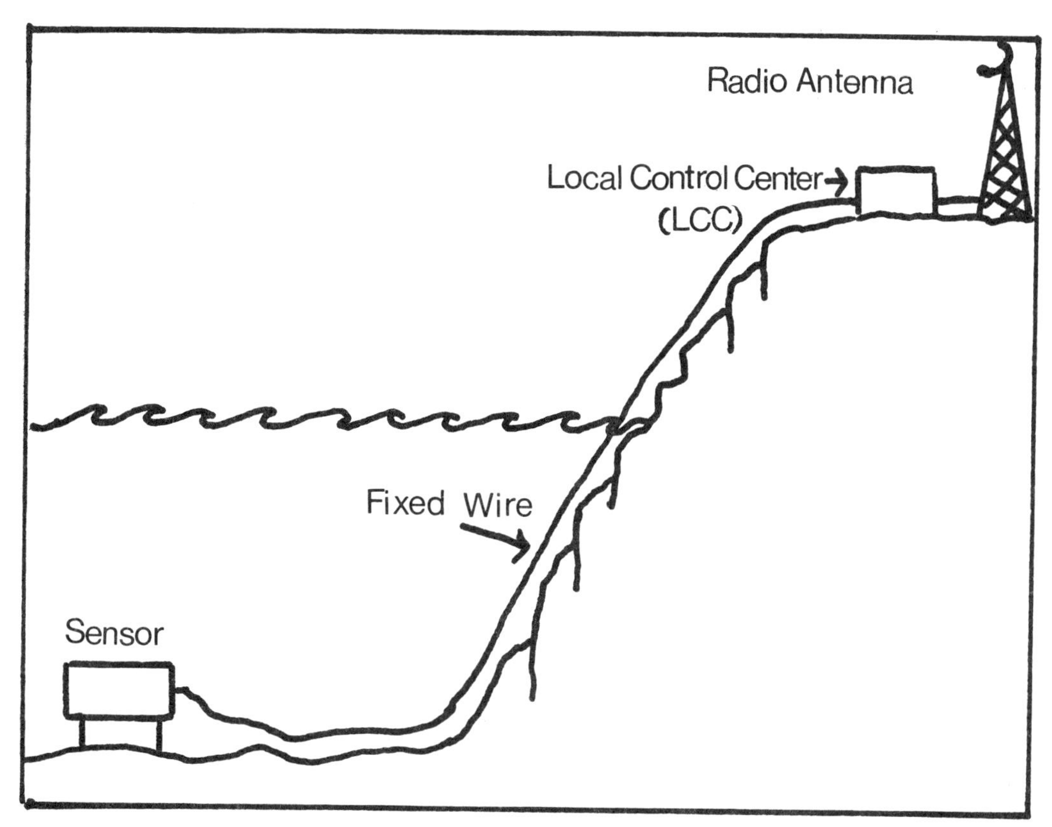

SENSORS: 1. Active sonar
2. Passive sonar

COMMUNICATION LINKS:

1. Fixed wire--From sensors to Local Control Centers (LCC).
2. Radio--MF, HF, VLF between LCC and Ocean Area Control Center (OACC).

Figure 5-15

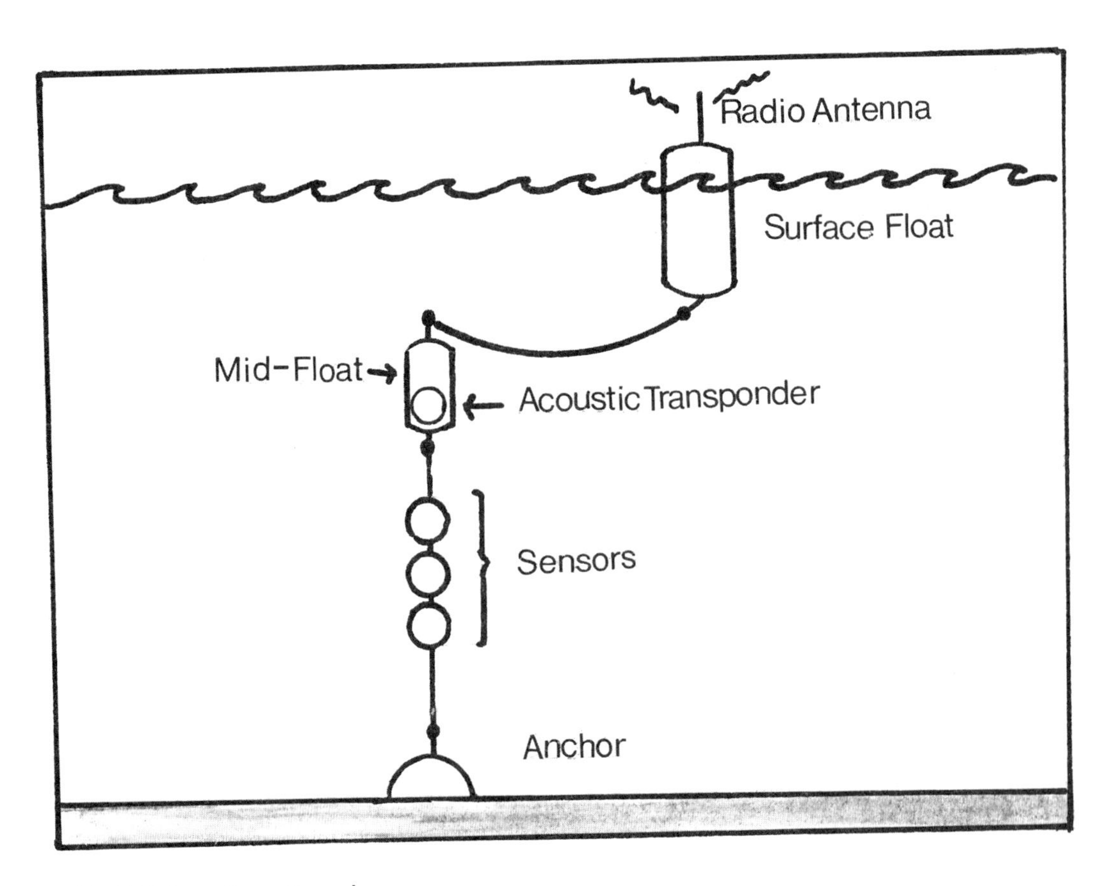

SENSORS: 1. Passive sonar
2. Magnetic Anomaly Detection (MAD)
3. Ocean Environmental Data (SVP)

COMMUNICATION LINKS:

1. Radio: UHF telemetry via antenna on surface float to:
(a) OACC via satellite
(b) Monitoring manned surface or airborne platforms within LOS

2. Acoustic: Encoded telemetry to interrogating submarine

Figure 5-16

NON-MOORED SURFACE BUOYS

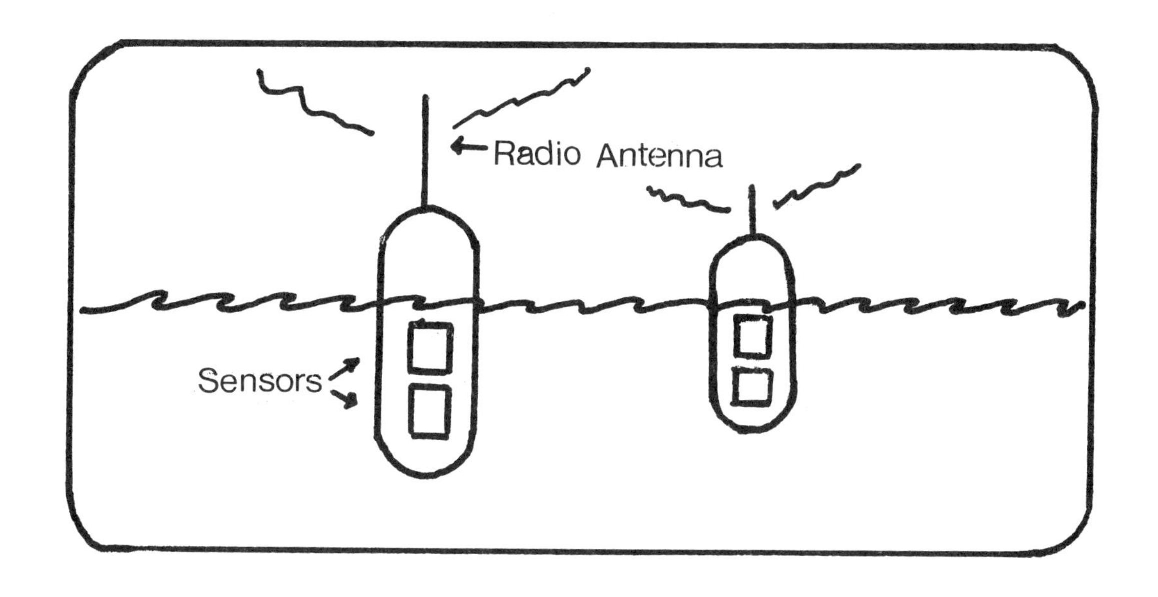

SENSORS: 1. Passive sonar
2. Magnetic (MAD)

COMMUNICATION LINKS:

1. Radio: UHF to monitoring airborne manned platforms

Figure 5-17

UNMANNED SURFACE PLATFORMS

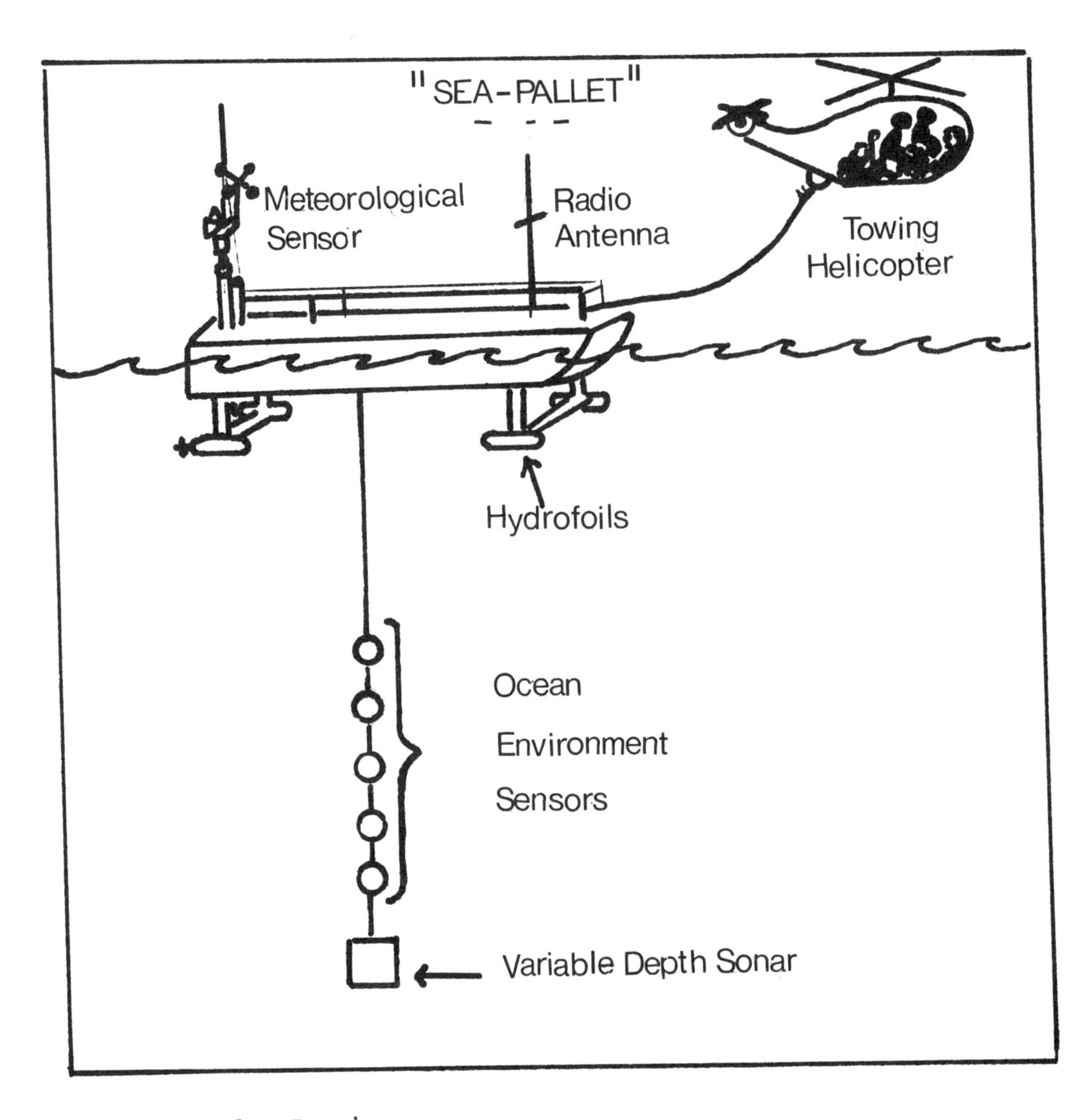

SENSORS:
1. Passive sonar
2. Active sonar
3. MAD (Magnetic)
4. Meteorological
5. Ocean Environment (SVP)

COMMUNICATION LINKS:

1. Radio: UHF Telemetry via antenna on platform to
   (a) Monitoring airborne platform
   (b) OACC via satellite relay

Figure 5-18

SURFACE SHIPS

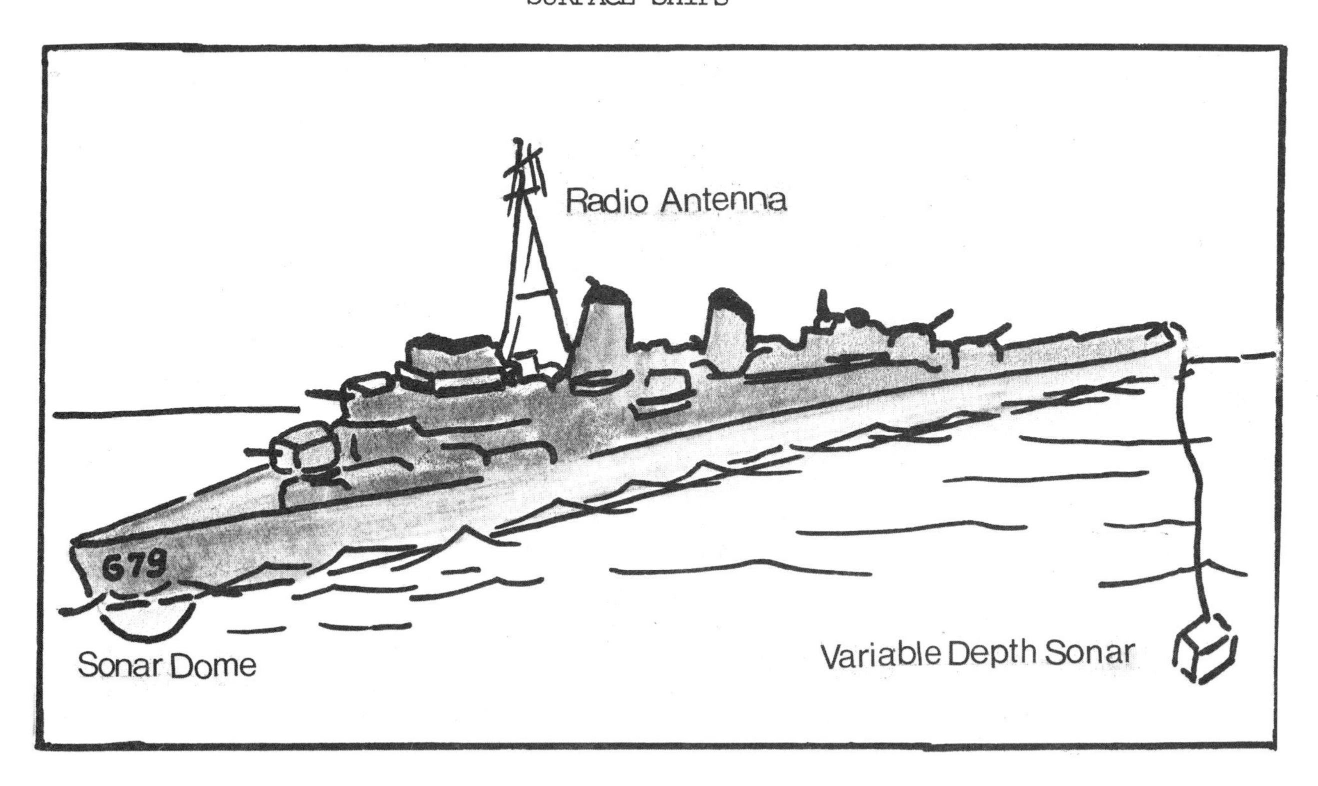

SENSORS: 1. Passive sonar
2. Active sonar
3. Optical

COMMUNICATION LINKS:

1. Radio
   (a) MF/HF to and from OACC
   (b) VLF from OACC
   (c) UHF to and from other manned platforms
   (d) UHF from unmanned platforms, satellite

2. Acoustic: U/W telephone to and from submarines

3. Optical:
   (a) Flashing light to and from other manned platforms
   (b) Flags and semaphore to and from other manned platforms

ADDITIONAL SENSORS:

1. Ocean environment (SVP)

2. Meteorological (wind velocity and direction, temperature, barometric pressure, humidity)

Figure 5-19

Figure 5-20

SUBMARINES

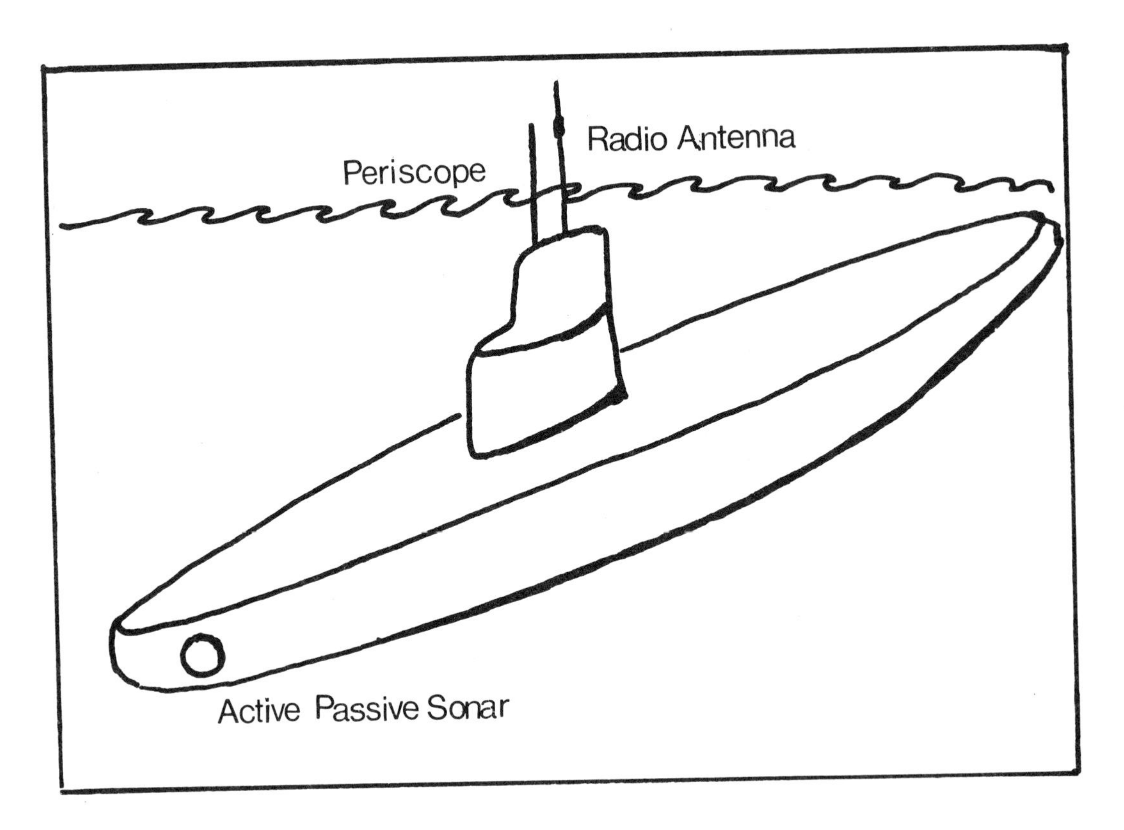

SENSORS:
1. Active sonar
2. Passive sonar
3. Optical via periscope

COMMUNICATION LINKS:

1. Radio:
   (a) Below periscope depth: receive VLF from OACC at shallow depths

   (b) At periscope depth or on surface:
   UHF--to and from other manner platforms
   MF, HF--to and from other manned platforms and OACC
   UHF--from satellite

2. Acoustic:
   (a) U/W telephone to and from other submarines or surface ships

   (b) Telemetered acoustic data from moored buoys upon encoded interrogation

3. Optical: Via periscope

ADDITIONAL SENSORS:

1. Ocean environment (SVP)
2. Depth

MANNED AIRBORNE PLATFORMS

TYPES: Helicopters and fixed-wing ASW aircraft operating from land bases and aircraft carriers

SENSORS:
1. Optical
2. Monitoring of acoustic and magnetic sensors

COMMUNICATION LINKS:

1. Radio: UHF to and from other manned platforms
   MF, HF to and from other manned platforms and OACC

2. Optical: From manned surface platforms

SATELLITES

SENSORS: None--pending future development

COMMUNICATION LINKS:

1. Radio:
   (a) UHF telemetry relay between surface platforms and OACC
   (b) UHF navigational and meteorological information to manned surface platforms and submarines at periscope depth

IV. BIBLIOGRAPHY

1. Anderson, V. C., "Acoustic Communications is Better than None," 1969 NEREM Record, IEEE Northeast Electronics Research & Engineering Meeting, Nov. 1969, Vol. 11.

2. Brahtz, J. F. (editor), Ocean Engineering, John Wiley & Sons, New York, 1968.

3. Craven, J. P., "Ocean Technology and Submarine Warfare," Astronautics & Aeronautics, April 1969, pp. 66-70.

4. Craven, J. P. and Searle, W. R., The Engineering of Sea Systems.

5. Hall, A. D., A Methodology for Systems Engineering, D. Van Nostrand Co., Princeton, J.J., 1962.

6. Hamsher, D. H. (editor), Communication Systems Engineering Handbook, McGraw-Hill, New York, 1967.

7. Kinsler, L. E. and Frey, A. R., Fundamentals of Acoustics, John Wiley & Sons, New York, 1962.

8. Knowlton, A. E. (editor), Standard Handbook for Electrical Engineers, McGraw-Hill, New York, 1941.

9. Metherell, A. F., "Seeing with Sound," Oceanology International, July/Aug. 1968, pp. 22-25.

10. Myers, J. J., Holm, C. H., McAllister, R. F. (ed.), Handbook of Ocean and Underwater Engineering, McGraw-Hill, New York, 1969.

11. Painter, D. W., "Underwater Acoustic Communications," Oceanology International, July-Aug. 1968, pp. 43-45.

12. Saltonstall, R., The Commercial Development and Application of Laser Technology, Hobbs, Dorman & Co., New York, 1965.

13. Reference Data for Radio Engineers, 5th edition, H. W. Sams & Co., (ITT), Indianapolis, 1969.

14. Our Nation and the Sea, A Plan for National Action, Report of the Commission on Marine Science, Engineering and Resources, U.S. Govt. Printing Office, Washington, D.C., Jan. 1969.

15. Principles and Applications of Underwater Sound, U.S. Dept. of Commerce, Office of Technical Services, Washington, D.C., 1946.

CHAPTER VI

NAVIGATION SUBSYSTEM

EPIGRAMS

Navigation on a rotating sphere is easier than navigation on a nonrotating sphere.

Latitude and north-south are concepts of a rotating sphere.

Longitude determination requires a clock.

Stellar navigation is impossible without one earth-oriented vector.

Local navigation is most easily accomplished by local bench marks.

For most sea systems the function of the navigation subsystem is to provide a means by which the collection of objects in the system can maintain position with respect to each other in such a manner that the system mission is accomplished. Utilizing this definition, the relative and absolute position determinations of elements within the system will be very much a function of the mission profile and the final navigation suit which is chosen will again be very much a function of the mission profile. For example, a merchant ship cargo transfer system is primarily concerned with extremely accurate navigation only during the entry into port and harbor. Its accuracy with respect to the pier at which it makes its final landing must be a few feet. Its ability to maneuver in the channel as it comes in must be accurate within a hundred yards or less. Its ability to find the entrance approaches to the channel must be accurate to within about one mile. Its ability to enter into offshore navigational networks and aids need not be much better than three to five miles. And its ability to navigate in the open ocean need not be any better than ten to fifteen miles or even twenty miles in the position accuracy. On the other hand, ballistic missiles systems must be precisely accurate at the time of missile launch, independently of the position in the ocean. Systems which are designed for object-finding and retrieval or for examining specific anomalies on the sea floor need not have any accuracy with respect to the absolute position on the earth, but should merely be able to identify the local topography and terrain in order to set up a local navigation grid which will permit accomplishment of the objectives. The designer of the navigation subsystems is therefore faced with a rather complex time-varying requirement on precision of navigation and a time-varying requirement for navigation on the terrestrial sphere vis-à-vis navigation with respect to local bench marks.

Three generically different types of navigation can be distinguished:

a) Navigation through observation of celestial objects,

b) Navigation by inertial measurement of the rotation of the earth, and

c) Navigation by distance measurements from known fixed objects.

The most ancient and traditional form of navigation is the first of these in which observations are made of the stars and the planets. The direction of north can be obtained from the Pole Star or, in principle, can be inferred from the apparent rotation of the other stars and the latitude can be determined from the declination of the Pole Star or of other stars. Thus, latitude and north can be determined without the use of a clock. Longitude, however, requires a precise clock in order to determine the hour angle between the observer and some arbitrarily- but nevertheless precisely-determined reference position on which the star ephemeris is based. In order to determine position by this manner, some vector which is oriented to the earth must be known with adequate precision. In general, and in particular on land, the vector which has been quite naturally chosen is the direction of local vertical, which for most navigational purposes may be presumed as the direction in which a plumb bob points. For extremely precise navigation, such a vector is inadequate, since the direction that a plumb bob points is a function of the local distribution of mass. Deflections in the earth from the true vertical resulting from such local distributions can be as much as one minute of arc or greater in many parts of the world. An alternative to the local vertical, which is extensively used at sea, is the use of the horizon. The appropriate observation of the horizon is again extremely difficult to accomplish with an accuracy which admits of precision, not only because of the optical problems associated with the atmosphere, but because the horizon is generally chosen because the local vertical is not easy to ascertain because of motions of the platform. Thus, the accuracy of the hand-held sextant is from two to five miles depending upon the skill of the observer and the state of the sea. The availability of stars is, of course, diurnal unless highly-collimated, extremely sensitive photometers or radiometric sensors are employed. Fortunately, the development of man-made satellites has resulted in the deployment of navigational aids (Navigation Satellite) which are sufficiently close to the earth that they can be tracked for a considerable distance. The satellite's positional ephemeris with respect to the earth can be determined at least to the extent that the earth's geoid is known and to the extent that the trajectory of the satellite can be extrapolated when it is moving over hitherto unsurveyed portions of the earch. With a precisely-known ephemeris and with the satellite transmitting not only its own identifying signals but transmitting its ephemeris and its own clock time, then the receiving ship need do no more than follow the trajectory of the satellite for a period of time which is long enough to create a sufficiently large baseline. That baseline then becomes the earth-oriented vector from which the position of the platform can be determined. Utilizing such systems, positional accuracies on an absolute basis in the middle of the ocean of one-tenth of a nautical mile can be achieved. The equipment for receiving these signals and computing position is now commercially available and will undoubtedly result in the superseding of the more ancient and classical methods of precise navigation at sea. The interval between fixes is, of course, a function of the satellite constellation and the satellite period. At the present time, fixes every three hours or so are generally available to platforms at sea. It should be noted that stellar navigation and satellite navigation are available only to craft at the free surface or above and that fixes of this kind, when taken by submersibles, are made at the expense of an excursion to the free surface by the submersible or by some craft vectored therefrom.

Between such stellar or satellite fixes and if required by the mission, some form of dead reckoning generally takes place. For systems where precision is not of great significance, this may be accomplished merely by steering a compass course, measuring water velocity by a pitometer, or electromagnetic flow meter, and estimating the correction due to wind and current. On the broad ocean and in the absence of fixed references, this can result in large integrated errors, even if the ability to measure flow velocity is precise. This is so because there is no absolute way of determining which part of the velocity vector is due to the motion of the ship through the water and which part of the velocity vector arises from the water current. Quite obviously, estimates of current as small as one knot result in errors of one mile for every hour between fixes. For ships or craft operating at or near the bottom, an attractive alternative is available in the form of Doppler sonar. In this instance, three orthogonal beams of acoustic energy are directed toward the bottom and reflected therefrom and, provided that the velocity of sound of the medium remains constant during the trajectory of the sound pulse, the shift in frequency will give an absolute measurement of the velocity in the direction of the acoustic beam. Accuracies of considerably less than 1% of the distance traveled have been attained with this type of technique, i.e., less than one mile error for each 100 miles which have been traversed. Unfortunately, the frequencies which are required of Doppler sonars are fairly high and the signal is attenuated beyond measurement when operating at considerable distances from the bottom. Some success has been achieved by reflection from the scattering layer or from sharp thermoclines in the water, but such layers are not always available and there is error which also may arise from motion of the scattering layer itself.

The recent development of inertial systems provides a solution for the dead-reckoning problem where Doppler sonar is not available and where continuous precise position is a requirement. The inertial platform in general consists of three orthogonally-mounted accelerometers and three gyros. When freely gimbaled and regardless of initial orientation and when held in a fixed location, such a set of gyros will be acted upon by the earth's rotation and the earth gravity field so that they will be oriented in a fixed position with respect to the gravitational vector and the direction of the earth's rotation. This orientation process which is referred to as gyro compassing will provide the direction of north and latitude for a fixed platform without the need for any external references. There will, of course, be small oscillations about these positions, resulting from the behavior of the gyroscopics system as pendulum. In most inertial systems, the orientation of the reference gyros are maintained at the stable orientation for the particular position on the earth through appropriate torques which are applied to the gyros as the ship system moves through space. The values of the torques which are to be applied are determined by the accelerometers which measure the accelerations of the platforms in the direction of the orientation.

A moment's reflection will indicate that, in order to achieve this torquing of the inertial platform, any horizontal acceleration across the earth must be accompanied by a proportional rotational acceleration as a result of the curvature of the earth. The inertial platform must therefore be "tuned" as a pendulum of appropriate apparent length in order to achieve this rotational-translational relationship. This tuning is known as Schuler tuning and results in an oscillation about the stable configuration having a period of approximately 84 minutes. There will, of course, be many uncertainty torques associated with the manufacture of any inertial system, as well as uncertainties in the scale factors of the accelerometers, all of which will tend to produce errors in the inertial system which will accumulate with time. Indeed, the impracticability of inertial systems prior to 1959 was due to the fact that it was not possible to manufacture gyros and accelerometers with sufficient precision and lack of uncertainty so that the inertial system would be able to travel significant times between fixes and still maintain a small error. At the present time, extremely high (and therefore militarily classified) accuracies can be achieved with inertial systems, albeit at the expense of high cost, so that they are adequate for ballistic missile systems and many hours can transpire before a fix is required. Systems which are commercially available or near-commercially available are now employed in aircraft where the time of flight is measured in the order of five to seven hours and on small submersibles, such as the United States Navy's deep-submergence rescue vehicle where the cost of high accuracy is warranted by the nature of the mission. The configuration of gyros which has been cited here is, of course, not the only configuration that is or has been available. For ships which are attached to the free surface, the need for a vertical accelerometer is, of course, not present, since the ship's vertical position will always integrate out to sea level. In other applications, preferred axes in which highly-sophisticated gyros and accelerometers are employed can be identified for particular missions and application. The great advantage of the inertial system is the fact that, after the initial position is determined, no further reference to surface navigational aids is required and, thus, such systems are highly useful, if not critical, to submarine system design.

Other inertial systems which have been considered include the electrostatically-supported spheres which are rotated at high speed in a vacuum. These spheres can be conceived of as a star in a bottle and, in principle and in fact, the uncertainty torques associated with such electrostatically-supported spheres is very small indeed. Although successful ESG platforms have been built, to the author's knowledge no electrostatically-supported gyro inertial system is currently available either commercially or developmentally. Nevertheless the system designer should know that the potential for accuracies in excess of those now experienced with ballistic missile systems has not yet been fully realized.

Even with the best of stellar and satellite fixes and with the best of inertial navigation systems, the accuracy that can be achieved will be measured in the order of magnitude of one-tenth of a nautical mile, i.e., 200 yards. For maneuvering in a confined harbor or in making up to a pier, such accuracy is not adequate. The third generic class of navigation must then be employed which consists of orientation with respect to fixed markers or bench marks on the surface of the earth or on the sea bed. These markers are sighted either optically, electromagnetically or acoustically by the device which desires to navigate itself. Every small boat sailor is, of course, familiar with the network of lights, buoys, and topographic charts which are available to the mariner for near-shore and inshore pilotage. In fair weather and on the surface, full reliance can be placed on these aids. Beyond sight of navigation aids or where optical visibility is limited, reliance can be placed on a number of electromagnetic navigation schemes. As in the case of communication, extreme precision can be achieved, utilizing high-frequency electromagnetic emissions. Accuracy is again achieved at the expense of the range at which the transmissions can be received. Extremely high-accuracy, short-baseline systems like High-Fix are available, in which two electromagnetic signal generators are located on the shore at a known fixed distance with respect to each other. The signals which are transmitted from them are time-coordinated. The phase of arrival thus determines a parabolic line along which the ship or platform is located. With a third station a second parabolic line can be determined, the intercept of those two lines being the location of the ship or platform. There is an ambiguity in this relationship, since the use of a CW (continuous wave) signal results in a whole series of parabolic lines each of which has the same signal phase relationship. Each of these lines is designated as a lane and, for extremely precise navigation systems, it is important that the platform have a means for counting the number of lanes from some known position point in order to ascertain without ambiguity its actual location. Where lower-frequency systems are used, the difference between lanes becomes quite large, and the ambiguity can be resolved by other more approximate navigational techniques. Quite obviously, these systems are difficult to utilize in the vicinity of one of the transmitters or at large ranges where the reception is poor, difficult, or where the reception has been interfered with by reflected waves from the ionosphere. Systems like High-Fix can have precisions as great as 10 to 20 feet under ideal or favorable conditions. For navigation further out to sea, i.e., several hundred miles out to sea, systems such as Loran A or Loran C may be available which, while having less precision, have a greater range. For full ocean application, systems like Omega are increasingly available and this system, as does Loran, as do most of the electromagnetic navigation systems, operates on a very similar principle. The advantage of these electromagnetic schemes is that continuous tracking capability is available with a precision which ranges from a mile or more on the high seas to a few feet for near-shore installations. As in the case of stellar and satellite systems, electromagnetic systems are not available in the submarine environment. Local navigation for submarine systems must therefore be accomplished acoustically. The use of acoustic transponder beacons in the frequency range of about 7 to 17,000 Hertz have proven satisfactory for navigation over areas of several miles. In these schemes the platform sends out acoustic pulse at one frequency, which causes the transponder beacon to respond at its own characteristic frequency. The time of travel for the pulse and the transponded signal is then used for a range determination and the position is determined by a range measurement from three or more bottom-mounted beacons. The reader will note the similarities between the problems of navigation by use of fixed references and that of communication. Both navigation and communication are in this sense just a subset of environmental sensing. In navigation, that aspect of the environment to be sensed is, on one hand, the rotation of the earth and, on the other, the position of the fixed bench marks either on the shore or in the ocean environment itself. These bench marks must, of course, communicate their position by means of optical, electromagnetic, acoustic or physical delivery techniques. The attenuation coefficients and range tradeoffs for this form of communication are the same as those indicated in the communication chapter. In communication, the environment is similarly employed, but the scope of the message is much wider.

CASE STUDY I

SEA-BASED ABM SYSTEM - NAVIGATION SUBSYSTEM

Millard S. Firebaugh, author of the "Sea-Based ABM System" report is presently an Ocean Engineering student at M.I.T., as well as a Lieutenant Commander in the U.S. Navy. His education includes: in 1966 an S.M. and the Naval Engineer's degree (M.I.T.), and in 1961 an S.B. (M.I.T.). Between 1961 and the present his naval activities have included a staff position on the CTG 126.1 in which capacity he worked in navigation during the search for the THRESHER. From 1966 to 1969 Mr. Firebaugh was a ship superintendent at the Portsmouth Naval Shipyard and was engaged in the overhaul of the U.S.S. SAM HOUSTON.

With such a comprehensive background, then, it is to be expected that Mr. Firebaugh's report is well thought out and grounded in the realities of modern navigation.

CASE STUDY I

SEA-BASED ABM SYSTEM - NAVIGATION SUBSYSTEM

Mr. Firebaugh's presentation of the various navigational hardware choices fits into the framework of a report dealing with a sea-based ABM system. The first part of Mr. Firebaugh's system report estimates the magnitude of the necessary ABM force and other external characteristics, based on present estimates of foreign powers' offensive missile strengths. Through three iterations he quantifies the sea-based ABM system so that it will fulfill the subjective human purpose: "to deter the USSR or the CPR from the launching a first strike of nuclear armed missiles by demonstrating a capability to intercept and destroy strike missiles before they reach targets in the continental United States or Hawaii. Damage to the United States from the supposed attack is to be limited to a level which allows essentially all important human activities to continue, including the United States' missile strike capability."

By the third iteration of the mission analysis and profile, Mr. Firebaugh has decided that enemy missiles must be intercepted in space, and has decided how many missiles will be needed to counter the supposed strike. He has also decided to subdivide his total ABM force into operational units which will consist of several missile platforms accompanying one large "flip"-type platform which will house radar, computation, command and control, and life support for the operational unit.

Having quantified the system and defined the platforms involved, Mr. Firebaugh defines the functions which the navigational subsystem must fulfill. They are:

1. To navigate the missile platforms and radar and computation platform out of confined waters and into the deep ocean;
2. To navigate this same operational unit from one port to another or to a specified station at sea;
3. To point the radar in the right direction and hold this orientation;
4. To calculate the trajectory of radar-detected targets relative to the earth, and to decide whether or not they will strike credible targets;
5. To navigate ABM missiles so that they intercept incoming missiles;
6. To avoid other ships, submarines, submerged objects and obstacles; and
7. To be able to look for targets based on advice from central command and control relative to a frame of reference other than the radar and computation platform.

Having defined the functions of the navigation subsystem, there must be a quantification as was done on a larger scale for the overall system. Some of the critical questions posed and answered by Mr. Firebaugh are:

1. At what range can an ABM effectively destroy an incoming missile?
2. How accurately must the radar know its own position in order to evaluate correctly the target of an incoming missile?
3. What coupling must there be between the radar and computation platform and the missile platform or missile in order to properly navigate the ABM?

In orders of magnitude these questions are answered as follows:

1. From general information on the yields of other ABM's we can assign a yield of about 0.5 megaton to our missile. It will detonate in space (above 100 km), and so the destructive force will appear in the x-ray output of the ABM. Since this radiation intensity falls off as $(\text{range})^3$, we must be rather close, within one kilometer.
2. Assuming that the radar is about halfway between the incoming missile and its supposed target, any uncertainty in the position of the radar will show up as twice that uncertainty in the missile's supposed target. So, as an order of magnitude estimate, the radar should know its own position to within one nautical mile.
3. Rather than trying to keep the missile platform solidly coupled with the radar platform, we need only provide the missile itself with a frame of reference that can be coupled with the radar platform. Using inertial guidance for the missile will make this link feasible, and will permit the radar and computation facilities to guide the ABM relative to its target. This, however, appears to be the most critical navigational problem which will be faced.

What follows is part of Mr. Firebaugh's report. Two of his comments, "definition" and "subprocesses," are included before the presentation of navigational hardware alternatives. After this, we will return to a final discussion of the navigation subsystem.

Definition: Navigation is the process of determining the observer's position, velocity, and acceleration relative to a coordinate system attached to some object of interest. Navigation subsystems are usually strongly interfaced with environmental sensing and command and control subsystems. In fact, navigation subsystems are so strongly interfaced with environmental sensing that they may frequently absorb the environmental sensing problem as part of the navigation subsystem. In many instances, the navigation interface with command and control is carried out through a human intermediary so more often command and control are not considered in the same subsystem. The breakdown between command and control and navigation may not be made, however, in highly automatic systems.

Sub-Processes: Draper makes a useful analysis of the navigation subsystem by observing that it is an informational process and, therefore, must have its own sub-processes of the following types:

1. Sensor Process - Sensors receive physical quantities and generate signals representing these quantities.
2. Transmission Process - Signals are transmitted to various components.
3. Data Processing - Sensor signals accepted and conclusions derived in forms compatible with signals from other subsystems.
4. Evaluation Process - Compare derived conclusions with other signals derived from stored information on programs and plans.
5. Programs and Plans Storage - Self-explanatory.
6. Control Order Generating Process - Control orders generated

to correct deviations from planned conditions.[1]

This analysis is biased toward the control interfaces but is still quite useful. Processes 1-3 are common to all navigation systems. Processes 4-6 may or may not be included, depending on the definition of the command and control subsystem.

PART II

A. Astro-ref Systems

Relates observer's position to fixed stars. May be used anywhere on the surface of the geoid or in space.

1. Hand-held Sextant and Chronometer

   Sensor - human eye and sextant sense elevation of known heavenly body above natural horizon. Measure time with a chronometer.

   Data Processing - hand computations with almanac.

   Accuracy - 1 to 2 N. Miles.

   Frequency - visible light.

   Advantages - small space and weight, inexpensive, $500.

   Limitations - intermittent, no instantaneous velocity or acceleration, cannot be done under conditions of reduced visibility, accuracy considerably reduced at night.

   Reference - Dutton, Navigation and Nautical Astronomy, Annapolis, 1951.

---

[1] Draper, C.S., "Survey of Inertial Navigation Problems in Sea Air and Spaces Navigation," Papers of the International Congress Long-Range Sea, Air, and Space Navigation, 26-31 August, 1965 in München, Vol. I, Section 8, page 10.

2. Bubble Sextant

Sensor - human eye and sextant sense elevation of known heavenly body above artificial horizon created inertially by reference to a spent bubble.

Data Processing - hand computation can be with computer assist.

Accuracy - 5 to 10 N. Miles.

Frequency - visible light.

Advantages - small space and weight, frequently used in aircraft.

Limitations - depends on clear visibility, intermittent vs. instantaneous velocity or acceleration. Can be used at night.

Reference - same as for hand-held sextant.

3. Type II Startracker Periscope

Sensor - periscope eyepiece aimed by computer input from some other navigation system (usually SINS) sights on known heavenly body and notes sighting error.

Data Processing - computers.

Evaluation - NAVDAC computer translates sighting error into update information for SINS.

Accuracy - quite.

Frequency - visible light.

Advantages - can be used at periscope depth below surface and at night. Essentially continuous information if desired.

Limitations - depends on clear visibility. High cost.

Reference - Rowan, W. J., "Notes on Submarine Periscopes," Navigation, Journal of the Institute of Navigation, Vol. 14, No. 1, Page 3.

4. Radio Sextant - part of the SSCNS (Ships Self-Contained Navigation Systems)

   Sensor - small antenna, receives a radio frequency signal from the sun or moon. Senses local vertical from an inertial source.

   Data Processing - computer.

   Evaluation - similar to Type II periscope.

   Accuracy - 1 N. Mile.

   Frequency - 16 Ghz or 35 Ghz.

   Advantages - can provide continuous data. Unaffected by visibility. Global coverage with the important exception of very northern latitudes in winter when for some periods no sun or moon is visible.

   Limitations - mostly used in conjunction with some inertial systems. Cannot be used submerged. Can only be used when sun or moon above horizon.

   References - Marner, G. R., "The Use of Radio Sextants in Maritime Navigation," Papers of the International Congress Long-Range Sea, Air and Space Navigation, 26-31 August 1965, in München, Vol. 1, Section 17; Frye, Chadima, Wallace, and Marner, "The Radio Sextant," Proceedings of the ION National Marine Navigation Meeting, Oct. 1967, page 110 ff. (provides weight and volume information).

B. Geo-ref Systems

Observer relates his position to fixed or presumably fixed set of coordinates on the earth. These can be considered in several subdivisions:

1. Hyperbolic Radio Systems

Observer measures delay in arrival time of signals from two different sources at known coordinates and with known signal transmission times. Position must then be on intersection of a hyperboloid and the earth's surface where the hyperboloid is uniquely determined by the magnitude and sign of the time delay. Consideration of signals from two sets of sources gives a point, namely, the mutual intersection of two hyperboloids and the geoid. These systems use various frequencies and radiation patterns providing coverage in some regions but not all. From a military point of view most systems have the disadvantage that some transmission stations may be on foreign soil and, therefore, subject to compromise. Variation in the electrical properties of the atmosphere, between day and night usually results in increased errors at night.

2. Rho-Theta Radio Systems

Observer measures delay in arrival time of a signal transmitted at a known time from a single source to get range and uses as direction-finding antenna to measure angle, thus finding position on a polar plot centered at the transmission station. A variation of this employs two or more stations so that the position is determined by range alone from the two or more stations. Position is then a point at the intersection of spheres with the geoid. Another variation finds a point as the intersection of two lines of direction from two sources. These systems have most of the disadvantages of hyperbolic systems and usually greater errors.

3. Magnetic Systems

Observer measures various parameters of the earth's magnetic field. A single observation generally only results in a N-S determination. Subject to variation of earth's magnetic field. Generally used with other systems frequently as a low-cost emergency source of navigation information.

4. Satellite Systems

Observer tracks an artificial satellite and compares its computed orbital parameter to its known parameters and hence concludes where the observer must be on the earth's surface. High accuracy is achieved by having the satellite transmitting a signal on a known frequency and measuring the doppler shift in frequency to determine the satellite orbit. The only such system in use allows fixes at any point on the globe every two hours. It is clever in that the known orbit of the satellite is transmitted to the observer on the same frequency he uses to measure the doppler shift. Time is also transmitted.

5. Piloting

Observer measures range and angle to fixed landmarks on shore, such as towers, lighthouses, or prominent geographical formations. This can be done visually or with radar. When done visually, it may only be done during periods of high visibility. With radar it may be done in all weather. Both visual and radar methods are limited to the vicinity of coastline.

6. Bottom-Reference

Observer measures his location relative to objects fixed on the ocean bottom. This technique uses acoustic ranging to measure the distance to either known bottom topography or the known locations of transponders. Navigation by bottom topography requires that the bottom topography be well defined by topographical surveys. Unfortunately, most of the ocean bottom topography is not well known to the accuracy necessary for reasonable navigation. The xponder technique is frequently employed to provide a local navigation system. That is, the latitude and longitude of the xponder beacons may not be accurately known but the location of objects of interest and the location of the observer relative to the xponder beacons may be very well known.

Examples of systems in these categories will now be given:

1. Hyperbolic Radio

a. Loran A

Sensor - radio receiver.

Data Processing - time delay measured by pulse matching partly by electronics and partly by human. Human then plots on special H.O. charts.

Accuracy - 1.5 to 5 N. Miles depending on location and time of day.

Frequency - about 1900 khz.

Advantages - low-cost commercial equipment available. General coverage of high traffic North Atlantic and North Pacific areas.

Limitations - coverage is not universal, can only be received on surface, process is not continuous, accuracy less at night, accuracy at best is not as good as some other systems.

Cost per Unit - $2,000.

Reference - Dutton (same as for sextant).

b. Loran C

Sensor - radio receiver senses pulses like Loran A but also matches phase of the pulses carrier frequency for greater accuracy.

Data Processing - computers give continuous readout of position coordinates. Frequently tied automatically to a SINS computer.

Accuracy - 300 yards.

Frequency - 100 khz.

Advantages - very accurate, all-weather, automatic readout, well adapted to be a subsystem in a more complicated navigation system.

Limitations - area coverage not great. Cannot be received with submerged antenna, high cost.

Cost per Unit - $30,000.

Reference - Baudin, J., "Basis of the Loran C System and Technique of the Ground Stations"; Vogeler, W. K., "Experience in Loran C." Both of the above in Papers of the International Congress Long-Range Sea, Air, and Space Navigation, 26-31 August 1965 in München, Vol. I.

c. Decca

Sensor - radio receiver matches phase of signals.

Data Processing - automatic readout in DECCA coordinates. Translation to latitude-longitude done by human with a chart.

Accuracy - similar to Loran C or better very close to stations. Reports on baseline between stations of accuracy within 20 or 30 feet.

Frequency - 70 to 130 khz.

Advantages - very accurate, all-weather, continuous.

Limitations - coverage limited to North Atlantic coasts. Surfaced antenna only.

Cost per Unit - $3,000.

Reference - Dodington, "Ground-Based Radio Aids to Navigation," Navigation, Journal of the Institute of Navigation, Vol. 14, Number 4, 1968.

d. Omega

Sensor - radio receiver matches signal phase - also counts phase so that it can alternatively be used as a Rho-Theta System.

Data Processing - automatic readout of position coordinator.

Accuracy - .5 N. Miles.

Frequency - 10 to 14 khz.

Advantages - global coverage in all weather. VLF transmissions can be received in the near-surface water layer, fairly accurate.

Limitations - vulnerability of foreign stations to compromise. System not yet fully operational.

Cost per Unit - $30,000.

Reference - Brogden, J. W., "The Omega Navigation System," Proceedings of the ION Nation Marine Navigation Meeting, Annapolis, 1967; Wright J., "Accuracy of Omega/VLF Range Rate Measurement," Navigation, Journal of the Institute of Navigation, Vol. 16, Number 1, 1969; Tibbals, M. L., "Application of Omega Navigation System to Sea Transport," Papers of the International Congress et al.

2. Rho-Theta Radio System

The following table is from Dodington, "Ground Based Radio Aids to Navigation":

NAME	TYPE	FREQUENCY	RANGE N. MILES	COST $	ACCURACY	SHIP/AIRCRAFT
Consol	Theta	100-300 khz	700	200	Poor	Ship
Vehicular D/F	"	200-1600 khz	200	1000	Moderate	Both
ILS	"	108-110) 330-335) mhz	20	500	Excellent	Aircraft
VOR	"	108-118 mhz	200	1000	Excellent	Aircraft
UHF D/F	"	225-400 mhz	200	3000	Moderate	Aircraft
Tacan/DME	Rho-Theta	960-1215 mhz	200	2000-6000	Excellent	Aircraft

3. Magnetic System

a. Magnetic Compass

Sensor - permanent magnet senses the earth's magnetic field, aligns with it. Gives only relation of ship's head to magnetic north.

Accuracy - approximately 2°.

Advantages - global coverage except vicinity of magnetic poles. Works underwater.

Limitations - fairly inaccurate. Magnetic N moves about 50-mile radius circle. Usually a backup system.

b. Magnetic Gradient Measurement

Sensor - measure gradient of magnetic field.

Accuracy - .5 N. Miles.

Limitations - very experimental.

Reference - Beisner, "Arbitrary Path Magnetic Navigation by Recursive Non-Linear Estimation," Navigation, Vol. 16, No. 3.

4. Satellite Systems

a. Navy Navigation Satellite Systems - Transit

Sensor - radio receiver receives signal from satellite. Receiver measures doppler shift as satellite passes ship position. Signal also contains orbit parameters and time.

Data Processing - computer.

Accuracy - .1 to .02 N. Miles depending on input of ship's own motions during the satellite pass.

Frequency - 150 and 400 mhz.

Advantages - very accurate.

Limitations - not continuous - get a fix every two hours. Antenna must be surfaced. Rather expensive.

Cost per Unit - $55,000.

Reference - Stansell, T. A., Jr., "The Navy Navigation Satellite System: Description and Status."

5. Piloting

Mention here will only be made of equipment. Accuracy is generally greater the closer one is to shore. Finally, at the time of mooring, it is typically a matter of a few feet or inches. It is worth noting in passing that supertankers are getting so big that some ships are installing sensors to measure velocity at the extremities of the ship to assist the pilot in knowing what the bow and stern are doing when he is bringing the ship alongside.

a. Pelorus - used in conjunction with gyro compass to allow observer to sight a line of bearing to an object.

b. Stadimeter - manual-optical device used for measuring range to an object of known height.

c. Charts - detailed maps of coastal area.

d. Gyrocompass - see section on inertial navigation.

e. Radar - radio direction and ranging device subject of another lecture.

6. Bottom Reference

No system in general use exists. Technique is frequently used in special oceanographic or ocean search situations such as Thresher search, Alvin search. Most active navigator in this mode is C. L. Buchanan at Naval Research Laboratory using his 3D tracking sonar and xponder arrangement in USNS MIZAR. He often uses this equipment to navigate a submerged object relative to a ship on the surface as well as to an object on the bottom. Accuracy is typically approximately 50 feet. Limitations of the system are short range and short life of bottom xponders.

C. Inertial Reference Systems

These systems are the subject of a separate presentation. Operation will not be explained here but the principal systems available will be mentioned.

1. Gyrocompass

Sensor - pickups sense relative motion between a gyro and ship's head.

Data Processing - human observation.

Accuracy - error $< 1°$

2. Stable Element

Sensor - senses motion of ship in pitch, roll, and head relative to gyros and accelerometers.

Data Processing - fire control computer in usual military situation. Function is to provide local vertical for solution of ballistic problems.

3. SINS - Ship's Inertial Navigation System

Sensor - senses motion of a ship relative to an inertial reference frame fixed on the earth's surface by measuring acceleration felt by a gyro stabilized platform.

Accuracy - quite.

Advantages - completely self-contained in the vehicle. Very accurate. Continuous readout. Service various fire control and other systems requiring local vertical. Easily integrated into control systems. Schuler timing used to limit errors.

Limitations - some drift may accumulate with time. Require occasional updating so some other high accuracy system must be used to update in order to preserve system accuracy. High cost.

Cost per Unit - $500,000.

Reference - Schuler, M., "The Disturbance of Pendulum and Gyroscopic Apparatus by the Acceleration of the Vehicle," reprinted in Navigation, Vol. 14, No. 1, 1967; Stater, J. M., Inertial Guidance Sensors, New York, 1964.

4. SGN-10 - Commercial Inertial Navigator for Aircraft

   Accuracy - about 1.9 N. Miles/hr.

   Cost per Unit - $100,000 (approx.).

   Reference - Caligiuri, J. F., "SGN-10 First Commercial Inertial Navigator," Navigation, Vol. 14, No. 1, 1967.

5. LN-3 - Similar to SGN-10

   Reference - Holm, R. J., "LN-12 in Civil Air Transport Operation," Navigation, Vol. 14, No. 1, 1967.

6. Dead Reckoning

   This is more of a method than a commercial system. Principle is to measure vehicle velocity through the water with some type of velocity-measuring equipment and read ship's head from a gyro or magnetic compass. Then, armed with a knowledge of ocean currents gleaned from charts and the time of day, lay out as supposed track and hope that one is on it. Naturally errors grow with time. Nevertheless, most navigation is done this way and it is satisfactory for some unsophisticated purposes. It is included here under inertial because of the ship's head input.

Armed with a fairly comprehensive listing of available hardware, final choices can be made. These choices must be compatible with specified functions of the navigational subsystem. Let us examine them in the same order as the functions were presented.

Navigation in confined waters will be a case of short-range sensing and communication. In Section 5 of georeference systems (Piloting), Mr. Firebaugh lists five alternatives, Pelorus, Stadimeter, Charts, Gyrocompass and Radar, all of which could be used for the sea-based ABM system. Mr. Firebaugh provides room on the bridge for a commanding officer, a pilot, lookouts, and an officer on duty. They will be aided by a gyro repeater and bearing ring arrangement, as well as by a short-range optical arrangement and plotting tables.

Navigation of the operational unit from one port to another or to a station at sea will be done with a SINS and the appropriate updating equipment. Updating the SINS is a difficult procedure because of the necessarily high accuracy and wide coverage of the updating facility. There are, however, three available systems, all of which will be used. The major criteria for their choice was accuracy (equal to or better than SINS) and coverage (Arctic, North Atlantic and North Pacific). They are NNSS (Navy Navigation Satellite System), Omega, and the Type 11 periscope, all of which were discussed in Mr. Firebaugh's listing.

NNSS is very accurate, but fixes can at present only be taken every two hours. Should a shorter interval be needed, more satellites can be put up. Omega, on the other hand, can give continuous fixes, and can be used to back up SINS in case of failure. The Type 11 periscope can be used to back up NNSS. Between SINS, NNSS, Omega and the Type 11 periscope, navigation will be possible on a continuous basis, with accuracies within one nautical mile of true position.

Both coastal piloting and open-ocean navigation are parts of the navigational subsystem which interface strongly with the rest of the system. The next several functions of the navigation subsystem are confined, largely to an individual operational unit, and as long as the hardware works, they should have little system level control over them. There will, for instance, be a continual flow of information from each missile platform to the radar and computation platform so that the position of each missile platform may be known with respect to the SINS. Guidance of the missiles, too, will be coordinated by the radar and computation platform, and will be quite independent of other operational units. From a system's viewpoint, then, the navigation subsystem should be responsible for the choice of hardware, and should be given maximal flexibility within the functional and interface constraints. Broad outlines of these subsystem level choices are schematized below:

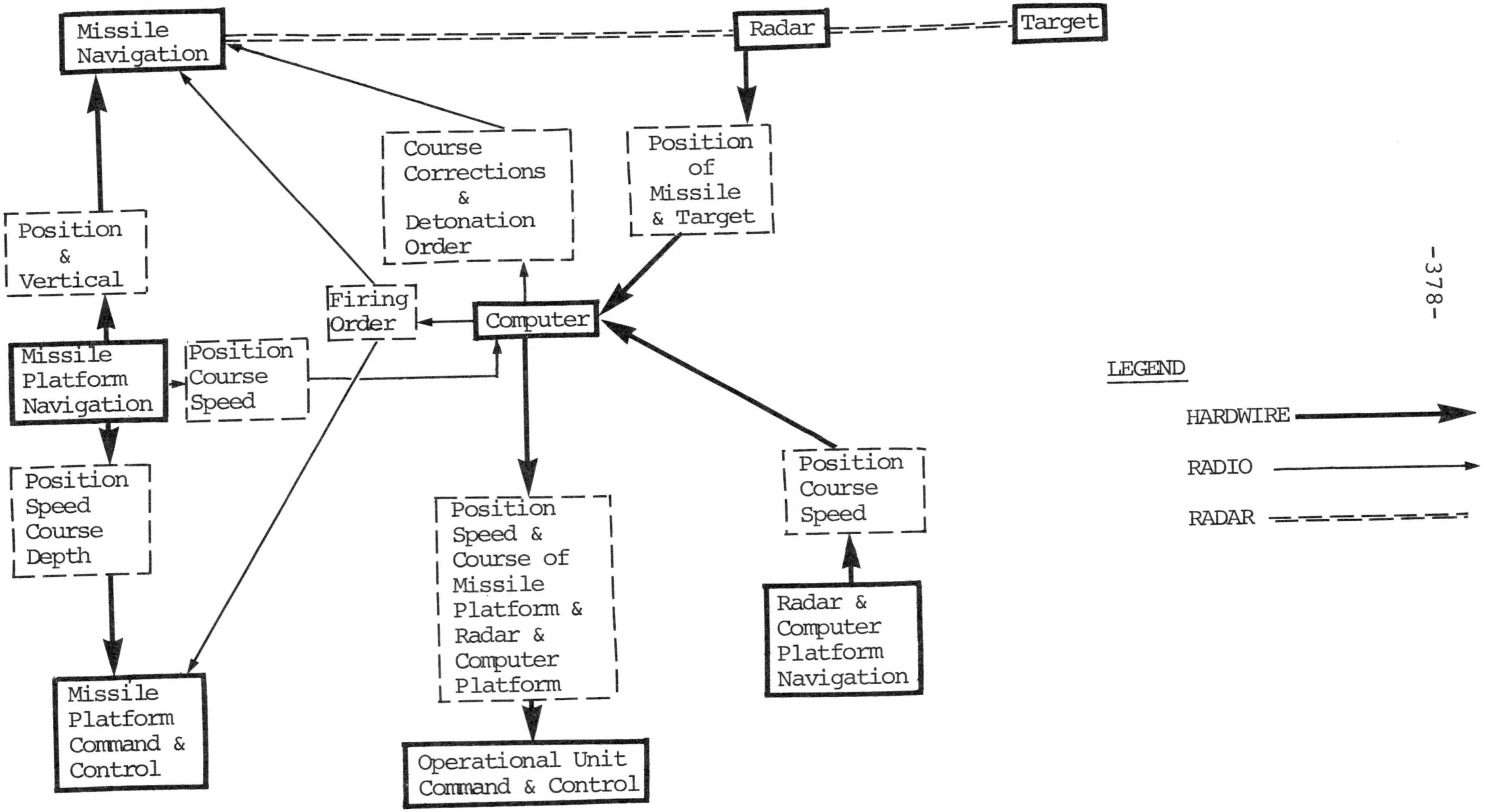

Missile Navigation
Radar
Target
Course Corrections & Detonation Order
Position of Missile & Target
Position & Vertical
Firing Order
Computer
Missile Platform Navigation
Position Course Speed
Position Speed Course Depth
Position Speed & Course of Missile Platform & Radar & Computer Platform
Position Course Speed
Radar & Computer Platform Navigation
Missile Platform Command & Control
Operational Unit Command & Control
LEGEND
HARDWIRE
RADIO
RADAR

Figure 6-1

CASE STUDY II

DEEP-OCEAN OBJECT RECOVERY SYSTEM--NAVIGATION SUBSYSTEM

William Hunter Key, Jr., author of the "Deep-Ocean Object Recovery System" report is a lieutenant in the U.S. Navy. His education includes a degree from the United States Naval Academy and the S.M. and the Naval Engineer's degrees (M.I.T.) in 1970. Lieutenant Key has served on the destroyers U.S.S. BENJAMIN STODDERT DDG-22 and the U.S.S. LOESER DE-680. He is also qualified as a U.S. Navy salvage diver.

With a background in practical Navy salvage operations and operations at sea, Lieutenant Key has had the appropriate experience for a realistic and pragmatic set of choices in his case study.

CASE STUDY II

DEEP-OCEAN OBJECT RECOVERY SYSTEM--NAVIGATION SUBSYSTEM

Problem Statement

Develop a system which will recover from the ocean objects of various size and density from both surface and subsurface regimes. The system will utilize existing state of the art as available with the idea of providing the system service at minimum cost. The system will have minimum search capabilities, that being the task of a separate system.

Mission Analysis

a. Quantity and type of objects to be recovered:

Single objects of varying density with maximum weight and volume that of a Saturn V booster rocket.

b. Location of objects:

The objects or object must be located prior to the arrival of the salvage system (i.e., search operations must be completed).

c. Constraints on retrieval:

Objects to be recovered will be of varying density with low-density objects unable to tolerate high surface loads; the object must be recoverable in 20,000 feet of water.

d. Obstacles to retrieval:

Recovery must be possible in state 6 seas.

Station must be maintained at the site 200 feet 2 σ_{rms}.

e. Time limit on retrieval:

Dependent on mission requirements.

f. Type of processing:

Object must be protected and preserved from corrosive action of sea water exposure from recovery to delivery to the specified port.

g. Limit on processing:

To commence upon recovery and to be finished within six hours.

h. Constraints on processing:

Object must not be damaged further in any way.

i. Distance:

4000 N.M. maximum after recovery of the object.

j. Time constraint on transportation:

None considering object properly preserved.

k. Obstacles to transportation:

Sea states great enough to endanger the recovered object.

Water depth at delivery port.

l. Obstacles to delivery:

Handling equipment in delivery port.

Tidal range in delivery port.

Labor problems.

m. Probability of system success:

80% success.

60% confidence level.

Figure 6-2

Mission Profile

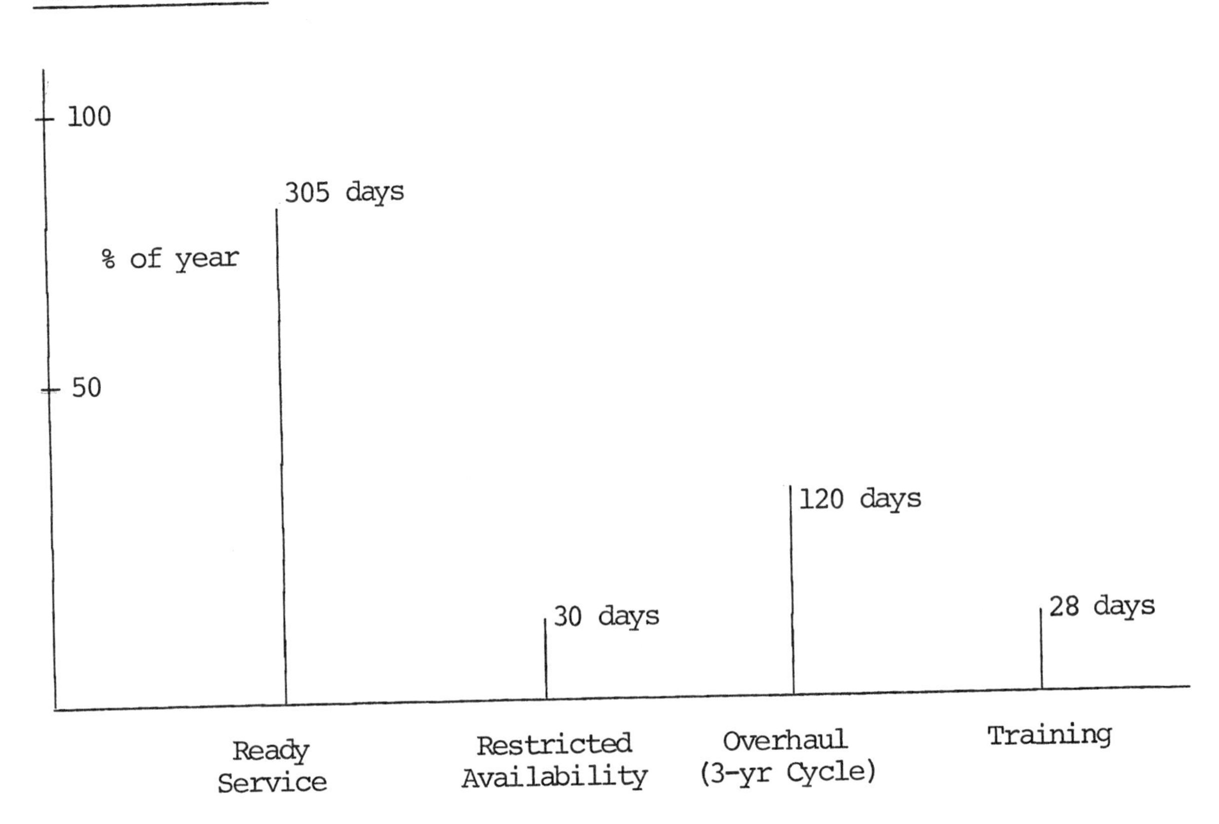

Profile Scenario

The system is in ready service with crew in an available status at a predesignated port. An object is lost which is deemed worthwhile to recover. The system proceeds to the site, makes the recovery and returns the recovered object to the designated port.

The recovery system will not be available for 30 days per annual cycle due to upkeep requirements which will consist of one two-week period every 180 days.

The system will also not be available during training periods of seven days each quarter for a total of 28 days per year.

Every three years, the system will receive a major overhaul and will not be available for 120 days.

Subsystems

The subsystems were divided by areas rather than by subject matter as the system manager feels that better continuity is achieved by this method.

1. Platform

   a. Hull and structure

   b. Powering

   c. Life support

   d. Maneuvering

2. Object retrieval

   a. Recovery

   b. Preservation

   c. Transportation

   d. Delivery

4. Command and control

   a. Navigation

   b. Communication

5. Terminal

   a. Handling facilities

   b. Preservation facilities

Command and Control Subsystem

I. Navigation

A. Problem Statement

Upon departure from a standby port, be capable of navigating to a given scene with an accuracy of 5 n.m. 1 σ_{rms}. Once on the scene, be capable of navigating to a position directly over the lost object with an accuracy of 200 ft 2 σ_{rms}. Have the capability aboard the submersible work boat to navigate to within 20 ft 2 σ_{rms} of the lost object and return.

B. Mission Analysis

The analysis will be considered in three parts, each pertaining to a certain phase of the navigation problem.

NAV I port to scene

NAV II scene to site

NAV III submersible guidance

1. NAV I

a. Precision of location:

5 n.m. 1 σ_{rms} in 4000 n.m.

b. Constraints:

1) Must be capable of operation in state 9 seas.

2) Must be capable of operation in overcast conditions (cloud cover, fog).

3) Must be passive at receiver (clandestine recoveries).

4) Must have no foreign ground stations.

5) Must be capable of being operated with a minimum amount of maintenance and training.

6) Must be capable of being maintained shipboard.

7) Must have minimum down time on equipment.

8) Must be operational regardless of ambient ocean or atmospheric conditions

temperature

pressure

interference

c. Shipboard obstacles to installation

1) Shipboard-generated electromagnetic interference

2) Restricted antenna or transducer location

3) Restricted antenna size and weight

4) Restricted unit size and weight

5) Shipboard-generated noise

6) Restricted unit location (bilge, mast, etc.)

7) Shipboard-generated heat

8) Equipment-generated heat

9) Shipboard-generated vibrations

d. Ambient obstacles to installation

1) Motion and G loadings

2) High humidity

3) Salt environment

4) Corrosion

5) Electromagnetic interference

6) Wave noise

7) Noise created by ambient life (shrimp, birds)

e. Distance capability

Global

f. Time limit on output

20 minutes

g. Probability of success

90% success

95% confidence

2. NAV II

a. Precision of location

200 ft 2 σ_{rms}

b. Constraints

1) Must be capable of operation in state 6 seas.

2) Must be capable of operation in overcast conditions.

3) Must have no foreign ground stations.

4) Must be capable of being operated with a minimum amount of maintenance and training.

5) Maintenance must be done shipboard.

6) Must be a minimum down time on equipment

7) Must be operational regardless of ambient atmospheric or ocean conditions: Temperature
Pressure
Interference

c. Shipboard obstacles to installation

1) Shipboard-generated noise

2) Shipboard-generated electromagnetic interference

3) Restricted unit weight and size

4) Restricted location requirements of unit (bilge, mast, etc.)

5) Restricted antenna or transducer size and weight

6) Shipboard-generated heat

7) Equipment-generated heat

8) Shipboard-generated vibrations

d. Ambient obstacles to installation

1) Motion and G loading in 6° of freedom

2) High humidity

3) Salt environment

4) Electromagnetic interference (natural to environment)

5) Wave noise

6) Corrosion

7) Noise created by ambient life (birds, shrimp)

e. Distance capability

15 n.m.

f. Time limit on output

30 seconds

g. Probability of success

98% success levels
95% confidence

3. NAV III

a. Precision of location

20 ft 2 σ_{rms} in 20,000 ft

b. Constraints

1) Must be capable of operation in sea state 6.

2) Must be capable of operation beneath surface of ocean.

3) Must have no foreign ground stations.

4) Must have takeout capability for maintenance.

5) Must be capable of being operated with a minimum amount of training.

6) Must be capable of operations regardless of ambient ocean conditions: temperature
pressure
interference

c. Obstacles to installation (onboard vehicle)

1) Vehicle-generated noise

2) Shipboard-generated noise

3) Shipboard-generated electromagnetic interference

4) Restricted weight and size of unit

5) Restricted antenna or transducer size and weight

6) Vehicle power limitations

7) Vehicle-generated heat

8) Vehicle-generated vibrations

9) Equipment-generated heat

d. Ambient obstacles to installation

1) Motion and G loading (at interface)

2) High humidity

3) Salt environment

4) Electromagnetic interference (natural)

5) Noise created by waves

6) Noise created by ambient life (fish, shrimp)

7) Corrosion

e. Distance capability

10 n.m.

f. Time limit on output

10 seconds

g. Probability of success

98% success level
95% confidence

C. Mission Profile

Scenario

Ship is given coordinates of lost object and is required to steam to the location within 5 N.M. 1 σ_{rms}. Once on this scene, the ship must position herself over the lost object. Once in position, may or may not direct submersible to object.

While in port in ready standby, the navigation equipment must be at all times capable of operation within the time required for the ship to get underway.

Repair and maintenance will be carried out by use of modules, thereby reducing equipment down time. There is no need for routine overhaul schedule as equipment will be repaired as required.

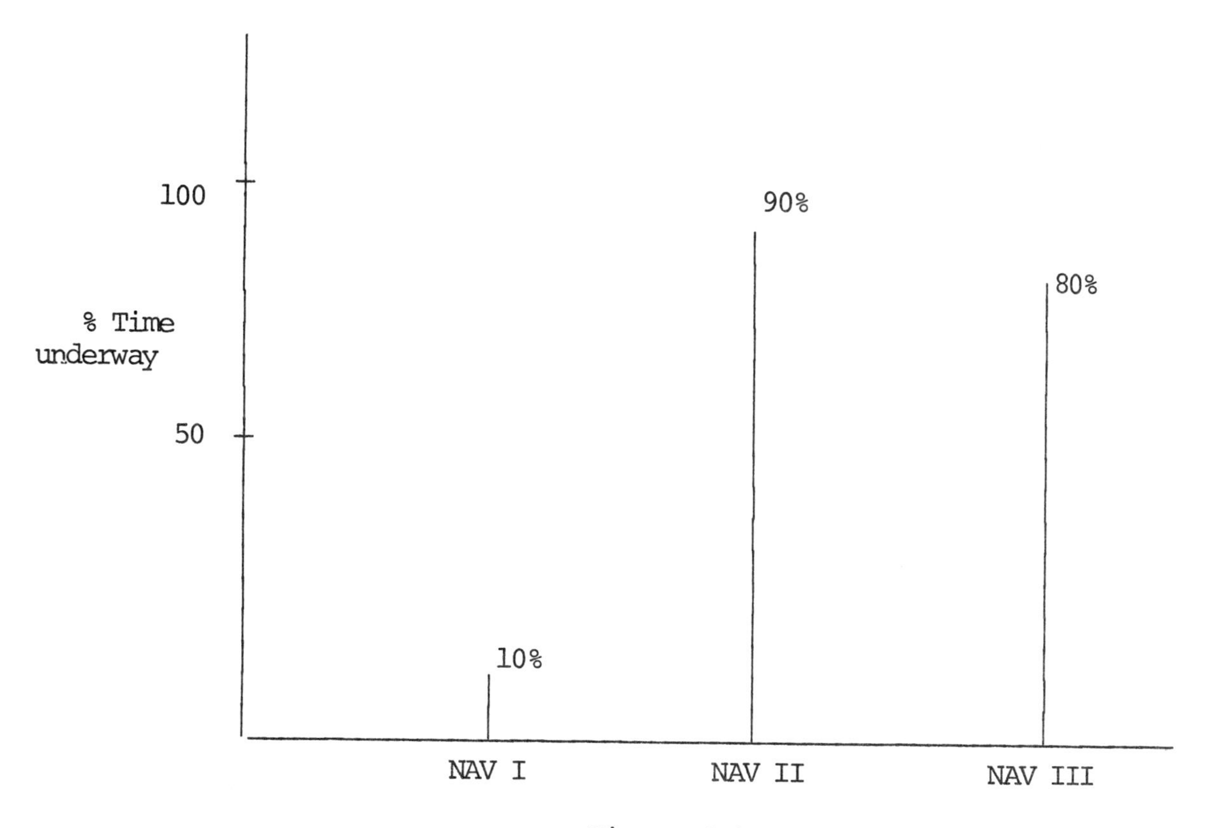

Figure 6-3

D. System Interfaces

1. Hull

   a. Equipment volume maximum

   b. Equipment weight maximum

   c. Antenna or transducer (transponder) volume

   d. Antenna or transducer (transponder) weight

   e. Antenna or transducer (transponder) location

   f. Antenna or transducer (transponder) configuration

   g. Equipment configuration

2. Powering

   a. Voltage stability

   b. Cycle stability

   c. Continuity of power supply

   d. Mechanical vibrations

3. Life support

   a. Human factors: operation
maintenance

4. Maneuvering

   a. Output suitable for use in maneuvering system

E. Component Selection

1. Explanation of definitions: In view of the fact that a system requirement of ± x miles is not meaningful, it is suggested to use the probability that the position will fall within a circle of given radius:

$$\sigma_{rms} = \sqrt{\sigma_x^2 + \sigma_y^2}$$

where $1\ \sigma_{rms}$ = .632 probability of success
$2\ \sigma_{rms}$ = .982 probability of success

2. NAV I (port to scene)

   a. Systems available with long-range capability

      1) Celestial-optical

      2) Radio-celestial

      3) Omega

      4) LORAN C

      5) Transit satellite

      6) Angle-measuring satellite

      7) Range-measuring system

      8) Doppler

      9) Inertial

   b. Selection process

      A matrix was used listing components and component constraints with failures to meet constraints marked. In this manner, all component systems which do not fulfill constraints, the remaining components are examined more carefully and final selection is made by subsystem manager.

3. Final selection of NAV I

   Transit satellite

NAV I

CONSTRAINTS

Component	State 9 Sea Operation	All Weather Capability	Passive at Receiver	No Foreign Ground Stations	Minimum Training & Main-tenance	Shipboard Main-tenance	Min Down Time	Ambient Conditions	Accuracy	Wt & Vol Consider-ations
Celestial Opt.		X								
Celestial Radio									X	
Omega				X						
Loran C				X						
Transit Sat.										
Angle-Meas. Sat.										X
Range-Meas. Sat.			X							
Doppler									X	
Inertial					X					

Figure 6-4

Celestial Optical Navigation System

Current status:	Operational
Capability:	Provide positional fix and azimuth reference
Accuracy:	4.0 to -.0 N.M. 1 σ_{rms}
Range:	Optical line of sight
Coverage:	Global
Availability:	Continuous
User/Type:	Unlimited; ships, a/c, and ground stations
Active/Passive:	Entirely passive
Weather cap:	Limited by cloud coverage and fog at low altitude
Weight/Volume:	5-75 lbs. 2 ft^3
Major Advantages:	1. No shore facilities
	2. Reference reliability ~ 100%
	3. Simplicity of sight reduction
	4. Low cost
Major Limitations:	1. Cloud cover, fog, and precipitation Obscure references at low altitude
	2. Running fix accuracy dependent on accuracy of D. R. system

Radio-Celestial Reference System

Current status:	Experimental
Capability:	Provide geographical fix and azimuth reference
Accuracy:	Current operational 21.0 N.M. (1 σ_{rms}) Future 2.3 N.M. (1 σ_{rms})
Range:	Radio line of sight
Coverage:	Global (except at high latitudes, 80°-90°)
Availability:	Poor at high latitudes; inadequate 45 N.S. in winter
User/Type:	Unlimited: ships, aircraft (limited), and ground stations
Active/Passive:	Entirely passive
All Weather Capability:	Yes in 8.7 M.M. to 1.9 C.M. wavelength band
Weight/Volume:	Weight 225-300 lb. Vol. 15 to 20 ft^3
Major Advantages:	1. Requires no supporting shore facilities 2. Reference reliability approximately 100% 3. Comparative simplicity of sight reduction and plotting
Major Limitations:	1. Dependent on accurate determination of the local vertical 2. Availability of two-source positions data very limited 3. Limited applications to s/c 4. Running fix accuracy dependent on accuracy of D.R. system
Special Requirements:	1. Accurate vertical reference system 2. For running fixes, very accurate dead-reckoning system

<u>Omega Navigation System</u>

Current Status	Operational
Accuracy:	Daytime = .5 N.M. 1 σ_{rms} Nighttime = 1.0 to 1.2 N.M. 1 σ_{rms}
Range:	Each station 5000 to 6000 N.M.
Coverage:	Global with eight stations
Availability:	Continuous
User/Type:	Unlimited; ships, submersibles, S/C
Active/Passive:	Passive at receiver
All Weather Capability:	Yes (ionospheric disturbances affect accuracy)
Weight/Volume:	Shipboard weight 75 lbs. Volume 1.5 ft^3
Major Advantages:	1. Provides worldwide coverage 2. Position data available to submerged research vehicles 3. High relative position accuracy (2.3 N.M. possible)
Major Limitations:	1. Special charts are required to plot lines of position 2. Anomalous propagation degrades system accuracy 3. Without instantaneous readout counter; loss of signal or SID can result in improper lane count 4. Requires fixed ground stations on foreign soil

LORAN C Navigation System

Current Status:	Operational
Capability:	Provides a position fix (hyperbolic coordinates associated computer can transform data to geographical reference)
Accuracy:	Sq. stations configuration ~ 70 ft 1 σ_{rms}
Range:	Groundwave range varies from 1000 to 1500 nautical miles
Coverage:	North Atlantic, Mediterranean Sea, Norwegian Sea, Bering Sea, Hawaii
Availability:	Continuous within limits of groundwave range
User/Type:	Unlimited; ships, aircraft, ground stations
Active/Passive:	Passive at the receiver
All Weather Capability:	Yes; major environmental factor is atmospheric noise
Weight/Volume:	Weight 75-100 lbs. Volume 1.75 ft^2
Major Advantages:	1. Position fix is independent of any assumption concerning the direction of the local vertical 2. Currently operational in U.S., European and Hawaiian areas. 3. Standard deviations of propagation well established
Major Limitations:	1. Requires fixed transmitter stations located on foreign soil 2. Special hyperbolic charts required for plotting LOP's. 3. Cannot be used by research submersibles without surfacing

TRANSIT Satellite Navigation System

Current Status:	Operational
Capability:	Provides geographical fix; aximuth can be determined by interferometer techniques
Accuracy:	~0.3 N.M. 1 σ_{rms} (high accuracy equipment) ~1.0 N.M. 1 σ_{rms} (single frequency)
Range:	Radio line of sight from satellites at 600 miles altitude
Coverage:	Global with best coverage at high latitudes
Availability:	Average, every 110 minutes
User/Type:	Unlimited: ships, ground stations, some aircraft
Active/Passive:	Passive at receiver
All Weather Capability:	Yes, ionospheric refraction only
Weight/Volume:	Wt. 100-125 lbs; volume 3-5 cubic feet
Major Advantages	1. Requires no transmitters on foreign soil 2. Provides worldwide reference system for other systems 3. Position fix is independent of any assumption concerning the direction of the local vertical
Major Limitations:	1. Reliability of satellite equipment must be very high 2. Any loss of radio contact invalidates position data 3. Position data not continuously available 4. Accuracy decreases during 12-hour intervals 5. User equipment is bulky and heavy

Angle-Measuring Satellite Navigation System

Current Status:	Proposal only
Capability:	Would provide geographical position and azimuth references
Accuracy:	1.0 N.M. 1 σ_{rms}
Range:	Radio line of sight from satellites orbiting at 4000 miles altitude
Coverage:	Global with best coverage at higher latitudes
Availability:	96% continuous; average gaps of 28 minutes
User/Type:	Unlimited; ships, aircraft (limited), ground stations
Active/Passive:	Passive at receiver
All Weather Capability:	Yes, for radio frequencies between 16 KMC and 35 KMC
Weight/Volume:	Weight 225-300 lbs; volume 15-20 cubic feet (excluding vertical reference)
Major Advantages:	1. Would require no fixed transmitting stations 2. Backup would be provided by the sun 3. No immediate requirement for computation and distribution of satellite 4. Compilation of position fix is simple and well understood
Major Limitations:	1. Accuracy of position fix would be dependent on accurate determination of the local vertical 2. Accuracy of predicted satellite positions would decrease during prediction intervals 3. User equipment is bulky and heavy

Range-Measuring Satellite Navigation System

Current Status:	Proposal only
Capability:	Would provide geographic fix, also have capability of displaying user's position to other agencies
Accuracy:	~1.0 N.M. 2 σ_{rms} (moderate accuracy mode) ~2.1 N.M. 2 σ_{rms} (high accuracy)
Range:	Radio line of sight from satellites at 5600 miles altitude
Coverage:	Global except at latitudes > 62° N.S.
Availability:	Virtually continuous gaps of 30-40 minutes in worst cases
User-Type:	Unlimited; ships, aircraft, unmanned buoys and ground stations
Active/Passive	Active except in passive hyperbolic mode
All Weather Capability:	Yes, ionospheric and tropospheric refraction errors < 100 feet
Weight/Volume	Weight 35-40 lbs; volume ~1.5 cubic feet
Major Advantages:	1. Position fix independent of any assumption about direction of local vertical 2. User position could be displayed at other stations 3. User relieved of navigation function and complex equipment 4. Could provide worldwide reference and calibration system
Major Limitations:	1. Could not be used by research submersibles without surfacing 2. Stations required on foreign soil 3. Satellite reliability requirements would be stringent 4. User is active except in hyperbolic mode

Doppler Navigational System

Current Status:	Operational
Capability:	Provide DR navigational data
Accuracy:	Ships and submarines ~0.1% of distance traveled 1 σ_{rms}
Range:	Unlimited with decreasing accuracy
Coverage:	Aircraft global, ships limited by bottom terrain and depth
Availability:	Continuous
User/Type:	Doppler sonar ships, submersibles Doppler radar aircraft
Active/Passive:	Active
All Weather Capability:	Yes
Weight/Volume:	Weight ~100 lbs; volume 2.5 to 3 cubic feet
Major Advantages:	1. Completely self-contained 2. Can provide precise value inputs to a north-seeking gyrocompass 3. Provide a means of measuring winds and currents
Major Limitations:	1. Accuracy decreases as function of distance traveled 2. Ships and submersibles: accuracy degraded by rugged bottom terrain; inadequate depth control; and depth capability of Doppler sensor

3. NAV II (scene to site- and station-keeping)

   a. Systems available with medium-range capability:

SHORAN	DECCA NAVIGATOR
EPI	SURVEY DECCA
Two-range DECCA	DECCA MINIFIX
DECCA LAMBDA	DECCA SEAFIX
DECCA HIFIX	LORAC A
RAYFLEX	LORAC B
AERIS II	RANA
SHIRAN	DOPPLER
MORAN SURVEY	RARIE
HYDRODIST	RAFOS
DM RAYDIST	MARS
LORAN A	STAR
LORAN B	

   b. Selection

   The majority of the above systems violate the constraint of no shore station. There are, however, several of the above which have buoy-mounted stations and therefore are not disqualified.

DECCA SEAFIX	(Hyperbolic)
RAYFLEX	(Electromagnetic ranging)
MARS	(Acoustic)
STAR	(Acoustic)

   c. Considerations

   There now consists two systems, surface and subsurface which have the following disadvantages:

Surface	Subsurface
Buoy mooring and placement	Environmental interference
Buoy position maintenance	Shipborne noise interference
Electromagnetic interference	Placement
Environmental interference	

NAV II

Figure 6-5

CONSTRAINTS

Component	State 6 Sea Operation	All Weather Capability	No For'n Ground Stations	Minimum Training & Main-tenance	Shipb'd Main-tenance	Min Down Time	Ambient Conditions	Accuracy	Wt & Vol Consider-ations	Range
Shoran			X							
EPI			X							
Decca (2 Range)			X							
Decca Lambda			X							
Decca HiFix			X							
RayFlex										
Aeris II			X							
Shiran			X							
Moran Survey			X							
Hydrodist			X							
DM Raydist			X							
Loran A			X					X		

NAV II (Continued)

CONSTRAINTS

Component	State 6 Sea Operation	All Weather Capability	No For'n Ground Stations	Minimum Training & Maintenance	Shipb/d Maintenance	Min Down Time	Ambient Conditions	Accuracy	Wt & Vol Considerations	Range	
Loran B			X					X			
Decca Navigator			X								
Decca Survey			X								
Decca Minifix			X								
Decca Sea Fix											
Lorac-A			X								
Lorac-B			X								
Rana			X								
Doppler							X	X			
Rarie								X			
Rafos			X								
Mars											
Star											

Electromagnetic Ranging Systems

System	Range	Accuracy Optimum	Maximum Range	No. of Users	Shore Sites
SHORAN	30-40	30'	50'	4 to 6	2
ELECTRON POSITION INDICATOR	400	195'	1520'	1 or 2	2
TWO-RANGE DECCA	150	30'	~50'	1 or 2	2
DECCA Lambda	425	25'	250'	1 or 2	2
DECCA Hi-fix	30	12'	100'	1 or 2	2
RAYFLEX	40	~2'		1	2
Cubic AERIS II	54	~2'	~6'	1	2
Cubic SHIRAN	450	10'		1	4
MORAN Survey	15-30	20'	50'	1	2
HYDRODIST	25	12'	100'	1	2
DM RAYDIST	200	12'	100'	2	2

Medium-Range Hyperbolic Navigation Systems

System	Range	Accuracy Optimum	Maximum Range	No. of Users	Shore Sites
LORAN A (standard)	750	0.25 n.m.	5 n.m.	Unlimited	3
LORAN B	250	45'	300'	Unlimited	3
DECCA Navigator	250	1.25 n.m.	2 n.m.	Unlimited	3
DECCA Survey	200	25'	300'	Unlimited	3
DECCA Hi-fix	40	25'	300'	Unlimited	3
DECTRA	2,000	~5 mi. ($2=\sigma_{rms}$)	~10 mi. ($2=\sigma_{rms}$)	Unlimited	3
DECCA Minifix	25	25'	300'	Unlimited	3
DECCA Seafix	25	25'	300'	Unlimited	3
LORAC A	200	15'	400'	Unlimited	3
LORAC B	300	15'	400'	Unlimited	4
RANA	~100	~50'	~100'	Unlimited	3

Underwater Acoustic Systems

System	Range	Accuracy Optimum	Accuracy Maximum Range	No. of Users	Shore Sites
SOFAR	~3,000	Classified	Classified	Unlimited	
RARIE	60-100	0.1 n.m.	Not established	1	
RAFOS	Classified	Classified		1	2-4
MARS	5-10	0.1% of range + 20'		Multi	
STAR	15	0.1% of range + 20'		Multi	

d. Final selection

In view of the fact that the submersible will require a NAV system, a composite is selected:

RAYFLEX on mooring buoys with transponders as anchors;

MARS system to utilize the transponders in buoys.

4. NAV III (submarine guidance)

a. The following can be considered for submarine guidance:

STAR

MARS

DOPPLER

Inertial

b. Selection

As before, except that in this case none of the component systems considered fulfilled requirements.

Further evaluation required:

Inertial	Dependent on initial position; must be used with additional system input for accuracy over long distances
DOPPLER	Dependent on initial position input and ambient ocean conditions
MARS	Dependent on ambient ocean conditions 16 KC frequency
STAR	Dependent on ambient ocean conditions 1.6 - 2.1 KC

c. Selection of NAV III system

A composite system was selected due to the fact that one system cannot provide the desired probability of sucess.

Composite system

MARS with an input to a backup inertial system.

Figure 6-6

CONSTRAINTS

NAV III

Component	State 6 Sea Operation	Subsurface Operation	No Foreign Ground Station	Take-Out Maintenance Capability	Min Training for Operation	Ambient Conditions	Accuracy	Range	
Inertial							X		
Doppler						X	X		
Star						X			
Mars						X			

E. Summary

NAV I	TRANSITE Satellite
NAV II	RAYFLEX and MARS
NAV III	MARS and INERTIAL

CHAPTER VII

ENVIRONMENTAL SENSING AND CONTROL SUBSYSTEM

EPIGRAMS

Most of the world's environment is uncontrolled. Every system eventually interfaces with the uncontrolled environment. The best way to regard the environment is as a subsystem whose manager will not listen to reason and will set his own constraints on the interfaces.

The environment is not an infinite sink.

Neglect of the environment can result in sudden catastrophe; neglect of the environment can result in slow but sure catastrophe.

The understood environment is a valuable system component. The understood and controlled environment can be a valuable subsystem.

The environment is physical, chemical, biological, and psychological.

General Purpose

The general purpose of an environmental sensing and control subsystem is to provide information to the rest of the system concerning the environment in order that the system may modify its relationship with the environment, if necessary, in order to achieve the system's objectives. The system modifications may be required in real time or what might be termed in operational time, or they may be long-term considerations requiring system adaptation at periodic intervals which are not related to specific mission operations. Ship routing to optimize weather conditions is an example of the operational type of adaptation, whereas harbor dredging or the construction of breakwaters and jetties are examples of periodic systems modifications which may be required to adapt the system to the environment. Systems managers have frequently neglected to include environmental sensing and control as a separate and distinct subsystem, or they have included it without being truly aware that such a subsystem has been identified. For example, in antisubmarine warfare systems, the sonar systems which are used for detection purposes are nothing more than environmental sensors and any sonar development group is no more and no less than an environmental sensing group.

The very recent concern on the part of the society with problems of pollution stems from the fact that many long-term interactions between systems and the environment have not been appropriately considered by the systems designer and developer and a very considerable economic and social penalty is now being paid for this lack of consideration. That the environment must be treated in future systems design is quite obvious, but many would differ on the manner in which it should be treated. For example, one of the students assigned environmental sensing and control as a special area of investigation for his project insisted, not without cause, that each subsystem manager should have responsibility for all his environmental interactions and the need for control of the environment which might stem therefrom. It is the author's view, however, that such a viewpoint can be taken only to the extent that the environment can be treated by the subsystem manager as an infinite sink and that he can be assured his interactions with the environment will not, in turn, change the environment in a way which will upset the anticipations of other subsystem designers. A few examples in this regard are illustrative:

a. The designer of a Doppler sonar presumes that the velocity of sound in the medium will stay constant during the time of transit of the emitted acoustic pulse to the bottom and its return to the receiver.

This will require that bubbles of air not be entrained in the water and pass over the face of his sonar dome during the measurement process. The designer of the ship's hull is highly desirous of maintaining optimum hydrodynamic shape and this will result in the entrainment of air in the bow wave and the sweeping of this air under the hull of the ship well within the vicinity in which any contemplated sonar dome might be operated. Without an environmental manager who will have established the functional requirements of the environment, such a problem might not be exposed until too late.

b. The designer of the classification object identification subsystem of an ocean search system desires to maintain optical clarity in the water along the optical path of his sensors. The designer of the propulsion system and propulsion control system desires to have the most efficient configuration of control jets in order to minimize power and optimize control ability. This configuration may turn out to be one which will stir the silt or create a great deal of turbulence within the optical path. Again, a manager, who is concerned entirely with the environment and its measurement and control, will have been able to identify this problem at an early stage and have aided in destroying environmental interfaces.

Some feeling for the type and variety of environmental parameters which might need to be measured or controlled in a system is highly desirable and therefore a brief listing of categorization is presented here. Major physical phenomena include the sea surface and wave spectra, the wind velocities and spectra, the temperature, the pressure, the internal wave structure, the acoustic spectrum, the optical clarity, the electromagnetic environment, the physical characteristics of the bottom and bottom sediments, etc. The major chemical parameters are the salinity, the dissolved salts, the dissolved oxygen, the free oxygen, dissolved or other free gases, the precipitated minerals, etc. The major biological phenomena are the crustacea and mollusks, the pelagic fish, the demersal fish, the sea mammals, the euphasides, plankton, phytoplankton, and algae, the scattering layer, the bioluminescent organisms, the sea fauna, barnacles, worms, and biological fouling organisms, etc. The major psychological phenomena are odor, taste, optical distortions, both above and below the free surface, humidity, which affect the human perception of orientation, etc. Many hundreds of possible parameters for measurement can be enumerated and for each there will be a temporal and spatial distribution. The systems manager must concern himself with a determination of those parameters and the time scale thereof which will in fact affect his systems operation. Figure 7-1 shows a plot of some 18 major ocean phenomena as a function of the wavelength and the frequency in cycles per year. This figure, which is taken from reference (8) is one of a series of figures which have been drawn for specific major oceanic parameters. Figures 7-2, 7-3 and 7-4 are examples for salinity, transparency, and wind.

A careful review of these factors and their temporal and spatial distribution in the areas in which the system will operate will permit determination of the uncontrolled environment. An assessment should then be made of the ways in which the environment can help accomplish the mission and those parameters which must be sensed in order to so utilize the environment. This will then determine the functional characteristics

1 = Large-Scale Wind-Driven Gyres
3 = Surface Current Meanders
4 = Internal Current Meanders
5 = Surface Topographic Eddies
6 = Submarine Topographic Eddies
7 = Semi-Diurnal, Diurnal, Semi-Annual Tides (as indicated)
8 = Inertial Currents
9 = Internal Waves
10 = Small-Scale Convective Eddies
11 = Surface Gravity Waves
12 = Rossby Waves
13 = Deep Scattering Layer
14 = Plankton and/or Algai Bloom
15 = Thermocline
16 = Turbidity
17 = Surges
18 = Tsunami

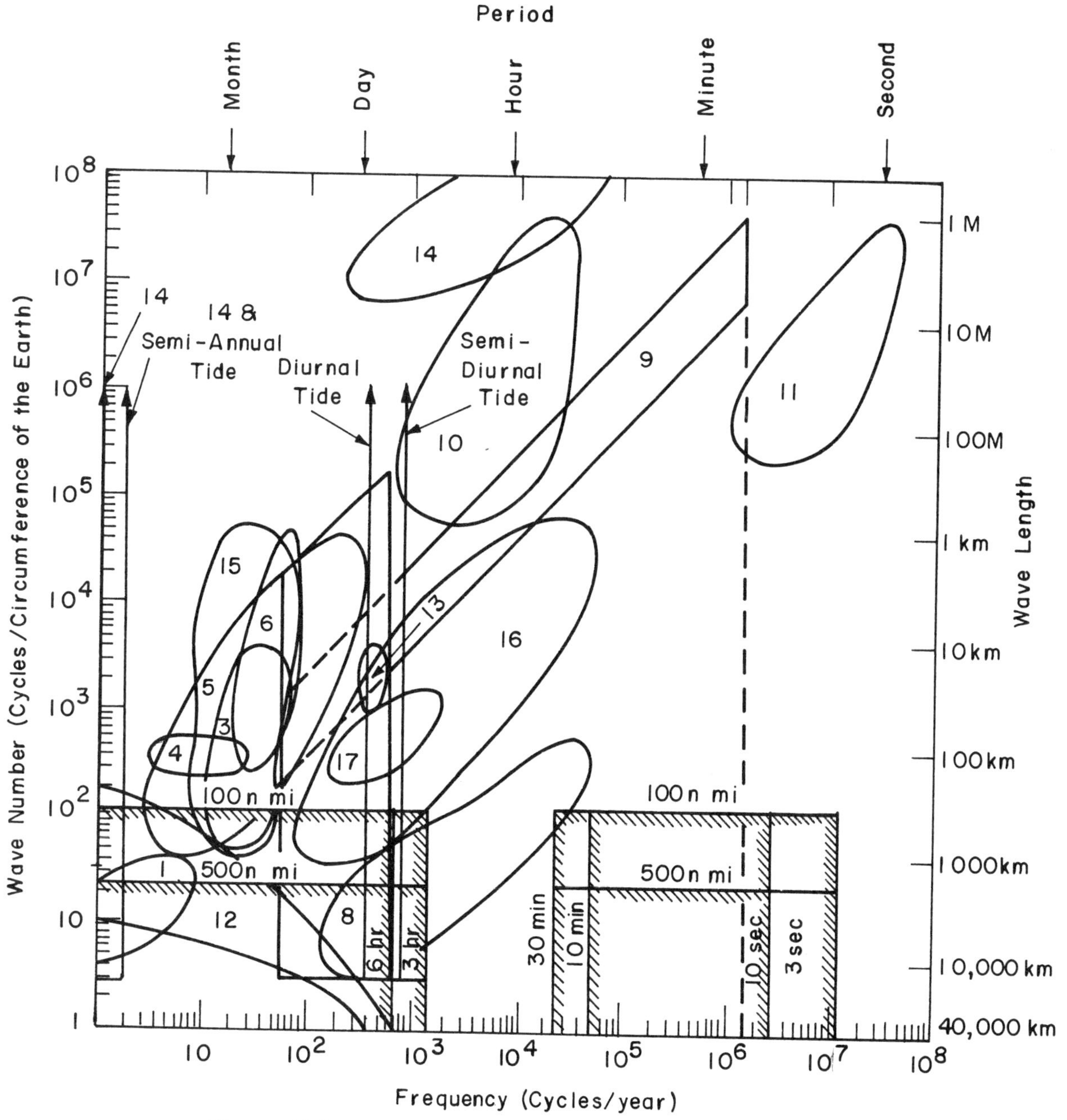

Figure 7-1. Major Oceanic Phenomena

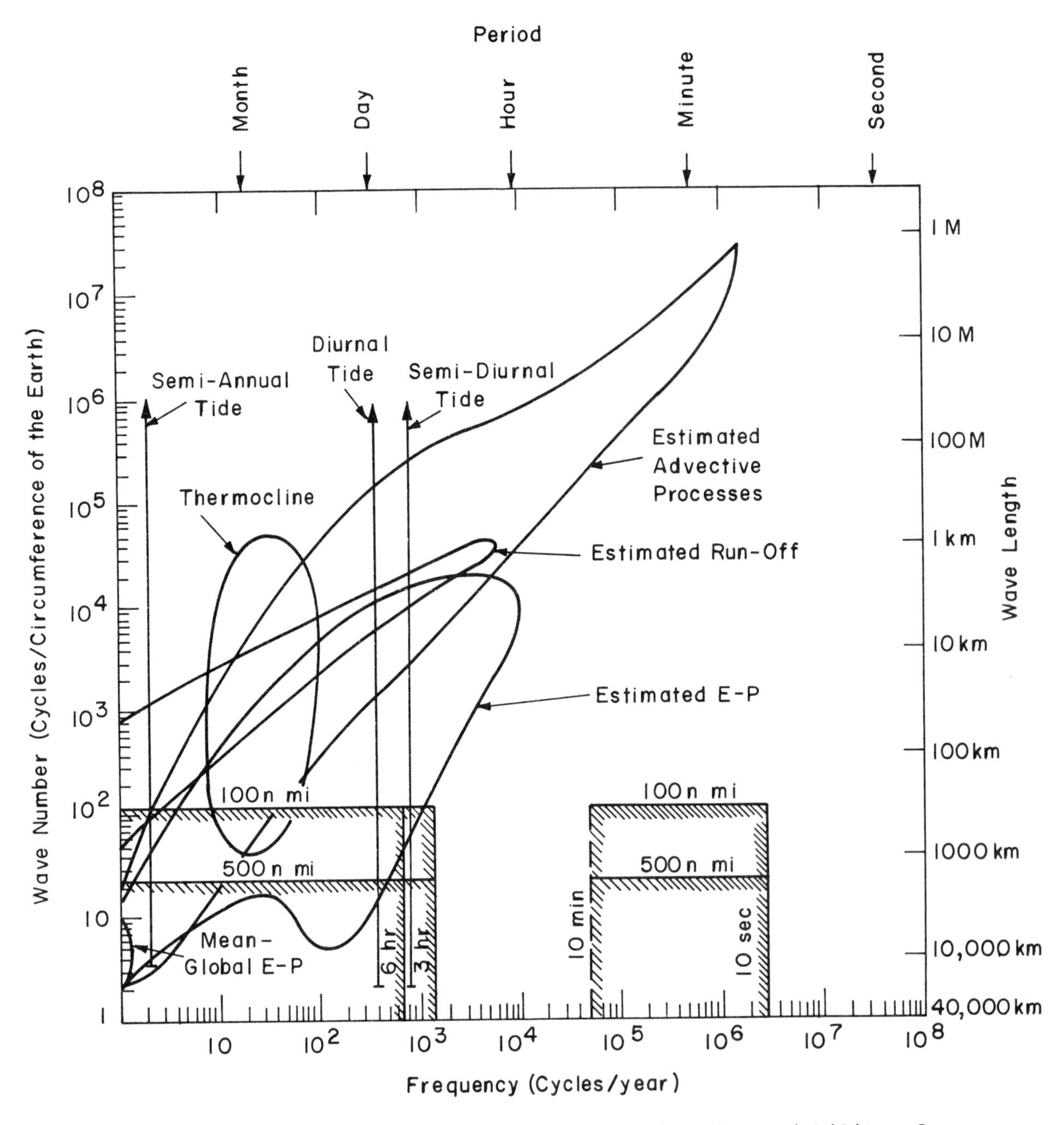

Figure 7-2. Phenomena and Processes Affecting the Variability of Salinity

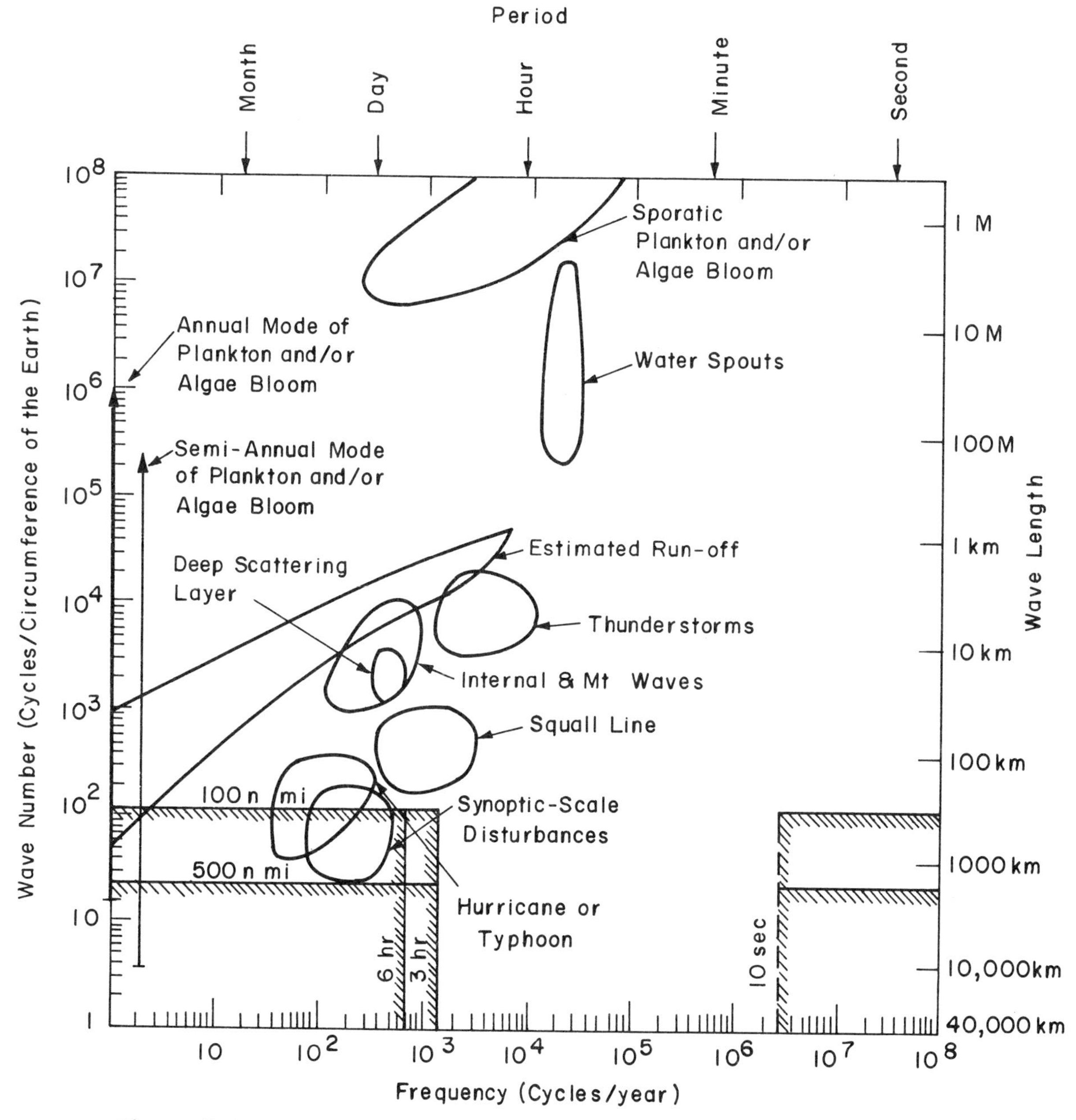

Figure 7-3. Phenomena and Processes Affecting the Variability of Transparency

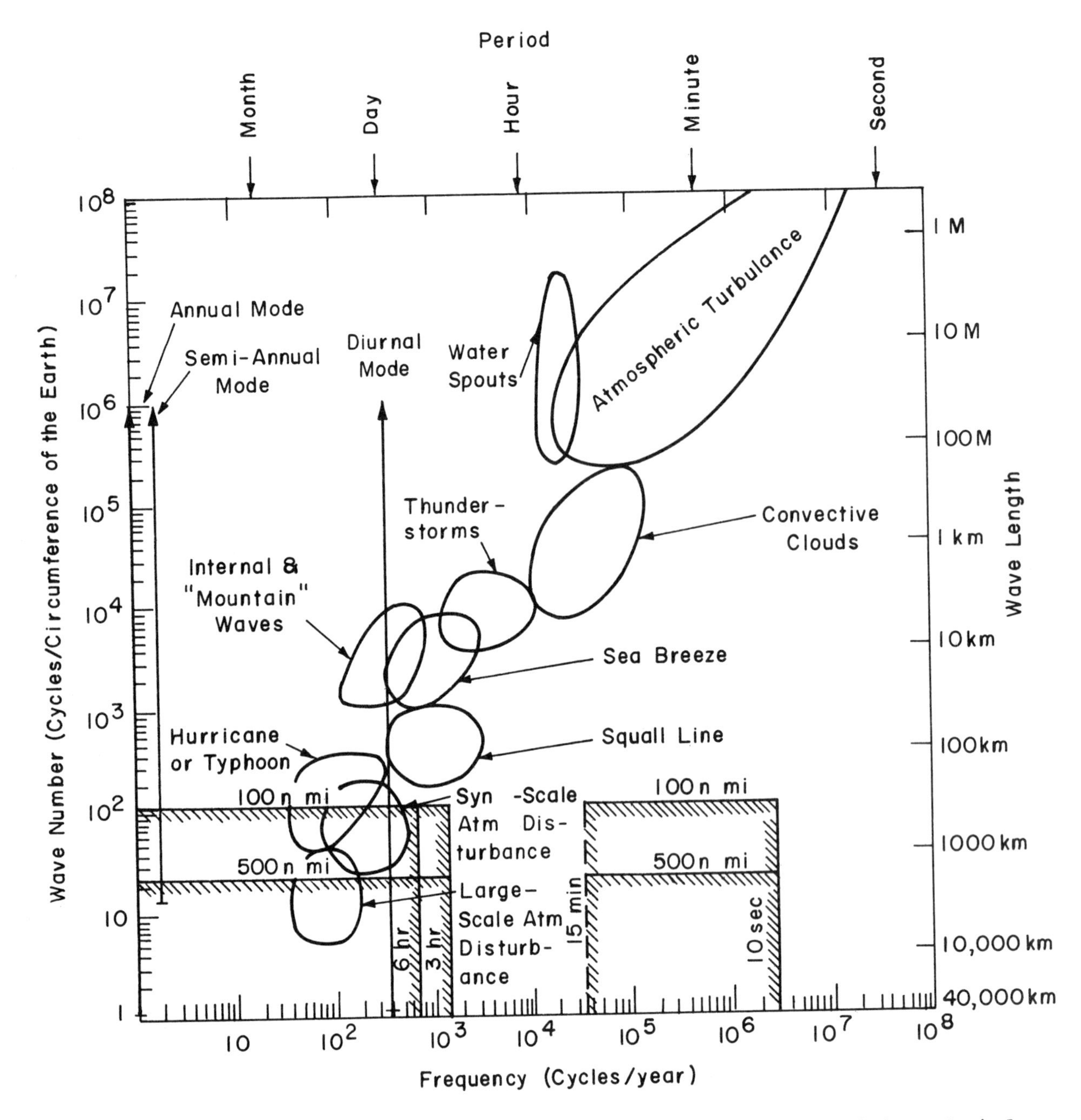

Figure 7-4. Phenomena and Processes Affecting the Variability of Wind

of the environment as a subsystem. At the same time an assessment must be made of the ways in which the environment will impede the accomplishment of the mission and a determination as to whether this impediment can be overcome most easily by an environmental control subsystem. If such is determined to be the case, then this will set the function of the environmental control subsystem. Similar reviews should be made of the interfaces, noting again that this situation is unique in that the environmental subsystem will impose interface requirements which the system manager may be powerless to alter, and to that extent the environmental subsystem should be conceived of as a system whose manager will not listen to reason.

Within these philosophic constraints, the environmental sensing and control subsystem can be handled in a manner very similar to that of other subsystems and, in particular, in a manner similar to that of communication and navigation are a form of environmental sensing. The restraints on measurement of the environment for those functional requirements will be the same as those for environmental sensing systems and the illustrative charts and graphs which were prepared for them. Those chapters are therefore directly applicable here.

In order to protect a shoreline, it is necessary to quantify the forces which act upon it, the materials of which it is composed, and to know how these materials react to various forces. With this knowledge, structures or programs may be implemented to modify the environment either by attenuating or augmenting the forces, or by modifying the materials, or both. All of these functions are part of the environmental sensing and control subsystem, although this subsystem will further be broken up into other subsystems.

Two problems which pose major threats to the success of the system are:

1. Because the shoreline will suffer most where the natural forces are greatest (the surf zone), it is imperative that this region be well instrumented and quantified. But accuracy will have to be compromised for ruggedness if the instruments are to survive in this region, unless measurements can be made from safe vantage points. The problem of instrumenting the surf zone is therefore crucial.

2. If predications are to be made concerning the shoreline configuration, care must be taken to be able to handle catastrophic storms which will appear as discontinuities in the time record of shoreline topography.

Each of these two problems belongs to a class of problems which frequently occur; the first problem is that of adapting the instruments to the environment in which they will operate, and the second is getting the instruments to give relevant information that can be used to control the environment or to serve as input for another subsystem.

CASE STUDY

SHORE PROTECTION SYSTEM--ENVIRONMENTAL SENSING AND CONTROL

"Shore Protection System" was written by John R. Mittleman and, at the time it was written, Mr. Mittleman's background in Coastal Engineering was purely academic. However, his experience in certain phases of instrumentation, particularly in materials analysis and in photogrammetry, is strongly supported by field work. In 1967 Mr. Mittleman worked for the Nassau County Museum of Natural History at Garvis Point, Long Island, researching the sedimentary geology of Long Island's north shore, and in 1969 he developed a methodology for the use of terrestrial stereometric photography in hydraulics research. Therefore, those parts of this case study which relate to the acquisition of data will be the strongest and the most detailed, and the focus of other parts will be shifted away from detailing hardware and toward resolving system problems.

MISSION ANALYSIS

Shore Protection: Preservation of the existing shorelines which are subject to destruction by noncatastrophic forces (i.e., waves, wind, littoral transport, etc.) and by human influence (i.e., stripping of vegetation, building projects, "shore protection" projects, etc.).

I would like to consider this systems problem, the subjective human purpose of which is to keep the shoreline where it is, from the beginning (related fluid mechanics research into the causes of shoreline destruction) through to the logical end, which may be at least an evaluation of the practical methods of coping with the destructive forces at hand.

The following quantification must take place before the broad outlines of the shore protection system can be specified:

I. Information

a. The objective which requires that information be obtained.

b. The required success probability and confidence limits in meeting the objective.

c. The temporal requirements for the synthesized information.

d. The parameters which must be measured within the present state of the art in order to synthesize information required for the objective.

e. The precision and accuracy with which these parameters must be measured.

f. The spectral and temporal location of the phenomena which are to be measured, and the spectral and temporal incentives which are permissible in parametric measurements.

g. Constraints on measurement.

h. Obstacles to measurement.

i. The requirements on information processing, within the limits of accuracy, precision, and specification of temporal nature.

j. The requirements on information processing imposed by the synthesis.

II. Interdiction and Diversion

a. The quantity and type of object to be handled.

b. The source of these objects or material, the precision of location, and the specifics of location, if this interdiction is to take place at the source area.

c. Requirements which must be met in order to interdict or divert at the source area.

d. Obstacles to interdiction or diversion at the source.

e. Time limits on location and interdiction.

f. Modes of transportation of the material away from the source.

g. The time of dissemination and transport.

h. Requirements which must be met to interdict or divert during the transport phase.

i. Constraints on interdiction during transport phase.

j. Obstacles to interdiction during transport phase.

k. Time limits on interdiction during transport phase.

l. Quantity and types of material in the final delivery.

m. Location of final delivery, precision of location, and time of delivery.

n. Requirements which must be met to interdict delivery.

o. Constraints on interdiction of delivery.

p. Obstacles to interdiction of delivery.

q. Time limits on interdiction of delivery.

r. Probability of success, and confidence level of success.

III. Transportation

a. Quantity and type of material to be collected.

b. Location of such material, and the precision of location, and the type of location which is required before retrieval is possible.

c. Constraints on retrieval.

d. Obstacles to retrieval.

e. Time limits on retrieval.

f. Distance of transport.

g. Time constraints.

h. Obstacles to transportation.

i. Required intermediate stops.

j. Quantity and type of object or material to be delivered.

k. Location of delivery, and precision of location.

l. Constraints on delivery.

m. Obstacles to delivery.

n. Time limits on delivery.

o. Probability of successful delivery, and confidence limits.

Having thus decided that the shore protection system is a hybrid system, I will attack it in pieces. The first will be instrumentation, information-gathering and surveillance, this being necessary before one could possibly decide how to protect the shoreline, or if indeed it needed any protection at all. The second piece will be a simultaneous consideration of solutions to the problem of shoreline destruction; that is, transportation and interdiction will both be considered separately, but at the same time. Transportation as a solution will involve such things as beach nourishment, whereas interdiction will treat such things as jetties and groin systems.

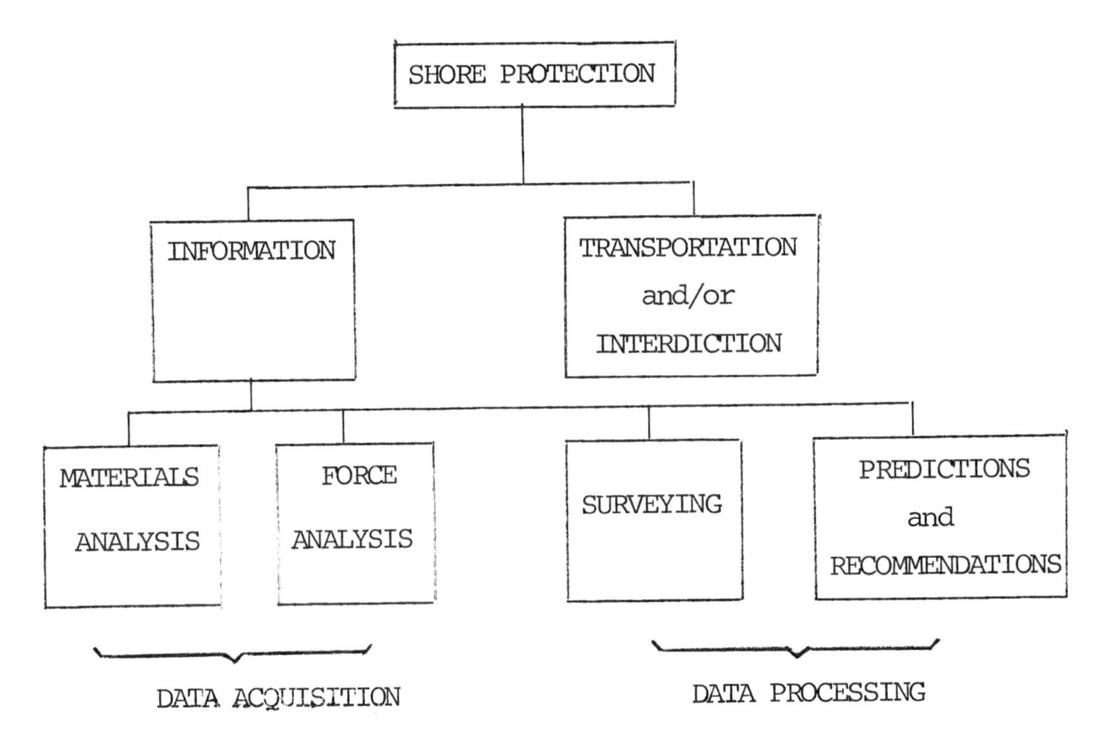

Figure 7-5

I. Quantification of the Information Subsystem

a) The information to be obtained will indicate the nature and extent of damage to the shoreline to be protected so that a decision may be made as to the necessity for action at that location.

This statement defines the nature of the output of the information subsystem, treating it at the interface between the information subsystem and the transportation subsystem. Later this interface will become a decision-making body, as implicitly stated, and the rest of system will get little more than an "Action/No Action" decision.

b) This decision will not specify the extent or the nature of the action called for, except in terms of recommendations to the subsystem responsible for the shoreline's repair or protection. What the information subsystem will furnish to the other subsystems is an "Action/No Action" decision and, in the event of an "Action" decision, all of the processed data concerning the forces and materials in play at the location to be treated. The decision as to whether or not action is needed is to have a 75% probability of being correct 50% of the time and, in view of these odds, the same buffer zone is to be considered at least three times before it would be destroyed, based on the information used in making the "Action/No Action" decision.

Separating out the decision-making body, the criteria used in deciding are defined here. Data will be acquired and processed, predictions will be made and, using these predictions, the "Action/No Action" decision will be made. To the rest of the system, the information subsystem will be almost like a black box in an "on" or "off" condition.

The difference between the information subsystem and an "on" or "off" switch is that data from anywhere within the box may be conjured up as it is needed for purposes other than setting the "on"/"off" switch.

c) Each shoreline sector will pass before the decision-making body in turn and, at each pass, an "Action/No Action" directive will be issued along with the allowable time limits for the repair of the sector.

d) The parameters which will be important in making the "Action/No Action" decision are:

1) The size of the buffer zone between the sea and the land which is to be protected from destruction by the sea.
2) The rate of destruction of the buffer zone (accretion or depletion of material from the buffer zone).
3) The nature of the material transported to or carried from the buffer zone under construction.
4) The nature and strength of the forces responsible for the buffer zone's destruction.
5) Temporal variations of I.d.3 and I.d.4.

These natural phenomena will be quantified, and will act as a forcing function on the "Action/No Action" decision. At this point it might be useful to schematize the information subsystem as follows, for purposes of visualization:

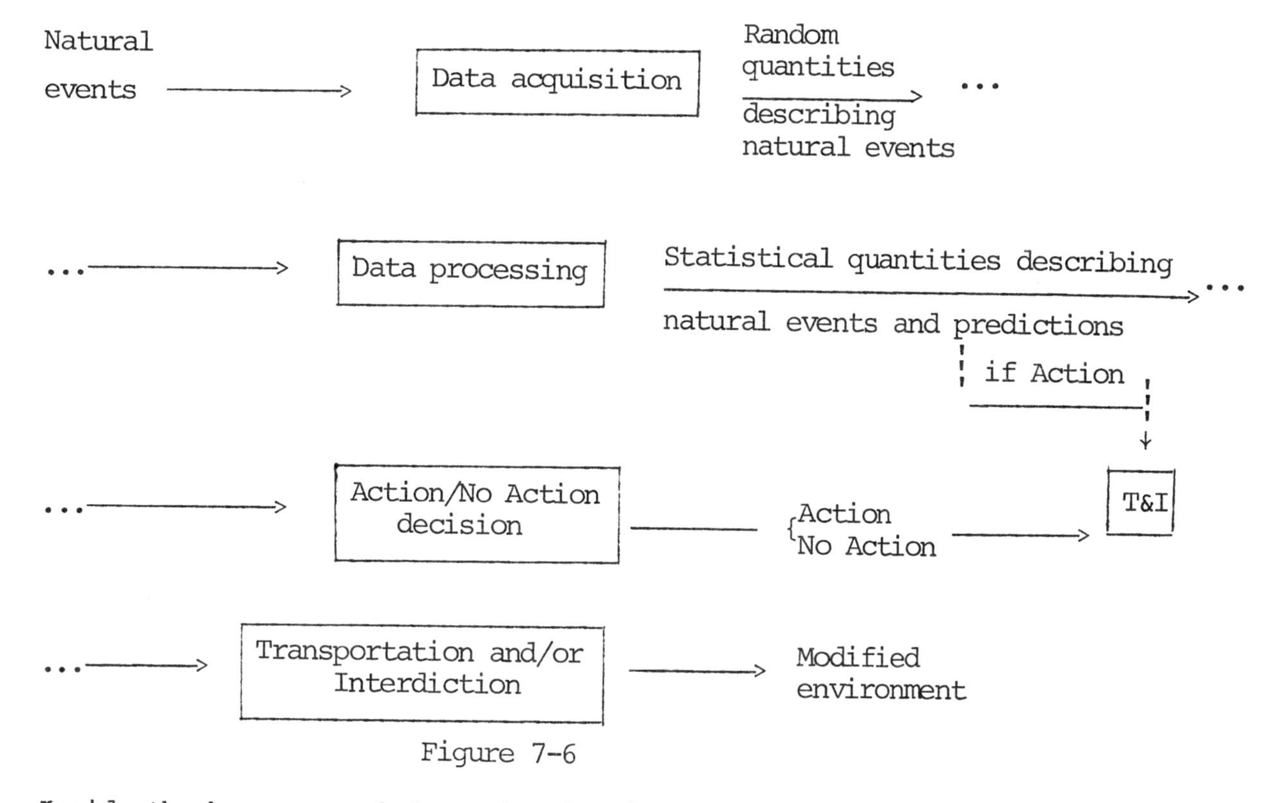

Figure 7-6

Inside the boxes are what may be thought of as transfer functions, and outside the boxes response functions which are, in turn, forcing functions.

e) The precision and accuracy with which each of these parameters is to be measured will vary with the individual location and with the method used to measure each parameter. Whenever possible and feasible, cross-checks should be made by different measuring techniques.

1) The buffer zone size must be known to within 10 feet for purposes of making the "Action/No Action" decision and may be found more closely on request.

2) The rate of destruction of the buffer zone must be known to within 0.1 of its width per year.

3) The nature of the materials must be known statistically, with not more than 20% error in the volume of material

involved in the net littoral transport, with not more than 10% in error in the contribution of each size unit of material.

4. The strength of the forces involved in the destruction of a buffer zone must be known to within 10% of their peak value.
5. The temporal variations of material size and force strength must be known to within 20% of the average value of each, respectively.

These baselines for hardware performance help define the data acquisition transfer function. They constitute an interface constraint placed on acquisition by data processing, and in turn by the overall system's reliability criterion and confidence level. On the other hand, the fact that the uncertainties are not smaller reflects a constraint placed on the system by the data acquisition people, and in turn by the state of the art of instrumentation.

f) The phenomena of transportation of material to or away from each location are expected to vary in a seasonal manner, without considering catastrophic events, such as hurricanes and tsunami which will be treated separately. Hence, the parameters to be measured must be measured at intervals or continuously throughout the year. The spatial extent of measurements will be at least large enough to enclose the entire buffer zone--that is, landward as far as the sea can be allowed to encroach without destroying developed areas, and seaward as far as accumulation of material would destroy developed areas.

In order to measure natural events accurately and meaningfully there must be a degree of compatibility between the time constants of natural events and those of the information subsystem. This idea, expressed above, generally finds its way into every system.

g) Constraints to measurement:
1) Reliability of data
2) Legal constraints above the high water mark
3) Time limits imposed by the seasons

h) Obstacles to measurement:
1) Weather
2) Sea state
3) Equipment failure
4) Random nature of events to be measured
5) Limited accessibility of certain shorelines

i) The processing of information must keep pace with the inflow of raw data and must furnish a basis for an "Action/No Action" decision at least three times before the buffer zone would be destroyed on the basis of information used in making the first "Action/No Action" decision, as discussed in Section I.b.

j) The synthesis of processed data is, essentially, the "Action/No Action" decision. The processed data must, therefore, present a prediction of the future state of the buffer zone, a statement of the nature of the materials and of the forces involved, and a forecasting of the effects of any proposed changes or developments on the future state of the buffer zone.

Mission Analysis from the Information Subsystem's Point of View

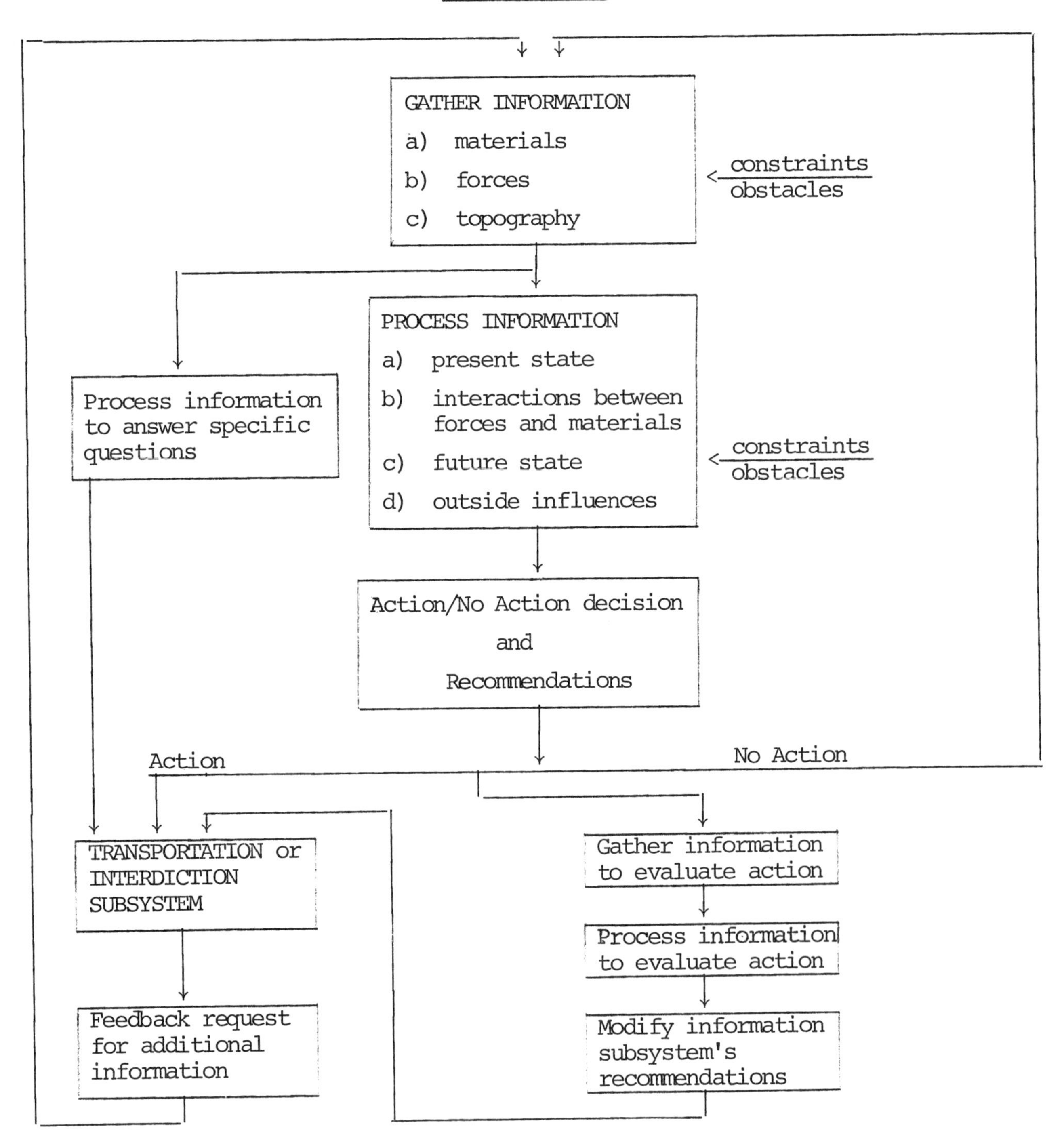

Figure 7-7

This flow chart presents the dynamic interactions between the subsystems which comprise and interface with the information subsystem. The interfaces are not detailed, but their presence is acknowledged so that, when the most rudimentary hardware choices are made, these interfaces will be considered. The division of the information subsystem into data acquisition, data processing, and the "Action/No Action" decision is based on the relevant technologies (broadly speaking, these are instrumentation, data processing, and judgment). In terms of the information budget, data acquisition and processing generate class B information, and the "Action/No Action" decision generates class A information.

II. Mission Analysis

A. Data Acquisition

1) The information to be gathered falls into three major generic groups, these being:

a) The materials involved in the processes of buffer zone destruction;

b) The forces responsible for the transport of material;

c) The topography of the buffer zone.

Having gathered the above information, one ingredient is still necessary, and that is a means of predicting the future of the buffer zone. Normally, this will be a relationship between forces, materials, topography, and transport. Ideally, one should be able to feed the former three in one end of a box representing this relationship, and come out with the future state of the buffer zone. This objective will rarely be reached analytically, but will be closely approached by empirical relationships and by accumulated experience.

2) The quantity of information to be gathered will depend heavily on the magnitude of the net mass transport to or away from each buffer zone sector. Where a fairly stable configuration is found, a winter profile and a summer profile of the sector will suffice, but where there is more activity in terms of mass transport, a more nearly continuous record will be necessary. At this point it is necessary to recognize a major systems problem which must be dealt with. That is, in the event of a catastrophic event, there will be a discontinuity in the records of buffer zone size, and the rate of change of buffer zone size cannot be assumed to be the same as it was before the event, owing to a change in the topography of the buffer zone. There will inevitably have to be a tradeoff between the cost of allowing catastrophic events to approach the buffer zone unimpeded and the cost of building protection specifically against catastrophic events.

3) The time at which information is gathered will be identified by seasons and by the sea state relative to long-term averages. In this way, a judgment can be made as to how representative the data is with respect to average conditions. Again, care must be taken to recognize that the largest amount of damage will be done during periods of above-average forces.

4) The location at which information is gathered will be identified not only topographically, but also with respect to the generic class of the buffer zone. That is, straight barrier beaches will be differentiated from coves and from headlands, etc., and sandy beaches

from rocky and pebbly, etc.

5) The constraints and obstacles to measurement are listed in Sections I.g and I.h, but at this point it is important to recognize that the major obstacle to measurement has systems importance; that obstacle is the local magnitude of the forces to be measured. The most important transport of mass will occur where the forces are strongest, probably in the breaker zone, and so it is at this location that it is most important to place instruments. The instruments placed in the surf must be extremely rugged and may, therefore, have limited accuracy. In light of this problem, it would be advantageous to be able to infer quantities such as current strength, etc., from measurements not taken in the breakers.

At this point, the two major systems problems have been defined in II.A.2 and II.A.5, but neither has been resolved.

B. Data Processing

1) The processing of information is effected in order to arrive at the following results:

a) The present state of the buffer zone, including a description of its topography and a statistical description of its constituent materials and the forces which act upon it.

b) A description of the interactions between the forces and the materials, and of the net effect of these interactions, that is, the rate of change of the buffer zone state.

c) A prediction of the future state of the buffer zone state.

d) Recommendations for preserving the shoreline and an analysis of the effects of a shore protection structure or process on the state of the buffer zone.

2) The constraints upon the data processing subsystem will stem primarily from the need to keep pace with the flow of incoming raw measurements and to be able to use the data which is provided by the data acquisition subsystem. This is essentially an interface constraint which can be expressed as the need for a functional relationship between the transport of material and the active forces.

3) The obstacles to the processing of information will stem mainly from the incompleteness of the acquired data and by the present state of the art in matters of shore protection structures.

C. The Action/No Action Decisions and Recommendations

In the Action/No Action decision there will be needed, in addition to the processed information, a great deal of judgment, for the decision will be severely constrained by such factors as:

1) The impossibility of undertaking too many projects simultaneously, due to a finite number of organizations capable of carrying out any given recommendation.

2) Priorities based on:

a) Political or civic pressure;

b) Sequencing of equipment;

c) Urgency of action.

3) The rule of thumb to be used is that action should be taken if the buffer zone will change by its own size in ten years, that is, if it will either disappear or become twice as large as it is presently

if accretion of material is undesirable.

In order for this scheme to work with the given reliability and confidence limits, as stated in Section I.b, at least three sets of processed data must be considered in ten years, thereby setting the upper limit of the time allowed for data acquisition and processing at three years. Since the forces involved are expected to have a roughly annual cycle, no set of data can be complete with less than a year of data acquisition. Hence, each shoreline sector will be considered once every one to three years. The actual amount will vary with the rate of destruction of the buffer zone so as to keep adequate track of rapidly-changing buffer zones.

By considering long-term effects and averages, the mathematical problems arising from discontinuous records are neatly solved. However, it should be apparent that such a maneuver in no way eases the effects of catastrophic events.

III. Hardware Alternatives for the Data Acquisition Subsystem

A. Materials

1) Standard sieve set separation of size groups

2) Hydrodynamic separation of size groups by settling rate

3) Chemical or thin-section analysis of mineralogy

4) Visual identification of minerals

5) Electron microscopic analysis of surface characteristics

B. Forces

1) Fixed current meters

2) Dye markers, marked particles and free buoys

3) Wave meters, partially or fully submerged

4) Doppler radar wave analysis

5) Aerial photography of wave patterns

6) Tide gauges

7) Wind meters

8) Analysis of evidence of forces

C. Topography

1) Aerial photography

2) Terrestrial photography

3) Rod and transit

4) Lead line

5) Sonar and side-looking sonar

This listing is not complete, but it will be sufficient for the purposes of the Shore Protection System. Among the important assets that it has is the avoidance of techniques which push the state of the art too much. It is much more important to get usable results consistently than to get extremely accurate results only part of the time. This all goes back to the philosophy which ranks a good system that works above a great system that doesn't work.

IIIa. Discussion of Hardware Alternatives for Data Acquisition

A.1. Standard sieve separation of size groups is an inexpensive, rapid operation in which confidence built on years of experience may be placed. Physically, the particles are separated on the basis of the smallest mesh size that they will fit through. Then the amount of material caught in each sieve is weighed and this amount is expressed as a percentage of the total weight of the sample. The three major problems associated with the interpretation of these results are:

1) due to the sorting of sediment by natural forces, it is difficult to estimate the validity of any sample in terms of how well it represents the beach as a whole; 2) the sieve sorting process does not in any way take into account the shape of the particles, making necessary a shape factor correction; 3) the sieve sorting process does not take into account the densities of the particles, thereby making necessary a correlation with standard quartz grain sizes.

Once all of the above corrections have been made, the relationship between the forces and materials can be expressed by curves similar to Hjulstrom's curves, which indicate whether the particles will be eroded, transported, or will settle out in the presence of a given current strength near the bottom.

A.2. Hydrodynamic sorting of size groups is based on the settling rates of the particles and is, in this respect, more organically tied to the problem. However, there is still the problem of getting a representative material sample and, in addition, the equipment used is far more complicated and delicate than the standard sieves. In cases where a sample can be processed in the laboratory as opposed to on the beach or at sea, hydrodynamic sorting can be used effectively. Furthermore, for a given beach, by which a given statistical size distribution is meant, there is a correspondence between the sieve distribution and the hydrodynamic distribution.

A.3. Chemical and thin-section analysis of the mineralogy of a beach's material is a fairly complicated process although it is well within the state of the art. In cases where the subtle differences in mineralogy are important for establishing the source area of a particle, then analysis of this sort may be necessary.

A.4. Visual identification of minerals is by far and away the most simple method where precise mineralogy need not be known. The hardware here is actually a trained sedimentologist. Such a specialist would be needed to interpret the results of chemical or thin-section analysis anyway, and he would also be capable of judging whether or not such analysis was necessary.

A.5. Electron microscope analysis of surface characteristics of particles is especially useful when it is necessary to know whether or not a given body of material is active in the coastal processes being considered. In cases where a source of material is needed to nourish a buffer zone, it is important to choose one which is not active in the net transport of material. Electron microscopy can identify fossilized dunes and differentiate between material which has been transported subaerially, glacially and submarine.

B.1. Current meters are an excellent way to measure currents (as contrasted with dyes or buoys, or other tracked material) because they can give an extensive record of current strength and direction as a function of time for a given location. Two major disadvantages, however, should be noted:

a) In order to establish the velocity field in a given region, the number of meters necessary increases rapidly with the desire for more accuracy. The exact location of the meter may also be of prime importance, especially when it is in close proximity to a small topographic feature. That is, the region measured may be on the order of inches, and topography of the same order of size may therefore affect the local velocity field as measured by the meter and, consequently, give the data processing subsystem a measurement irrelevant to the overall process of movement of materials. If, in order to avoid this obstacle, the meters are placed a small distance above the microtopography, then a relationship between the currents measured and the actual bottom currents will be necessary. One might at this point consider mapping the microtopography in the immediate vicinity of each meter at the time it is installed.

b) It is immediately obvious that the largest part of the process of movement of materials will occur in the roughest zones--that is, the breaker zones. It is then clear that the most important measurements will be taken in the most difficult zones from the point of view of instrumentation. It is therefore essential that a careful estimate of the destructive forces at each prospective instrument site be made and that the instrument be built to withstand these forces while operating. It might well be possible to establish a relationship between the currents in the breaker zone and the currents just outside this zone so that measurements may be taken elsewhere than in the breaker zones. This work would be carried out by the research facilities of the data processing subsystems.

Current meters are generally divided into four generic categories:

1) Resistive, in which a fixed surface experiences drag forces;

2) Impeller, in which water is trapped and its kinetic energy is used to give high rotational velocities to the meter's rotating parts;

3) Propeller, in which the current stream produces torque on an assemblage of foils;

4) Acoustic, in which the differential celerity of sound is measured in opposite directions or in which particle velocities are measured by the doppler shift.

The most important task of the data acquisition subsystem manager is to choose the category of current meter to be used on the basis of its applicability to the particular location at which it is to be placed. The choice of the particular meter to be used will be left up to the component engineer.

B.2. Dye-marking the water offers a way to estimate surface currents, but tells little about the bottom currents and therefore has little bearing on the problem. Fluorescing of materials is a neat way to use marker particles and is very organically tied to the problem of determining rates of transport because the particles used can be chosen to be exactly the same as the buffer zone materials, statistically. Buoys of controllable depth may also be useful when they are adjusted to stay very near the bottom and follow the bottom currents. In extremely rough areas, however, their usefulness will be limited by the destructive forces acting on them, and in shallow water their size might be too large with respect to their depth to consider that they are not interacting bilaterally with the flow.

B.3. Wave-measuring devices are also available in a large variety of forms, the step resistance staff being the most common, followed by the capacitor type gauges. The choice between these and more exotic types (such as upward-looking sonar, terrestrial photogrammetry, pressure sensitive devices, etc.) will again be made on the basis of applicability to the location at which it is to be placed. Where frequent maintenance is possible, the simplest may be the resistive type, but where surface action is strong enough to cause destruction of the instrumentation, it would be wise to sense the waves from below, using perhaps a pressure sensitive meter. Again, the responsibility of the data acquisition subsystem manager is to choose the generic type of meter and to leave the particulars to the component engineer.

B.4. Doppler radar wave analysis affords a method of measuring waves without being in the waves at all. Observations may be made from land where power maintenance, life support, etc., are less of a problem than they are at sea or in the buffer zone. In return for these conveniences, the system pays the price of having to correlate the radar output with the wave system it saw. More accuracy may be lost than can be afforded, but this depends on the particular case being considered.

B.5. Aerial photography gives as a gratuitous offshoot of terrestrial topography a generally clear picture of the trends in wave patterns. Such information can be used to spot trouble areas where wave energy is concentrated or to get an idea of the effect of a structure such as a jetty on the wave pattern, etc. Dominant wave formations can be spotted and their wavelength can be measured. In using this information, the quality of the wave measurements can be improved and their relevance can be better estimated.

B.6. Tide gauges are very much like wave gauges with the difference being that tide gauges need not respond to high frequency waves.

There are several ways of cutting out the high frequency waves and recording only the tide, but most of these methods rely on the response time of the instrument being used. In many cases, it will suffice to enclose a wave meter in a tube whose bottom is pierced by a very small hole; then the response time of the water level in the tube will depend directly on the ratio of the tube's diameter and the hole's diameter. Another form of tide gauge that should be considered is the pressure-sensitive transducer which could be placed in a protected location, for instance, on the bottom as opposed to the free surface.

B.7. There is little point in worrying about wind meters in view of their great importance in aviation. There seems to be little choice of wind-metering instruments, but the anemometer and vane should be entirely sufficient. The most difficult job will be the interpretation of wind data, for it is often difficult to estimate the fetch, the effects of tide and topography, etc. An attempt will be made to correlate wind data with wave data in each sector, but this will be carried out in the data-processing subsystem.

B.8. Analysis of the evidence of forces would include such things as analysis of ripple marks, natural sorting of materials, accretion or depletion of material in the vicinity of existing structures, etc. Although quantitative analysis of this sort of evidence is very difficult to use in forecasting, it can be extremely useful in improving the intuition of those working on shore protection projects. It is interesting to note that the supply of examples of the effects of natural forces on poorly-designed jetties, groin systems, and channels is almost unending, and much can be learned from this ample supply of past mistakes.

C.1. Aerial photography provides large amounts of information per excursion and is particularly suited to studies which are to encompass large areas, such as shorelines which may be in the order of miles, tens of miles, or hundreds of miles long. The record that is obtained preserves the context of each location, and the photography can be assembled to give a continuous picture of the entire buffer zone from one end to the other. The average photography, as taken by the U.S. government, covers three miles square on a nine-by-nine-inch print. Lower altitude photography can be taken to give more detail with less coverage per print. That is, flying at about 6,000 feet will give coverage of one mile square using the standard nine-inch lens. Not only can photogrammetric measurements be taken from the aerial photography (giving topographic quantities such as buffer zone width, elevations, etc.), but also interpretation of the photography can reveal many interesting things, such as the nature of adjacent soils and bedrock, drainage patterns, the nature of the vegetation cover, land use in the buffer zone and adjacent areas, offshore formations, sources of pollution, and wave characteristics.

Due to the great expense involved in the formation of an in-house photogrammetric facility, the data acquisition subsystem manager must consider the feasibility of turning all or part of the operation over to a subcontractor. The technical aspects of obtaining photogrammetric results--that is, the flying of photography, the processing of films, and the plotting procedure--are all quite routine and objective and can therefore safely be handed out, but the interpretation of the photography is very much more subjective, and the data acquisition subsystem manager would be wise to keep this phase in-house where he has more control over it. In any case, the photography provides a permanent record of the state of the buffer zone and adjacent areas at the time the photography was taken.

C.2. Terrestrial photography can be taken with a matched pair of stereometric cameras and can subsequently be reduced to a three-dimensional model of the photographed location much like aerial photography. It is extremely well adapted to providing a permanent record of the state of a comparatively small sector of shoreline and to recording the evolution of such a sector in successive photographs. The two cameras used to obtain the photography are often mounted on a vehicle such as a small truck or jeep which can be taken right up to the location to be photographed. Terrestrial photogrammetry has all of the advantages of aerial photogrammetry when applied to a smaller scale and has, as well, the advantage of the possibility of in-house operation due to the reasonable cost of the equipment.

C.3. Rod and transit surveying has the great disadvantage of only being able to record one data point per measuring operation. That is, the amount of work in the field depends on how many points are to be measured, whereas with photogrammetric methods, one pair of plates are exposed in the field and as many data points as desired are later measured in the comfort of a plotting room. It is, however, a reasonable method for establishing the ground control points necessary for establishing the absolute orientation of the three-dimensional model provided by photogrammetric methods. Such control points are not always needed, however.

C.4. The lead line provides almost the only way to measure profiles of the submarine topography in areas where the surf activity is too strong to permit the approach of a sonar-equipped boat but, even so, it is a tedious procedure of limited accuracy suited only to the abovementioned extreme circumstances.

C.5. Sonar and side-looking sonar have recently become popular to the extent of being used for navigation based on the bottom's topography. Sonar can indeed provide very good information about the submarine topography, but is limited by the draft requirements of the platform. The most important application of this technique to the shore protection system will be in establishing the topography which controls the refraction of waves as they approach the shoreline from deep water. Due to the ease of operations and the good accuracy obtained, this method really does not have any serious competitors in the near shore (but not in the surf) zone.

IIIb. Hardware for the Data Acquisition Subsystem

A. Materials

For the analysis of beach materials the hardware that will be used will be the standard sieve set. Visual identification of minerals will be relied upon in most cases, but the possibility of chemical mineralogical analysis should be afforded to the sedimentologist responsible for materials analysis. Also, the possibility of electron microscopic analysis should be afforded this scientist in extreme cases. Neither the chemical nor microscopic facilities need necessarily be acquired by the shore protection system if it is possible to have the job done by another concern.

The interface constraints that are posed by the choice of the sieve set method of separation have only to do with the transfer of data from the data-acquisition subsystem to the data-processing subsystem. The quantities which must be presented are: 1) the percentage of the total sample occupied by each size group; 2) the beach sector on which the sample was taken; 3) the exact location of the sample site with respect to the high water mark. The time at which the sample was taken (by month or season) will also be presented. It is immediately clear that it will be advantageous to measure several quantities at each site at essentially the same time.

B. Forces

1) Currents. The hardware to be used for the measurement of currents is divided into two categories by the presence or absence of breaking waves over the instrument. In areas where there is relatively little danger of destruction to the instrument, a current meter of the resistive type will be used because of its suitability to areas where clogging by sediments would hamper the operation of meters having exposed moving parts. Also, the output, conveniently, is an electrical signal which can easily be converted into digital form.

The interface constraints posed by this choice are as follows:

a) The navigation system will have to be able to accurately place and retrieve the instrumentation.

b) There will have to be a data link between the instrument and the data-processing facilities, whether by the retrieval of an information storage block physically attached to the instrument, by a cable connecting the instrument to a remote data storage block, or by electromagnetic transmission of data from a buoy connected to the instrument by a cable link.

c) The maintenance time cycle may be controlled by the instrument's ability to store data or energy.

In areas of relatively active surf, where the possibility of setting out instruments is limited, the currents will be measured by dyes or by fluorescing of sand particles. There will be an attempt made by the data-processing subsystem to correlate the mass transport as indicated by the fluoresced particles with the longshore currents indicated by the water-carried dyes and both of these, in turn, with the offshore wave pattern. This will necessitate taking several different kinds of data at essentially the same time, as mentioned before in Section IIIb.A.

2) Waves. Again, the sensing of waves will be controlled by the presence or absence of breaking waves. In deep water a moored buoy will be used as a platform on which to mount wave-metering instrumentation. The choice of the specific instrument will be made by the component engineer, but simplicity can be called for in deep water where the waves are relatively calm.

In shallow water, where breaking waves are expected, it is important to keep the instrument from being destroyed. Thus, a pressure-sensitive device may be located on the bottom and, by recognizing the hydrodynamically-induced pressures in addition to the hydrostatic pressure, wave height can be computed.

For an overview of the wave patterns, aerial photography will be used. On the basis of this photography, the type of meter to be used will be determined and the location of the instrument will be chosen.

The possibility of using doppler radar is relatively new, but should be investigated more thoroughly by the data-processing subsystem's research group because it can give a record of the wave spectrum without immersed instruments.

The interface constraints imposed by the choice of buoy-mounted meters in deep water are:

a) same as IIIb.B.1.b

b) same as IIIb.B.1.c

c) the buoy's response to all forcing functions must be known and the maximum response may be limited by orientation limitations on the instrumentation.

The interface constraints imposed by the choice of a resistive type meter to be used in shallow water are:

a) same as IIIb.B.1.a

b) same as IIIb.B.1.b

c) same as IIIb.B.1.c

The interface constraints imposed by the choice of a non-immersed metering system, such as doppler radar, will have mainly to do with the specific requirements of the metering system, such as power supply, transportation along the shoreline and interpretation of resulting data.

3) Tides. Since the affect of the shoreline on the tidal range is rather important, it is important to place meters near enough to shore. They can be affixed to any structure, preferably in a protected orientation. The tide meter to be used will be a resistance

type staff with a time constant long enough to filter out the high frequency waves that arrive at the staff.

The interface constraints imposed by such a meter are the same as those stated for the other immersed meters.

4) Wind. An anemometer and directional vane will be used in conjunction with an electrical or digital output. The interface constraints posed by this choice are: 1) the maintenance time cycle may be controlled by the instrument's ability to store energy or data, and 2) there will have to be a data link between the instrument and the data-processing facilities either by cable, electromagnetic transmission, or recovery of an information storage block from the instrument.

C. Topography

1) Subaerial. Most of the surveying will be done by aerial photogrammetry because this form of surveying is particularly suited to mapping large and extensive areas. In cases where more detailed information is called for, the tool to be used is terrestrial photogrammetry. The aerial section will be contracted out in order to avoid the purchase of expensive equipment that is already available, but the terrestrial section will be done, after all, in locations which have been designated as interesting by the data-processing subsystem.

The interface constraints that are posed by the choice of aerial photogrammetric surveying are: a) the subcontractor, who is part of the data-acquisition subsystem, will fly coverage requested by the data-processing subsystem, and so there must be adequate feedback possible in matters of flight height, time of flying, film types and flying pattern. All of these are important to both the data-processing subsystem and the photogrammetric concern, but none of them have any real systems implication outside of this, if they are handled correctly; b) the final output of the plotting of the photography must be in a form which can be handled by the data-processing subsystem and, if this is to be digital, then the plotting equipment must be equipped with shaft digitizers and suitable printout equipment.

2) Submarine. Sonar soundings will be used wherever possible. The interface constraints imposed by this choice are: 1) adequate navigation is needed to locate the position of each sounding; 2) the platform used must be particularly well suited to the near-shore environment, and its response to the appropriate spectrum of waves (which will, in general, be of shorter wavelength than would be found in deep water) must be known, and 3) the soundings must be presented to the data-processing subsystem in a convenient form, preferably digital.

It should be noted in each of the above decisions that the component level engineer has been left a great deal of design freedom.

IVa. Discussion of Hardware for the Data-Processing Subsystem

A. Present State

1) Cartographic information is the direct output of photogrammetric plotters.

2) The size of the buffer zone is to be calculated from its landward limit to its seaward limit (as defined in Section I.f.), using

a) digital data processing

b) scale

B. Interactions between Forces and Materials

This will be a fairly independent group whose purpose it will be to establish relationships on which estimates may be based predicting the amount of material moved by given force conditions. They will use whatever modeling, testing, computational or observational facilities they can justify in terms of the pros of measuring one phenomenon to predict another and the cons of spending money to find out how to do it. They will work primarily to solve problems posed elsewhere in the Information Subsystem.

C. Future State

Having the data concerning materials, forces, and topography at each location and the relationships between these quantities and the amount of material transported, the calculation of the future state of the buffer zone will be done numerically. If direct measurements have been made by fluorescing particles and tracking them, then the two may be compared to arrive at a final transport rate and at an estimate of the future state of the buffer zone.

D. Outside Influences

The most important work that will be carried out here is, as stated in Section II.B.1.d, an analysis of the effects of the shore protection structures on the shoreline. As can be seen from most existing jetties, the anticipated effects are not always had. The hardware that might be used still depend upon the specific project, but in general one can count on the need for model-testing facilities, computational facilities, and all of the other processed data, especially from the work done on the interactions between forces and materials.

IVb. Hardware for the Data-Processing Subsystem

A. Present State of the Buffer Zone

1) Photogrammetric plotting facilities

2) Digital processing facilities

B. Interactions between Forces and Materials

1) Digital processing facilities

2) Research facilities

a) Wave tank

b) Model beach basin

c) Proposed instrumentation

d) Etcetera for specific research projects

C. Future State of the Buffer Zone

1) Digital data-processing facilities

D. Outside Influences

1) Digital data-processing facilities

V. The Action/No Action Decision and Recommendations

A. Action

1) Interdiction of the movement of materials is the first of two ways of affecting the nature of the buffer zone and this interdiction can be accomplished in many different ways. Each method of preventing the movement of materials to or away from a given beach sector is particularly well adapted to a more or less specific type of problem and less well suited to others. It will be the responsibility of this subsystem, which is actually a form of command and control (of the strategic genre), to suggest the type of structure or program to be followed, but it is not their responsibility to give any more

specific directives, for further directives would limit the freedom of the interdiction subsystem. The type of structure or program called for would be, perhaps, one of the following types:

a) Jetty

b) Groins

c) Revetement

d) Seawall

e) Storm fence

f) Vegetative cover

or it might be something else more suited to the particular problem at hand.

2) Transportation of material is the second way to affect the state of the buffer zone, but extreme care must be taken to be assured that it will be affected in the manner planned on. One of the most disastrous pitfalls encountered in this field is associated with the choice of a source of materials. If material is taken from an area in the buffer zone that is, itself, subject to littoral transport, then the rate of change of buffer zone size at this location will be affected just about as much as the area on the receiving end of the transportation process. If, on the other hand, materials are taken from a region which is fossilized with respect to the global balance of littoral materials, then it is possible to create another location which will begin to accumulate all of this material and grow rapidly and unexpectedly. It is therefore imperative that, in addition to a decision calling for a transportation process, such as nourishment by pipeline or barge, a decision be made indicating the acceptable source areas.

B. No Action will be the decision of the Action/No Action and Recommendations subsystem in the event that the buffer zone is not changing rapidly enough to warrant action. The data acquisition and processing will continue for such a sector at a relatively slow rate, that rate to be specified at the time the No Action decision is made.

VI. Complementary Facilities

Having made the choices of hardware to be used in each field imposes the need for complementary facilities that will be needed.

A. Transportation

1) Four-wheel drive vehicle for onshore collection of materials and as a platform for the stereometric cameras to be used for terrestrial photography. This same vehicle will also be the physical link between the laboratory and the buffer zone.

2) Displacement boat, possibly amphibious, for the offshore collection of materials, for submarine surveying and for the emplacement and recovery of offshore instruments.

3) Airplane for the acquisition of aerial photography of the buffer zone.

B. Laboratory, including facilities for

1) Materials analysis

2) Photographic processing

3) Photogrammetric plotting

4) Digital card punching and processing

5) Storage of materials, data and portable hardware

6) Research into the relationships between forces and materials

C. Buoys, Piers and Etcetera for Offshore Instruments

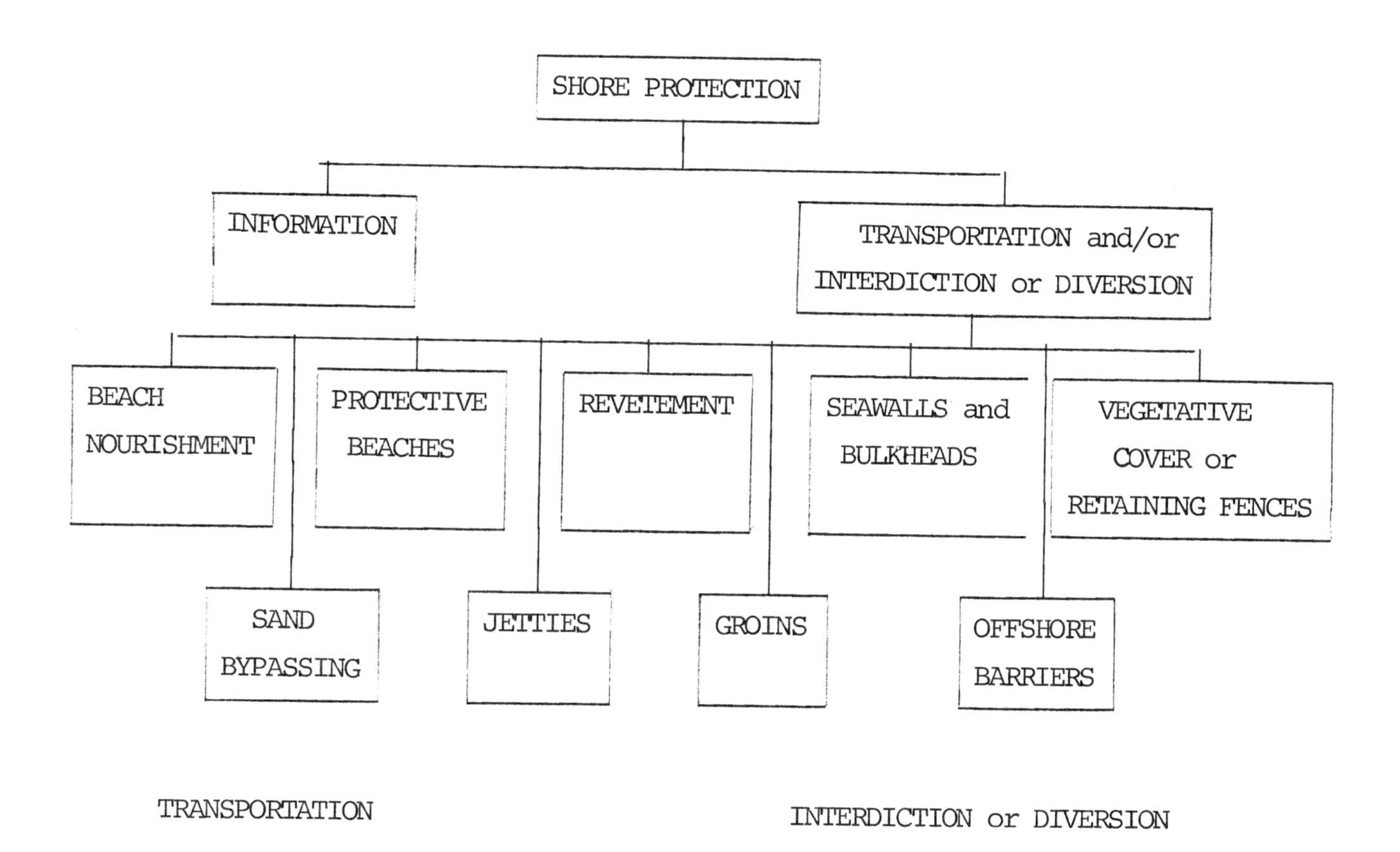

Figure 7-8

I. Quantification of the Transportation and/or Interdiction Subsystem

A. Interdiction or Diversion

a) The quantity of material which must be interdicted or diverted will depend entirely upon the beach sector under consideration, but before it becomes of any real consequence experience has shown that it will be in the order of hundreds of cubic yards per day except in cases of extremely narrow and critical buffer zones. The type of material to be handled will in general be sand and silt-sized particles.

b) The source of this material will be the beach sector under consideration and the location of this sector will be given to the nearest quarter mile unless larger lengths of buffer zone are being destroyed by the same mechanism.

c) The requirement that must be met is in all cases the preservation of the buffer zone, whether this is by stabilization of the existing materials or by reducing the net influx of material. This subsystem will neither create nor destroy material, but it may modify the material exposed to the destructive forces or modify the location at which these forces are dissipated.

d) The obstacles to interdiction or diversion are both the large quantities of material to be handled and the state of the art concerning effective methods of interdiction or diversion.

e) The time limits on location and interdiction will be governed by the rate of buffer zone destruction and will be the concern primarily of the information subsystem, although once the trouble is located it must be corrected before the buffer zone is destroyed.

f) The modes of transportation of the materials to be handled will be determined by the information subsystem and will play an important part in the choice of the generic type of structure or program to be used.

g) The time of transportation of materials will generally vary seasonally, but this information will be determined by the information subsystem as well. Except in the case of sand bypassing, which may be varied, and the infrequent case of variable groins, the time of transportation will be relatively unimportant.

h) In order to interdict the transportation of materials in their dissemination phase, the program will generically be one of stabilization aimed against erosion in the subaerial or submarine environment.

i) The constraints imposed upon interdiction in this phase are related to the allowable effects any program or structure may have on other beach sectors; that is, the interdiction of movement of materials in one sector must not be responsible for the destruction of another sector.

j) The obstacles to interdiction or diversion during this phase are both the large quantities of materials to be handled and the state of the art concerning effective methods of interdiction or diversion.

k) The time limits on interdiction or diversion in this phase are, again, imposed by the rate of destruction of the buffer zone as determined by the information subsystem.

l) The quantity of material which is deposited to the

detriment of the buffer zone will be a function of the location of the beach sector under consideration. The type of material will generally be sand or silt-sized particles.

m) The location at which this material is deposited will be found by the information subsystem and given to the nearest quarter mile or more.

n) The material which would otherwise be deposited to the detriment of a given beach sector will either be contained in a location other than in this sector or will be bypassed around this sector.

o) The constraints imposed upon the interdiction or diversion of material, which would otherwise be deposited in a given location, are the allowable effects that such action may have on other locations; that is, keeping material from being deposited in one location must not be responsible for the destruction of the buffer zone in another location.

p) The obstacles to diversion or interdiction at delivery are both the large quantities of material to be handled and the state of the art concerning effective methods of interdiction or diversion.

q) The time limits on interdiction or diversion at delivery will be determined by the rate of destruction of the buffer zone by unwanted deposition of materials and will be determined by the information subsystem.

r) Considering the grave implications of system failure, primarily of property damage, it would seem that a high reliability and confidence limit should be set, but, in terms of past experience and of the many partial failures evident in this field, it would appear to be unrealistic to quote figures too high. Therefore, the success of this subsystem will be a measure of how stabilizing the structure or program used is in terms of the decreasing rate of destruction of the buffer zone. For the effectiveness of this subsystem, the probability of success must be 80% with a confidence level of 75%.

B. Transportation

a) The quantity of material to be collected will be determined by the rate of change of buffer zone size in the beach sector to be treated. Transportation of material to a beach sector is known as beach nourishment and transportation away from a sector is known as dredging.

b) The location of the material to be transported will depend either on the location of excessive accretion, as determined by the information subsystem in the case of dredging, or upon the availability of material in the case of nourishment. The possibility of an operation analogous to cut and fill exists.

c) The constraints upon collection of the material will, in general, be determined by the ability of the hardware to handle the quantity of material needed and the ability to operate in the environment of the area of collection.

d) The obstacles to the collection of material will be:

1) sea state

2) seaway

3) state of the art concerning the transport of suspended materials

e) All time limits will be determined by the rate of

destruction of the buffer zone and by the seasonal changes in environment.

f) The distance over which transport will take place will, in general, be limited on the lower side by the location of suitable source or dump areas and on the upper side by the size of the shore region to be protected. County and state lines may in part limit the distance of transportation.

g) See I.B.e.

h) See I.B.d.

i) The material to be transported will need no processing, and no intermediate stops will be required by the mission.

j) The quantity and type of material to be delivered is the same as the material to be collected.

k) The location of delivery will, in the case of dredging, be determined by the possibility of dredged materials reentering the balance of materials transported by natural forces. In general, it will suffice to carry dredged materials out of the littoral zone except in the case that the same material is to be used elsewhere for beach nourishment. In the case of nourishment, the location of delivery will be determined by the information subsystem.

l) See I.B.c.

m) See I.B.d.

n) See I.B.e.

o) For the reasons found in the quantification of the interdiction and diversion subsystem, the probability of success will be set at 80% with a confidence level of 75%.

II. Mission Analysis and Profile

When the information subsystem determines the need for action in a specific location, it presents this decision to the transportation, interdiction and diversion subsystem along with all of the pertinent processed data. On the basis of the recommendations of the information subsystem and the experience of the transportation, interdiction and diversion subsystem, a generic choice will be made concerning the form of the action to be taken. Then a rough preliminary design will be drafted and modified on the basis of an evaluation from the information subsystem. The design will be modified as many times as is necessary by competing contractors based on both the additional information available from the information subsystem on request and on the evaluation from the information subsystem given gratuitously. From all of this will emerge a final design which will be built or implemented on the shoreline by the contractor chosen.

Mission Analysis from the Transportation or Interdiction or Diversion Subsystem's Point of View

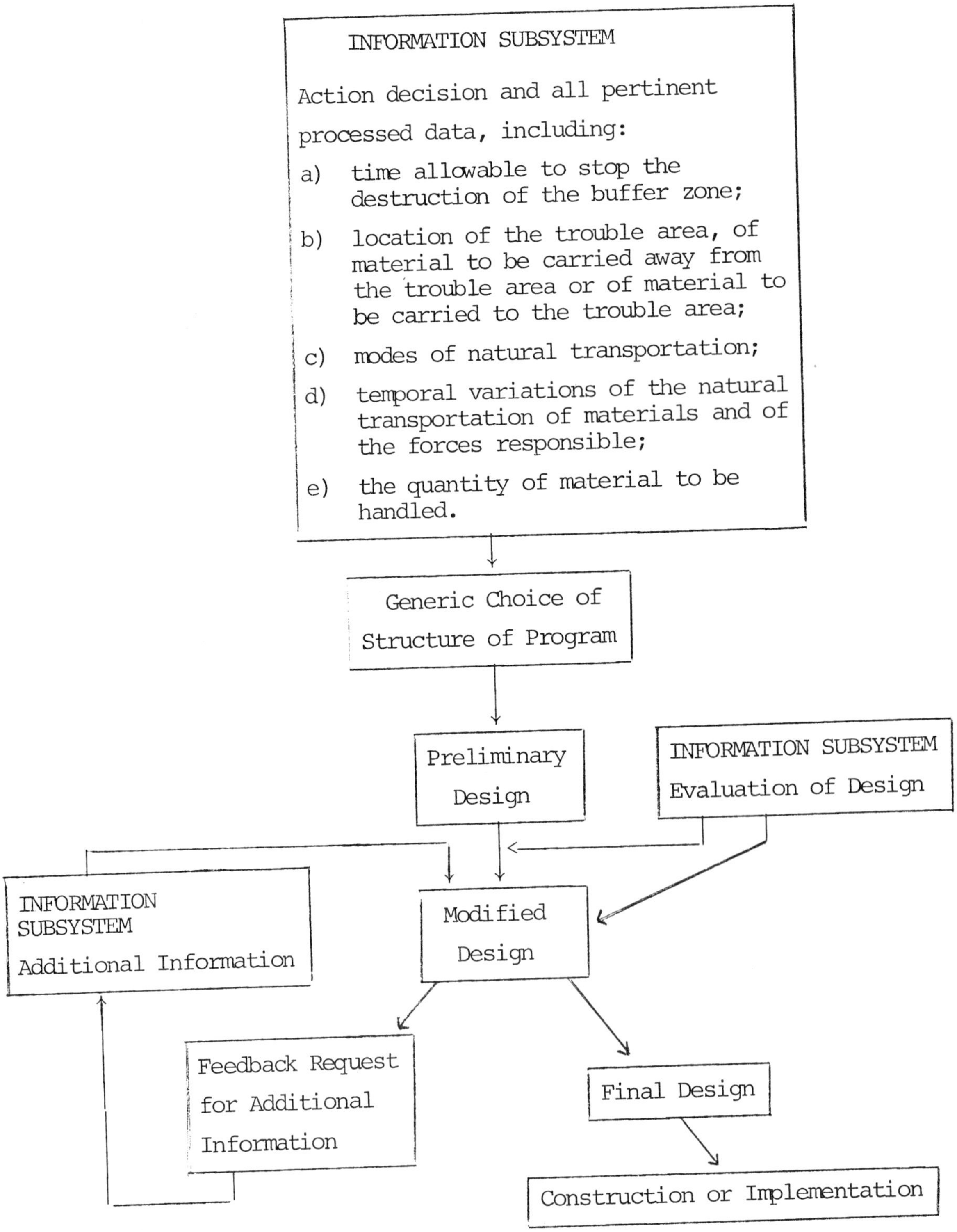

INFORMATION SUBSYSTEM

Action decision and all pertinent processed data, including:

a) time allowable to stop the destruction of the buffer zone;

b) location of the trouble area, of material to be carried away from the trouble area or of material to be carried to the trouble area;

c) modes of natural transportation;

d) temporal variations of the natural transportation of materials and of the forces responsible;

e) the quantity of material to be handled.

Generic Choice of Structure of Program

Preliminary Design

INFORMATION SUBSYSTEM
Evaluation of Design

INFORMATION SUBSYSTEM
Additional Information

Modified Design

Feedback Request for Additional Information

Final Design

Construction or Implementation

Figure 7-9

III. Discussion of the Various Types of Shore Protection Structures and Programs

A. Interdiction or Diversion

1) Vegetative cover and retaining fences are used to control subaerial erosion of relatively small-sized particles such as sand. There are, however, few instances when they can be effectively employed because of the nature of the erosional process and because of the circumstances under which subaerial erosion is important. Subaerial transport of particles is usually found to be predominantly a process of saltation where the particles actually are airborne for short times. Hence, there is a tendency for such particles, once initially eroded, to overtop fences of reasonable height (measured in terms of a few feet). If the fence does successfully stop particles, it can only do so by degrading its own height, for the particles so stopped will accumulate at the base of the fence and reduce the height of the top of the fence above the general beach surface. Continued arrival of particles can only result in the eventual overtopping of the fence. On the other hand, vegetative cover is aimed at the stabilization of the beach against the initial erosion. But those areas which erode most readily are, in fact, the least conducive to the support of plant cover. Beach grasses which can sustain the adverse moisture conditions cannot tolerate excessively high salinities and are found to thrive best only above the berm leaving an important fraction of the beach's profile exposed to subaerial erosion. Luckily, the problem of subaerial erosion is very much less important than that of submarine erosion.

2) Seawalls and bulkheads generally are designed to serve

two functions: retaining a land mass above the general beach level, often in conjunction with a road or causeway, and preventing destruction of a land mass by wave action. There are three subgroupings of seawalls and bulkheads--those being wall-faced, concave seaward and convex seaward (including stepped). Wall-faced seawalls or bulkheads are generally not built to withstand large wave forces, but rather to retain a land mass. Such structures are built in quiet locations, such as in protected harbors or bays or landward of a large, gently-sloping beach. Concave seaward seawalls are generally designed to prevent waves from overtopping the structure and are often built along long stretches of developed shoreline which are subject to intensive wave action. Convex seaward and stepped structures are less effective against overtopping by waves and are generally built against scour and run-up. In all cases, roughness may be added to dissipate energy and to prevent the undercutting of the seaward base of the structure. Also, hydrostatic pressure from the landward side must be considered and is often controlled by permitting seepage seaward through the wall at predetermined locations. The most unfortunate aspect of the use of seawalls in the ocean environment is the tendency to generate development of the shoreline within feet of the swash mark. It is almost axiomatic that such development will suffer losses periodically. In many cases, especially in the Gulf of Mexico, the design of the developed areas admits nothing less than catastrophic losses in severe weather. The shore protection system must eventually be able to exert the political and social pressure necessary to inhibit such forms of development.

3) Revetement is almost an extension of the seawall and bulkhead category in that its purpose is to prevent the erosion of the shoreline by wave action. The general form of revetement is either based on energy dissipation, as with rubble revetement, or energy reflection, as with smooth concrete revetement. Such revetement often causes the exhaustion of the supply of small-sized land-derived particles from a given location and leaves a cobble or shingle beach. These locations are also noted for the great disparity between winter and summer profiles for the same reasons.

4) Offshore barriers are designed to cast a wave shadow in which small craft and swimmers will be relatively safe. The attendant energy deficiency causes sedimentation shoreward of the barrier, thereby narrowing the straits between the mainland and the barrier. This, in turn, concentrates the longshore currents if the barrier is entirely outside the littoral zone, but causes the longshore current to be displaced seaward if the barrier is in the littoral zone. If properly designed, the increased current strength will maintain a free passage between the mainland and the barrier, but, if improperly designed, the result will be a land bridge between the barrier and the mainland, making passage impossible for boats. Furthermore, the accretion of material between the barrier and the mainland may cause the net erosion of downcurrent sectors and will move the littoral zone seaward, thereby causing the barrier to be no longer entirely outside of the littoral zone.

5) Groins are designed to trap or retard the particles involved in littoral drift and thereby to stabilize the shoreline.

They are a short-term solution to the problem because the material they trap serves to alter the profile of the shoreline in their immediate vicinity and thereby cause operation outside of the design conditions. What usually happens is that the upcurrent side of the groin is rapidly filled and the downcurrent side is scoured clean, producing a scalloped shoreline where there are several groins. After this stable configuration is reached, sand is effectively swept around the seaward end of the groin and continues just as it did before the construction of the groin, if the seaward end of the groin is in the littoral drift zone. Otherwise, material will be permanently lost. During the stabilization, there will be a general deficiency of material downcurrent of the groin corresponding to the volume of material caught by the groin. It should be noted, however, that, except for the accretion and scour localized at each groin, the shoreline is usually stabilized by the groins. In order to protect a length of shoreline miles long, however, it requires groins all along the length at spacings measured in the hundreds of yards or less. Groins can lead to trouble, especially when there is insufficient material to achieve the stable configuration without damaging the downcurrent beaches. Since the downcurrent side of the groin will be eroded, there can be no developed areas too close to the shoreline at the time of the groin's construction. There is also the possibility of flanking the groin if the groin is anchored seaward of the eventual position of the eroded downcurrent shoreline. As a last comment, the expense of groins is large enough to see competition from other solutions, namely, beach nourishment.

6) Jetties and breakwaters are generally of the same form; that is, they are shore-connected structures, extending seaward for the dual purpose of directing and confining both littoral particles and currents or waves. Jetties are often built at the entrance of harbors or channels in order to confine the tidal current and thereby increase its peak velocity, thus helping to scour the entrance way and maintain a desired channel depth. Particles moving along the shore are diverted on reaching such a structure and are either confined to the upcurrent side or are swept out to deeper regions where they may or may not shoal the channel. Jetties and breakwaters both serve to reduce wave action either by diffraction or by directly absorbing the wave's energy. As was the case with groins, a jetty acting as a barrier to littoral transportation can only function effectively until it reaches its capacity in terms of the trapped volume of material. Thereafter, material will accrete elsewhere--generally across the mouth of the channel in the case of jetties perpendicular to the shore, and off the tip of breakwaters built with the extreme seaward end more or less parallel to the shore. Hence, jetties and breakwaters may be considered short-term solutions to the problem of protection and will eventually require maintenance such as dredging or sand bypassing. Also, as was the case with groins, the downcurrent side of the jetty will suffer erosion due to the deficiency of sand created by the barrier to littoral transportation. A jetty or breakwater, therefore, cannot be built to function forever without maintenance except where there is no littoral transport of materials.

7) Protective beaches can be maintained seaward of a developed shoreline for the purpose of dissipating wave energy. In fact, a beach is the most natural dissipater of wave energy that exists, as evidenced by their worldwide presence. However, the problem of building a beach where natural conditions are not favorable (for, otherwise, there would be a natural beach) is quite difficult. A careful study of the existing forces must be made so that one can know how much maintenance will be necessary to stabilize such an artificial formation, and so that the effect of maintaining a sand buffer zone in one location can be estimated in other locations. Generally, the maintenance of a protective beach will help control erosion downcurrent of the beach, but this may cause problems if a channel is downcurrent of the beach, for it is generally desirable to keep materials from accreting in channels. If a jetty or sand-bypassing program exists at the time when a protective beach is built upcurrent of the jetty or bypassing, the balance of materials at the jetty or bypassing will be upset to the detriment of the channel. Hence, careful consideration must be given to the long-range effects of the construction and maintenance of a protective beach.

8) Sand bypassing is one solution to the problem of unwanted accretion of material in the vicinity of a channel whose depth must be maintained. As discussed in the preceding section, short-term solutions to the problem of littoral drift into channels must be supplemented by continual maintenance programs. In sand bypassing, material is taken from the area of accretion, upcurrent of the structure, and is transported to the downcurrent side artificially, thereby

preventing detrimental accretion upcurrent of the structure and also preventing excessive erosion downcurrent. The two most frequently used systems are classified as either land-based dredging or floating dredges. In both cases, suction-type dredges collect material and pipelines carry it past the channel's mouth and redeposit it in the downcurrent region. One of the biggest problems with both systems is that the amount of material arriving at the dredging site varies considerably during the year and so that dredge will operate under optimal load conditions infrequently. Land-based dredges are also plagued by the variances in the area of natural accretion and floating operations must be designed to survive near-shore wave action. In spite of these problems, sand bypassing is a long-term solution that has been used successfully in many locations where simple fixed structures have failed. In cases where the source and deposit areas are widely separated, mobile land- or sea-based vehicles have been used to transport the dredged material.

9) Beach nourishment is the replenishment of beach material from artificial sources by artificial means. One of the first things to consider is that the material used to maintain a given beach will inevitably be carried down the shoreline and will affect the balance of materials everywhere downcurrent of the nourished region. In cases where long stretches of shoreline are plagued with an insufficient supply of material, this method is extremely valuable. The second important thing to consider is the choice of the borrow area. Often there are sufficient onshore deposits of material to nourish a beach for many years, and in this case the problem of depletion in unexpected

areas due to the borrowing of material will not be present. In other cases, offshore formation can be used as the source of materials, but care must be taken to choose one which does not actively participate in the local littoral balance. For, if it does, feeding one beach may starve another. Such fossilized formations can often be identified by the surface characteristics of the particles, especially in the case of fossilized sand dunes and beach ridges, both of which display the pitted surface characteristic of particles subaerially eroded. Fossilized glacial features such as kames can be identified by the vitreous cracking of mechanically-eroded particles. The size characteristics of the chosen particles must conform to the prevalent forces in such a way as to be relatively stable after their emplacement. This is a long-term solution to the erosion of the shoreline and must be treated as such; that is, a continuous program must be implemented for the delivery of material to the beach.

IIIa. Resume of the Important Characteristics of the Various Shore Protection Structures and Programs

STRUCTURE or PROGRAM	DESTRUCTIVE MECHANISM to be CONTROLLED	STAGE OF DESTRUCTION ENCOUNTERED	REMARKS ON THE LONG-TERM EFFECTS OF THE STRUCTURE OR PROGRAM
Vegetative cover or retaining fences	Subaerial erosion and wind transport	Erosion of material from the buffer zone	Vegetative cover can be an extremely valuable technique for stabilizing the upland buffer zone especially where bluffs or unstable cliffs are present. Nearer the shoreline both methods become less valuable because of the gradual self-destruction by excessive accretion or by the difficulty of their implementation in high salinity regions.
Seawalls and bulkheads	Waves and scour	Erosion of material from the buffer zone	Effective method of creating a very narrow buffer zone. This is usually undesirable because of the potentially catastrophic situation it fosters. Outside constraints, usually social or political, may force this solution to be adopted in spite of its flaws.
Revetement	Waves and scour	Erosion of material from the buffer zone	Effectively cuts down on wave and scour destruction but causes the depletion of land-derived particles, thereby changing the particle size distribution everywhere downdrift of the revetement.
Offshore barriers	Waves	Erosion of material from the buffer zone	Short-term solution because of sedimentation landward of the barrier.
Groins	Longshore transport	Transportation of material away from the buffer zone	Short-term reduction of longshore drift accompanied by downdrift depletion of materials. Long-term stability thereafter.
Jetties and breakwaters	Longshore transport	Transportation of material into the buffer zone	Short-term reduction of longshore drift followed by resumed accretion of material in deeper water (usually in the channel) and accompanied by depletion of materials downdrift of the structure.
Protective beaches	Waves and scour	Erosion of material from the buffer zone	Long-term solution requiring continual maintenance. The only case in which protective beaches can be destructive is when downdrift accretion of materials is undesirable.
Sand bypassing	Sedimentation	Delivery of material into the buffer zone	Long-term solution requiring continual maintenance.
Beach nourishment	Waves and longshore transport	Erosion of material and transport of material away from the buffer zone	Long-term solution requiring continual maintenance, especially useful where long stretches of shoreline are deficient in materials.

Figure 7-10

PHOTOGRAPHS

1. Sand ripples can often be used to measure bottom currents and are especially interesting because of the strong correlation between the forces responsible for the formation of ripple marks and those responsible for the transport of materials in the buffer zone. A description of the statistical inferences that can be made on the basis of ripple marks is given by J. C. Harms in the G.S.A. Bulletin (3/69). By using such evidence as ripple marks in conjunction with measurements taken during their formation, much can be learned without instrumenting the surf zone.

2. The curved seawall is generally designed to reflect breaking waves back to sea. Notice also that the smooth seawall is used in conjunction with rubble whose function is to absorb wave energy by creating turbulence.

3. A close-up of the same seawall shows that the material of the seawall is readily eroded by particles in suspension. The turbulence created by the rubble also causes particles of appreciable size (as seen at the base of the seawall) to attack the seawall.

4. Straight seawalls are often used both as retaining walls and to reflect non-breaking waves. The step construction at the base of the seawall is an energy dissipater, but most of it is now covered by sand.

5. Stone revetement is used as an energy dissipater and does not reflect waves significantly. Many other forms of revetement are used including rubble and tetrapods.

6. This offshore breakwater was designed to cast a shadow which would protect the beach from waves. In fact, this is the last of four such breakwaters which run parallel to the shore. The longshore current runs from left to right in the picture and all of the other breakwaters are to the left of this one. They do indeed cast a shadow, and its effect is, unfortunately, to cause

all of the material which was in suspension to drop out. Hence, there is now a land bridge connecting the mainland to the jetty, thereby completely cutting off the longshore current in this area. The future of this area is continued deposition until all of the jetties are connected to the mainland.

Note also in this picture that the beach material is by no means homogeneous from land seaward. The problem of where to take a representative sample from or of how many samples are needed across the width of the buffer zone is not easy, and the degree of homogeneity of each sector must enter into the calculations strongly.

RALPH

SPEED
LIMIT
30

REFERENCES

1. U.S. Coastal Engineering Research Center, Shore Protection, Planning and Design, Third Edition, 1966.

2. Inman, D. L., Sorting of Sediments in Light of Fluid Mechanics, Journal of Sedimentary Petrology 19, 51-70 (1949).

3. Inman, D. L., Wave Generated Ripples in Nearshore Sand, Beach Erosion Board, Corps of Engineers Tech. Memo No. 100, 1957.

4. Myers, Holm and McAllister, Handbook of Ocean and Underwater Engineering, McGraw-Hill, 1969.

5. Weigel, R., Oceanographical Engineering, Prentice Hall Series in Fluid Mechanics, 1964.

6. Harms, J. C., Hydraulic Significance of Some Sand Ripples, Geological Society of America Bulletin 80, 363-396 (1969).

7. Morris, H. M., Applied Hydraulics in Engineering, Roland Press, 1963.

8. Jacobs, Clifford A., Pandolfo, Joseph P. and Aubert, E. J., "Characteristics of National Data Buoy Systems: Their Impact on Data Use and Measurement of Natural Phenomena," TRC Report 7493-334, prepared for the U.S. Coast Guard, The Travelers Research Center, Inc., Hartford, Conn., December 1968.

CHAPTER VIII

LOGISTIC SUPPORT AND MAINTENANCE

EPIGRAMS

Spare parts can sink a ship.

Systems move on supplies.

An uncoordinated supply system can lose parts faster than they can be made.

An unregulated supply system can create spare parts requirements that do not exist.

The magnitude of the supply problem is determined in the initial design.

The magnitude of the supply problem is determined in the maintenance and reliability philosophy.

The magnitude of the supply problem is determined in the return and repair philosophy.

The magnitude of the supply problem is determined by the failure and usage reporting philosophy.

The magnitude of the supply problem is determined by the depot, warehousing and tender capability.

The magnitude of the supply problem is determined by the communication capability.

The magnitude of the supply problem is usually underestimated.

The function of a logistic support and maintenance system is to ensure that malfunctioning or potentially malfunctioning components of the system are identified, that the malfunction is corrected by repair or replacement, that the potential malfunction is prevented by repair, replacement or preventive maintenance and that the level of repair, replacement and maintenance is such that the system is able to perform with its prescribed reliability and operational effectiveness.

There is probably no subsystem which is more frustrating to the systems manager, more difficult to quantify or more intractable in management than the logistic support and maintenance subsystem. In general, if the personnel who operate the system are psychologically satisfied with the subsystem performance, then it is probably true that too large an expenditure of resources has been used for this subsystem and many spare parts will have been purchased which are never needed. Conversely, if the system is adequately supported at a "cost-effective" economic level, then it is generally true that occasional operational aborts will be experienced due to lack of spare parts, fault identification or maintenance with a consequent high level of psychological dissatisfaction. If the criteria for subsystem performance, for accounting for trouble and future reporting, for size and scope of logistic pipeline, etc., are not precisely identified, it is easy for a major system to degenerate into a Babel of supply communication, a miasma of spare parts, an exponentiating number of requisitions, cannibalism, degradation of system performance and deterioration of personnel morale.

The systems manager must therefore delineate at the outset and continuously update the following functional and interface requirements which must be met by the logistics and maintenance subsystem.

a) The percentage of time that sea-based platforms of the system will have aborted their mission as a result of deficiencies in the L&M subsystem and confidence limits on that percentage.

b) The percentage of time that sea-based platforms of the system will be operating in defined degraded modes as a result of deficiencies in the L&M support and confidence limits on that percentage.

c) The percentage of time that sea-based platforms will be unable to leave the terminal in full operational status (or will be unable to have a support ship reach the platform) as a result of deficiencies in the L&M support and the confidence limits on that percentage.

d) Similar quantification for other elements of the total system.

The subsystem manager will be unable to utilize these apparently precise statements without a great deal of information and estimate about the characteristics of the system. Indeed, the very nature of the supply problem is determined in the initial design. At the very minimum, it is an informational requirement that each subsystem designer generate a spare parts and maintenance requirements estimate as a fundamental parameter of his component design. These estimates are critical since an underestimate can result in operational crises when the underestimate is suddenly discerned and the lead time for spare parts manufacture is long. When there is an overestimate (a much more trivial sin), there is a continuous generation of excess (and often costly) inventory. Some adaptive reporting procedure must therefore be generated to provide early and meaningful feedback for adjustment of the estimate. Several serious traps are inherent in this process. The first of these results if a feedback process based on "usage rule" is employed. Characteristically, when a part experiences trouble or the rumor of trouble, supply managers at all levels will tend to stockpile, hoard and overrequisition the suspect component. This will result in an apparent usage rate which is greatly in excess of the real requirement. A more meaningful reporting system is that of trouble and failure reporting where each failed component is required to be described in terms of mode of failure. Such a system must be highly disciplined or underestimate will occur due to failure of reporting. Weighted combinations of usage and TFR have been suggested and ought to help balance the swings of the pendulum. Even so, the additional phenomenon of "cure" may arise where either the manufacturer changes design as a result of the TFR or the operator modifies his operating mode to avoid the failure condition. In any event, the systems manager must adopt some adaptive mechanism for adjusting spare parts requirements as a result of operational experience.

Almost inevitably, the generation of spare parts requirements will result in a spare parts inventory which is too large or too expensive to be carried by each sea-based platform. In order to meet his functional requirement the L&M subsystem manager must have some estimate of the effect of spare parts outage on mission performance. Without this he will be unable even to establish load list priorities, much less demonstrate subsystem performance. This has led to the generation of systems for which each spare part is assigned a Mission Essentiality Code.

The most complex of these systems attempt to construct complete or nearly complete analytical models of the system and the effect on the system of component failure. As the reader might suspect, the author regards such attempts as informational folly. At the other end of the spectrum are subjectively determined priority ratings which are employed in quasi-empirical models with coefficients which are adaptively modified by operational experience. Again the reader will detect the author's preference.

At intervals, the sea-based platform visits a terminal or is visited by a support ship or tender. A basic decision must be made early in systems design on the extent to which one of these facilities is either a "warehouse" or a "garage." This, in turn, requires a basic decision on levels of repair and replacement. At one extreme is the high-maintenance philosophy in which the crew is sufficiently knowledgeable and skillful to diagnose and make repairs at the lowest component level (e.g., circuit boards, cabling, soldering, valve and pipe repair, brazing and welding, machinery repair, bearings, bushings, minor forge and foundry, minor machinery, etc.). At the other extreme is the throwaway module with replacement. It is the author's view that the latter approach results in cost savings and systems efficiencies which are well in excess of those calculable by any cost model. If the component designer has the notion that a highly-trained individual will be present to repair, conduct preventive maintenance and engage in occasional inspection and timing operations, he will have a tendency to accept such requirements even if they reduce component cost only slightly and without computing the cost of training and deploying the skilled individual. If, on the other hand, he has the notion that he is designing for a "wooden"-"nonmaintenance"-"throwaway" environment, he will introduce training and supply problems only to the extent that he fails in his design.

The extent of the L&M system or, in other words, the locale of interface between the logistics manager and other subsystems has many dilemmas built in. At one extreme the logistics manager is responsible for procurement and production of spare parts in conformance with performance and configuration specifications of the subsystem manager. He is responsible for repair and repair facilities and for parts and components substitution. Under this philosophy he is equivalent to "Manny, Moe and Jack," the Pep Boys of the automotive industry.

At the other extreme, the interface is such that he operates only as a supplier and distributor of spare parts and logistics which are designed, produced and repaired by the individual subsystem managers.

Many hybrid combinations can exist, such as subsystem manager responsibility for initial operating spaces and logistics, or subsystem manager having continuous design and qualification responsibility, etc. Rather delicate tradeoffs in system reliability and cost are inherent in the decision on the interface, but it cannot be overemphasized that an early and clear determination of the precise limits of responsibility is required.

Only through early establishment of logistic goals and interfaces can the supply system manager design his subsystem. Substantial hardware development and acquisition will be required. These may include:

On-board test equipments

On-board repair facilities

On-board computation and communication facilities

Tenders, supply ships, tankers, lighters, helicopters, and aircraft (or the assured services thereof)

Depots and warehouses

Computation, communication and accounting facilities

Handling facilities, roads, wharfs and piers, etc.

Quite obviously, for most major ocean systems, the logistics and maintenance subsystem will be a major system in its own right. Several major pitfalls in the design of the subsystem should be avoided, as follows:

a) Mismatch in Communication and Reporting Capability in the System

The sheer volume of accounting involved in most logistic systems is such that the designer is led to the use of digital computers and digitized communications. If all elements involved in the system are not capable of receiving and generating information in the same format and at an information rate which matches the system, then serious accounting problems can be created. This is most aggravated when some intermediate supply point (tender, supply ship) has a communication capability which is less than that of the central supply facility and also less than that of the user element. The supply system, accounting and requisition systems, do not merely reduce to the communication capability of the intermediate supply facility, but degrade even further due to the inability of the intermediate point to synopsize or place priority ordering on the incoming volume of information. This condition results from the inability to handle the incoming flow of information. For both user and supplier the net effect is the arbitrary discard of a certain percentage of the supply information. Therefore, a most important feature of a supply communication network is that it be balanced.

b) Mismatch in Handling and Transfer Capability.

Replenishment and particularly replenishment at sea can be a long, tedious and even dangerous process, even under good design, when no special consideration is given to the systems implication of the replenishment hardware. In that event transfer at sea may not be accomplished except in the calmest sea states and transfer in a port or harbor may require intermediate loading and unloading. In a well-designed system, transfer cranes and stowage locations on the supply ship will be matched with transfer facilities and stowage locations on the serviced ship or platform. Pier and terminal facilities will be similarly designed.

c) Excess Inventory in Pipeline

When the ocean system contains a substantial inventory of parts having a high unit cost or a high unit cost per volume, then amortization costs and spare part redundancy costs can become significant. Accessibility of terminals to major airports thus enters as an element in the spare-parts philosophy, since the use of airlift or airdrop can aid in this inventory problem.

d) Commonality, Single-Source Procurement and Obsolescence

A great deal of pressure is and should be placed on the subsystem designer to encourage the use of common modules or large blocks or standard mechanical components in his design. The presumption here is that the inventory of spare parts is greatly reduced. A number of subtle pitfalls are inherent in this generality. When a spare part is common to a mission essential item (e.g., the navigation system) and a nonessential item (e.g., the captain's record player), the nonessential item can and may deplete the spare parts inventory unless a portion of the inventory is dedicated to the high essentiality item. Such dedication (mandatory in terms of performance effectiveness) of course vitiates the advantage of commonality. Standardization also invites substitution of functionally equivalent components. The pitfall here is in the specification of equivalence. All too frequently "equivalent" spares, in the sense that they completely match written specifications, will fail or give degraded performance in specific applications.

On the other hand, spare parts uniqueness creates serious problems if the components thereof become obsolete or when the retooling for a new run of parts is deemed financially unattractive by an manufacturer.

Guaranteed single-source procurement is one technique for avoiding these pitfalls. This solution has itself many pitfalls, but these are chiefly monetary. Excess cost derives not only from lack of production competition, but also in the stimulation of design modification to increase reliability, commonality and the avoidance of obsolescence. The alternative to single-source procurement is a careful management of configuration control and qualification. The system manager's dilemma here relates to the assignment of responsibility for this function. The supply subsystem manager will usually desire that this function be his responsibility after the initial subsystem design is fully qualified. The subsystem manager will usually desire that this function remain his responsibility in order to insure subsystem performance. The systems manager's decision should probably be based on the expected life of the system, the criticality of spare parts characteristics with respect to system performance, the cost of qualifications, etc. For example, missile and aerospace components are extremely expensive to qualify, requiring as they usually do a flight test for assured performance. Inadequacy of spares will not be exposed until actual system performance with the resulting high cost of failure. In such instance, tight control by the missile subsystem manager is mandatory. Conversely, in a continuously mandatory redundant navigation system, a supply manager can make spare parts substitutions in relative safety, since performance will, in general, be degraded but not destroyed, the awareness of impairment will be quickly determined, and the inadequacy of the "equivalent" part quickly exposed.

LOGISTIC SUPPORT AND MAINTENANCE SUBSYSTEM CASE STUDY

The concept of undersea nuclear power plants for civilian power application is very effective from the standpoint of public safety and for minimizing pollution. Commander L. K. Donovan has a unique background as a Civil Engineer Corps officer in the United States Navy with nuclear power experience.

CASE STUDY
NONPOLLUTIVE POWER SYSTEM
LOGISTIC SUPPORT AND MAINTENANCE

by

L. K. Donovan

I. Subjective Human Purpose

Provide safe and reliable power and drinking water in quantities large enough to support East Coast port complexes in a manner that will significantly reduce air and water pollution at a cost which is economical compared to correcting the pollution problems with present-day plants.

II. Concept Formulation

A. Discussion of Subjective Human Purpose

1. Almost all large East Coast port complexes are located in areas of large population and large power utilization. Most of these areas have air and water pollution problems of varying degrees that are continually increasing. It is estimated in some East Coast metropolitan areas that the production of power alone contributes between 20 to 30% of the pollutants in the atmosphere. Also of concern, as more and larger generating stations are constructed to meet the electrical demand, is the thermal pollution of waterways and lakes caused by the dumping of excess cycle heat. This is even more severe with nuclear power stations which have a cycle efficiency of about half that of the conventional plant. The nuclear plant, however, contributes insignificantly to air pollution compared to the conventional plant. If the thermal effect of

the nuclear plant could be minimized or even made useful, the conditions of the subjective human purpose to minimize air and water pollution could best be met with nuclear power plants.

2. Assuming that no other major power source will be developed in the next ten years, the alternatives for minimizing the thermal effect from nuclear power plants are as follows:

a. Locate the plant near the largest body of water available, such as the ocean, and exhaust cooling water as far out to sea as possible where the thermal load would be absorbed with little noticeable change in the ecological balance.

b. Locate the plant near a body of water where the increase in temperature caused by the exhaust heat would serve a beneficial purpose.

c. Locate the plant in the ocean, offshore at a depth where the thermal effects would be minimized or useful.

3. The economics of the alternatives can be roughly estimated and it is certain that alternative c. would appear most expensive at first. Let us consider the advantages and disadvantages of each alternative.

a. The first alternative would require expensive coastal property and would still have siting problems due to population concentrations along the coast. Because of the population density, the AEC is requiring extensive safeguard systems and reviews which are the major cause today of the slowdown of the nuclear power industry. The alternative would use current technology and the only new technological problem is the long exhaust pipeline which, though not insignificant, should be easy to solve.

b. A proposal has been made to use the thermal discharge of a power plant to create a thermally-controlled environment to assist in the cultivation of lobsters in Maine.[1] Water temperatures in the area have dropped and the lobster population has diminished. Similar applications for the second alternative have one thing in common, that they are usually well distant from the centers of large power utilization; thus, distribution costs are very high. The number of beneficial applications also seems small compared to the East Coast power needs and the magnitude of the pollution problem.

c. The third alternative would involve many new technological problems and would be a very large initial capital expense for a large-scale power plant. It does provide the best solution to the pollution problem and has the potential of reducing the cost of expensive safeguard systems since the effects of the maximum hypothetical accident (MHA) on the population centers would be minimized. In addition, this alternative would satisfy the requirement by the Commission on Marine Science and Engineering[2] that the technology be developed and an experimental nuclear plant be constructed on the continental shelf to provide power for undersea exploration and exploitation.

B. Selection of a Concept

The undersea nuclear power station was selected as the project to be developed because it best satisfied the pollution problem and because of the need to develop the technology for future use of the undersea. The power output of the project was scaled down, however, and the purpose was revised to act as an experimental prototype of future large plants where new technology can be developed and the effects of the artificial thermal upwelling on the ecology and micrometeorology of the area can be studied.

C. General Criteria

1. Because of the potential military benefits of the experience gained from the construction and operation of this system, it is assumed that a high-level government decision was made that the Navy would develop and operate the prototype or prototypes in cooperation with the Atomic Energy Commission. It was further assumed that, since this plant will be located on the ocean bottom, the Naval Facilities Engineering Command would take the lead in the development for the Navy.

2. The potential capital savings from duplication of facilities would run about 20% for the second plant and 30%

for the third plant if bought under a single contract. Thus, the cost of three sister plants on a single contract would be approximately 2.5 times the cost of a single plant. It is also apparent that the Navy has potential use of this power around its major fleet and submarine bases. Thus, it is also assumed that Congress approved the construction of three such plants at New London, Connecticut, Norfolk, Virginia, and Charleston, South Carolina.

SUBSYSTEM LIST

TABLE	DESCRIPTION
A.	Platforms Including Transfer Vehicle
B.	Terminal
C.	Nuclear Power and Auxiliary Systems
D.	Construction and Recovery
E.	Life Support
F.	Communications, Navigation and Environmental Sensing
G.	Maintenance and Logistics
H.	Emergency Systems
I.	Test and Evaluation
J.	Electrical Distribution
K.	Crew and Training

MAINTENANCE AND LOGISTICS SUBSYSTEM

I. General Information Concerning Integrated Logistic Support

- A. Objectives
- B. Definitions
- C. Elements
- D. Relationships among Elements

II. Statistical Decision Theory for Logistics Planning

- A. General
- B. Bayesian Statistical Analysis
- C. Subjective Prior Distributions
- D. Loss Function

III. Maintenance Program

- A. Levels of Maintenance
- B. Types of Maintenance
- C. Maintenance Interfaces with Other Subsystems

IV. Logistics Program

- A. Types of Support
- B. Logistics Interfaces

I. General Information Concerning Integrated Logistics Support

A. Objective

1. The basic objective of an integrated logistic support (ILS) program is essentially twofold:[1]

   a. To recognize the implication of maintenance and logistics during design.

   b. To structure and provide the required support elements as dictated by the design.

Both of the above concepts, influencing the design of hardware by logistic-support considerations and designing an integrated support system, must commence coincident with project conceptual formulation.

2. The first objective will involve designing to ensure that alternatives are available so that systems and components can be selected with a balance between performance and logistics support of the item over its entire life. The second objective causes an integrated management and technical approach to be taken in structuring the logistics support system. This means that an engineered plan to maintain an engineered plan to logistically support must be developed concurrently with the design of new hardware. These plans must be managed from the time they are developed through the engineering development phase on through the operational phase to the end of the life of the project.

3. Maintenance is very closely related to ILS and as part of the plan, a baseline level of maintenance must be

determined during design for each hardware component prior to operation. This baseline is based on the best subjective and test information available and desired degree of reliability. As experience is gained in operation, the ILS must be able to adjust to the latest information from the using activity concerning usage data, maintenance, and reliability.

B. Definition

Integrated logistic support is a composite of the elements necessary to assure the effective and economical support of a system or equipment component at all levels of maintenance for its programmed life cycle. It is characterized by the harmony and coherence obtained between each of the elements and levels of maintenance.

C. Elements

The elements listed below will vary somewhat for other systems depending on the other system-related activity which plays a part in ILS for that system. The list is complete for this system and is divided into three parts. First are design features which have impact on the support activities; second are part of the developmental effort in general that must be applied to ILS or interface with it, and third are the logistic support activities and their components.

1. Design Features with ILS Impact

a. Maintainability (a characteristic of equipment/system design).

b. Reliability (mean time between failures).

c. Durability (capacity not to "wear out" before project end of life).

d. Standardization (common spares for as much equipment as possible).

2. Developmental Activities Applying to or Interfacing with ILS

a. Overall equipment/system design (the relation of the four design features above and some of the support activities below to the design as a whole).

b. Technical data and communications (feedback reports required by ILS and channels of information such as malfunction and failure reports).

c. Configuration control (control of equipment configuration and the plan to maintain it with respect to this interface with ILS).

3. Logistic Support Elements and Their Components

a. General planned maintenance (the philosophy, plan and procedures relative to the management and accomplishment and quality control of preventative and corrective maintenance at all levels of maintenance).

b. Specialized maintenance (major maintenance requiring system shutdown and requiring additional personnel with special training: examples: refueling, turbine overhaul, etc.).

c. Contractor maintenance (maintenance resource acquired on contract, including technical field representatives).

d. Spares and Repair Parts Provisioning

   1) Operational end-use items

      a) Special parts/spares

      b) Common parts/spares

   2) Nonoperational end-use items (housekeeping, etc.)

      a) Consumable

      b) Nonconsumable parts/spares

   3) Support and Test Equipment

      a) Special parts/spares

      b) Common parts/spares

e. Support personnel (the numbers and skills required in maintenance and logistics personnel and the training required, including courses and training aids).

f. Technical documentation (all data, including technical manuals, drawings, maintenance and inventory records, and feedback data required to perform maintenance or make logistics decisions).

g. Support and test equipment (all tools, test equipment, etc., used to facilitate or support maintenance and monitor the operational status of systems and equipment).

h. Support facilities for maintenance or logistics.

i. Modification and/or modernization (all equipments, tools, procedures, tests and services required).

j. Packaging and transportation

1) Operational end items (incoming, short shelf time)

2) End items to be repaired (outgoing)

3) Spares and repair parts (long shelf time)

k. Disposal operations (waste, radioactive material, excess equipment).

D. Relationship among Elements

1. Inherent in the design of an ILS system is a study of the relationship of one element on another. To assist in this task, the elements listed above are tabulated in matrix form with a checked pair indicating a significant potential for interaction or influence on one another.

	Maintainability	Reliability & Durability	Standardization	Overall Design	Technical Data & Comm.	Configuration Control	General Planned Maint.	Special Maint.	Contractor Maint.	Spare and Repair Parts	Support Personnel	Technical Documentation	Support and Test Equipment	Support Facilities	Modification/Modernization	Packaging & Transportation	Disposal
Maintainability				√	√		√	√	√			√					
Reliability & Durability							√	√		√	√						
Standardization				√						√		√			√		
Overall Design	√	√	√		√	√	√	√		√	√	√	√	√		√	√
Technical Data & Comm.		√					√	√		√		√	√	√			
Configuration Control							√	√		√	√						
General Planned Maint.	√				√			√	√	√	√	√	√	√		√	√
Special Maint.	√			√	√	√	√		√	√	√	√	√	√			√
Contractor Maint.					√						√						
Spare and Repair Parts	√		√	√								√		√			
Support Personnel	√						√	√				√	√	√	√		
Technical Documentation					√	√				√	√						
Support and Test Equipment					√		√			√	√	√				√	√
Support Facilities	√										√					√	
Modification/Modernization	√	√	√		√	√	√	√	√	√	√	√	√			√	
Packaging & Transportation													√	√			√
Disposal										√		√				√	

√ means element interacts with or influences one on top.

TABLE 8-1

II. Statistical Decision Theory for Logistics Planning

A. General

1. The concepts of general statistical discussion theory that apply to logistics planning can be described as having three stages:[2]

   a. The estimation of some future state, such as anticipated failure rate or personnel required.

   b. The evaluation of consequences (loss or disutility) which depends on the current decision and eventual true value of the estimated state.

   c. The systematic combination of these subjective measures to determine an "optimum" decision.

B. Bayesian Statistical Analysis

1. The estimation procedure makes use of subjective probability distributions and the Bayesian consideration of additional information. This decision will only consider models useful for this application. More precise details of the application of Bayes' Theorem is available in the literature.[3]

2. Consider that the ultimate consequence of a decision depends upon the eventually-known value of some state, Θ. Since it is unknown at the time of the decision it is a random variable, $\tilde{\Theta}$. Before any objective data is known,

we can represent all known subjective information as a prior probability density function, $f'(\tilde{\Theta})$. If we now run an experiment yielding a sample outcome, Z, it permits transformation of the prior probability density function into a posterior probability density function, $f''(\tilde{\Theta}|Z)$. The statistical experiment is such that the likelihood of particular sample outcomes is conditioned by the true value of Θ or a likelihood function, $\ell(Z|\Theta)$.

Then, from Bayes' Theorem,

$$f''(\tilde{\Theta}|Z) = f'(\tilde{\Theta})\ \ell(Z|\Theta)\ N(Z) \tag{1}$$

where N(Z) is a Normalization Function.

Now let Θ represent the unknown failure rate in an electronic system and let Z represent the experimental outcome of a test which yielded r failures in t hours and consider the failure mode as a Poisson process. The sample likelihood is given by the Poisson density function:

$$\ell(r,t|\Theta) = \frac{(\Theta t)^r e^{-\Theta t}}{r!} \tag{2}$$

Now consider the prior probability density of $\tilde{\Theta}$ to have a gamma distribution with parameters r' and t', the primes denoting prior. Then

$$f'(\tilde{\Theta}) = \frac{t'(\Theta t)^{r'-1} e^{-\Theta t'}}{(r' - 1)!} \tag{3}$$

Now combining eq. (2) and (3), the posterior distribution given this prior and another sample outcome of r and t is

$$f''(\tilde{\Theta}) = \frac{t'^{r'} t^{r} \Theta^{r+r'+1} e^{-\Theta(t+t')}}{r!(r'-1)!} N(r,t) \tag{4}$$

Now let the double primes denote the posterior so that $t''=t+t'$, and $r''=r+r'$ and with appropriate $N(r,t)$, then

$$f''(\tilde{\Theta}) = \frac{t''(\Theta t)^{r''-1} e^{-\Theta t''}}{(r''-1)!} \tag{5}$$

3. As an example of how this is used, suppose subjective prior statements about an unknown failure rate indicated an eq.(3) distribution with $r'=4$ and $t'=2000$. Say, then, information was received from test that $r=3$ and $t=1000$. Then these posterior feelings should be according to eq.(5) or with $r''=7$ and $t''=3000$. Another way of saying it is that the prior mean is $\tilde{\Theta}'=r'/t'=0.002$, the sample mean $\tilde{\Theta}=r/t=0.00333$ and the posterior mean should be $\tilde{\Theta}''=r''/t''=0.00266$ or 2.66 failures per 1000 hours.

C. Subjective Prior Distribution

1. The subjective prior distribution is very important if the system is new and little objective sample data are available. As can be seen by eq.(5), sample data are systematically factored in, as available, through recursive application and the subjective data become less and less influential. The art of relating subjective statements to probabilities is very difficult and imprecise at best. A similar method, using groupings of greater than/less than levels, is used here to determine the composite opinion of the experts available on what the value of the unknown quantity will be. Methods of deducing the parameters r' and t' from the subjective

information in a gamma distribution are complex and can be studied in detail in the references.[4,5,6,7]

D. Loss Function

1. The other consideration of this statistical model is the conditional loss due to a discrepancy between a planned-for state and eventual actual state. In logistics planning, this loss is the consequence of an eventual excess or deficiency of a logistic asset such as a spare part. The loss, then, is some measure of the total consequence as perceived by the decision-maker of such discrepancies. This loss function is very difficult to evaluate appropriately in relation to the organization as a whole, as the person who is affected most by the loss may tend to overinflate its value.

2. Let Θ' be the planned value for the value Θ, that is, selection of Θ' is the decision. Let K_1 = per unit disutility of excess and K_2 = per unit disutility of deficiency. Then, if $F(\Theta')$ = probability that $\Theta \leq \Theta'$, then

$$F(\Theta') = \frac{K_2}{K_1 + K_2} \quad \text{or} \quad \frac{F(\Theta')}{1 - F(\Theta')} = \frac{K_2}{K_1}$$

3. The above allows a comparison of the consequences of being one unit deficient versus being one unit in excess. Let's say that you need a 95% assurance that a stock is adequate. Then,

$$F(\Theta') = 0.95 = \frac{K_2}{K_1 + K_2} \quad \text{and} \quad \frac{K_2}{K_1} = \frac{0.95}{0.05} = 19$$

This would indicate that the per unit cost of deficiency is 19 times that of excess. This may not be true in all systems and may seem illogical to the decision-maker. The practice is to factor experience in at this point and the Θ' may be higher or lower than what is given by the above equations. Additional refinement of this technique is discussed in the references.[3,8]

III. Maintenance Program

A. Levels of Maintenance

1. In order to define levels of maintenance for systems, subsystems and components, it is first necessary to determine the consequence of a failure on the whole system. Thus, the following types of failures are defined:

   a. Class I. This class of failure results in the unplanned automatic shutdown of the nuclear power subsystem or creates a situation where a hazard to personnel or equipment is such that corrective action will require the shutdown of the nuclear power subsystem.

   b. Class II. This class of failure results in the activation of a redundant system or component without which the nuclear power subsystem would be automatically shut down or be required to shut down manually due to the hazard created.

   c. Class III. This class of failure does not result in shutdown but causes a reduction in nuclear power subsystem capability.

   d. Class IV. All other failures that do not cause reduction in plant capability.

2. Each of the classes of above failures will also have two types associated with each class.

a. Type I. The failure is annunciated by an alarm on the master control and monitoring panel due to direct monitoring of an associated parameter.

b. Type II. The failure is not annunciated.

3. The desired level of maintenance of the system is to have an optimization between reliability, durability and maintainability such that:

a. Only one annual shutdown maintenance period is required and the mean time between Class I failures is approximately 15 months. In addition, the design should insure that no Class I, Type II failures can occur at all with a 95% reliability and 95% confidence.

b. The modularized design configuration of the system is such that any Class II failure can be repaired within two hours for electronic failures and within 10 hours for all other types of Class II failures. The desired mean time between failures (MTBF) is 2000 hours.

c. Class III failures must be repairable within 48 hours with a desired MTBF of 2000 hours.

d. The failure rate determined by manufacturer's specification will be accepted for Class IV failures.

B. Types of Maintenance

1. Preventative maintenance (PM) includes all scheduled inspections and maintenance performed on a periodic basis, i.e., weekly, monthly, quarterly, annually.

2. Unscheduled maintenance includes any maintenance caused by malfunction and maintenance resulting from PM inspections which require more than 0.2 man hours to repair and is performed along with the PM inspection.

3. Scheduled maintenance - Unscheduled maintenance which can be deferred and scheduled at a later time.

4. Shutdown maintenance requires subsystem shutdown to complete maintenance.

5. Special maintenance - Projects large enough in scope to require additional personnel other than normal crew or that require special crew training or qualifications, such as refueling.

C. Maintenance Interfaces with Other Subsystems

1. The design target for overall plant reliability is an average of 92% availability over a three-year period. Availability is accumulated whenever the plant is exporting or capable of exporting power.

2. The refueling cycle is approximately three years at full power with 92% availability. Refueling will be accomplished at the terminal facility. The period of time allotted

to refueling from shutdown to startup is 35 days.

3. By meeting the availability goals above, only an average of 29.2 days per year or 87.6 days per 3-year cycle is available for shutdown, maintenance and testing. This time must include the refueling.

4. All non-shutdown maintenance will be performed by the operating crew augmented by one full-time maintenance man. Each watch will have a balance of maintenance specialties so that there is always at least one man trained in each of the maintenance specialties, i.e., electrical, mechanical, health, physics and process control, and instrumentation. The specialty of the full-time maintenance man will be rotated for accomplishment of all PM's and inspections in the various disciplines.

5. If a component can be removed and replaced by a spare, the down component will be repaired at the terminal or by contract, whenever possible.

6. All special maintenance and scheduled shutdown maintenance will be performed by maintenance teams from the parent organization.

7. Support shops at the underwater facility will be minimal to allow PM performance and minor unscheduled maintenance only.

8. All major component repair or replacement will be performed at the terminal facility, when required.

9. The terminal facility will contain adequate space and equipment for all envisioned maintenance, including final disassembly and disposal. Some major components will be replaced

at one-half project life (e.g., reactor coolant pumps, turbine rotor, etc.).

10. Maintainability will be considered in the design of all subsystems.

11. The transfer vehicle will be maintained by the terminal facility.

12. A maintenance control program having features of master scheduling, equipment history records, logistics records, interface and test equipment, calibration and repair will be instituted.

IV. Logistics Program

A. Types of Support

1. Spare/Repair Parts Procurement and Inventory Control

a. Mission Spares and Supplies. Supplies required in direct support of the operational mission and spares required to correct Classes I, II, and III failures.

b. Emergency Spares. Back-up items or duplicate major components that may be required for unplanned failures. These spares may be planned to be utilized at some time during the life cycle of the system.

c. Routine Spares. Spares required to correct Class IV failures.

d. Consumable Supplies. All consumables not required for direct mission support.

2. Storage

a. Storage at the plant is required for essential mission spares and at the terminal for supplies that are to be used in a planned fashion, routine spares and consumables and for items awaiting shipment for repair or for disposal.

3. Contract Repair Service. Maintenance contracts with outside firms for maintenance, repair and calibration of components and test equipment will be established to maintain system maintenance force at constant level.

4. Packaging and Transportation. Packaging of all items to be shipped from the terminal and arranging for transportation of outgoing and incoming shipments will be accomplished by the logistics work force.

5. Disposal. An organization for arranging for disposal of excess components, spare parts and supplies and for the collection and disposal of radioactive and nonradioactive wastes will be established.

B. Logistics Interfaces

1. Identify types of spares as defined above for each component.

2. Standardization will be accomplished to the maximum extent within the types of spares defined above. There will be no attempt to standardize between the mission and emergency spare types and the routine and consumable spare types. This will prevent essential mission spares being used for routine items.

3. At least one line item for each of the spares required to correct Class I failures will be stocked and more if maintenance history increases requirement.

4. Spare parts required to correct Class II and Class III failures will be stocked with 95% assurance.

5. Emergency spares will be identified during design and sufficient quantity will be ordered for use until system end of life.

6. Malfunction and equipment maintenance records will be reviewed whenever rate of usage exceeds original estimate and new usage rates are verified.

7. Spare parts required to support Class IV failures will be stocked based on best subjective and test information available in the amount of a 6-month supply.

8. A 3-month average supply of consumables will be maintained unless procurement lead time for an item is longer that three months.

9. High and low stocking levels will be determined by evaluation of the estimated failure rate of a component during the design phase as modified by later usage data. Usage data will be totally reviewed every six months.

10. When a component is removed from the system, all spares associated with that component will be removed from inventory and, if not cross-referenceable for use on some other component, will be disposed of.

11. All radioactive material in excess of releasable limits established by the AEC will be disposed of in accordance with AEC regulations by the terminal facility.

12. All nonradioactive waste will be disposed of by commercial contract.

13. Storage space for majority of inventory will be located at the terminal. Only essential mission spares and a 30-day supply of routine and consumable spares and supplies are to be stored at the plant.

REFERENCES

1. NCEL Report CR 68.012, "Concept Development of Manned Underwater Station," Vol. I, July 1968.

2. M.I.T. Instrumentation Laboratory Report E-1350, "Statistical Decision Theory for Logistics Planning," May 1963.

3. Raiffa, Howard and Robert Schlarfer, "Applied Statistical Decision Theory," Harvard Business School, Division of Research, Boston, 1961.

4. Wadsworth, George P. and Joseph G. Bryan, "Introduction to Probability and Random Variables," McGraw-Hill Book Company, New York, 1960.

5. Molina, E. C., "Poisson's Exponential Binomial Limit," D. Von Nostrand Company, Princeton, N.J., 1942.

6. Pearson, Karl, "Tables of the Incomplete Gamma Function," Biometrika Office, University College, London, 1934.

7. Lloyd, David K. and Myron Lipow, "Reliability: Management, Methods and Mathematics," Prentice-Hall, Inc., Englewood Cliffs, N.J., 1962.

8. Schlaifer, Robert, "Probability and Statistics for Business Decision," McGraw-Hill Book Company, New York, 1959.

INDEX